The Origin and Evolution of the Solar System

The Graduate Series in Astronomy

Series Editors: **M Elvis**, Harvard–Smithsonian Center for Astrophysics
A Natta, Osservatorio di Arcetri, Florence

The Graduate Series in Astronomy includes books on all aspects of theoretical and experimental astronomy and astrophysics. The books are written at a level suitable for senior undergraduate and graduate students, and will also be useful to practising astronomers who wish to refresh their knowledge of a particular field of research.

Other books in the series

Dust in the Galactic Environment
D C B Whittet

Observational Astrophysics
R E White (ed)

Stellar Astrophysics
R J Tayler (ed)

Dust and Chemistry in Astronomy
T J Millar and D A Williams (ed)

The Physics of the Interstellar Medium
J E Dyson and D A Williams

Forthcoming titles

The Isotropic Universe, 2nd edition
D Raine

Dust in the Galactic Environment, 2nd edition
D C B Whittet

The Graduate Series in Astronomy

The Origin and Evolution of the Solar System

M M Woolfson

Department of Physics
University of York, UK

CRC Press
Taylor & Francis Group
Boca Raton London New York

CRC Press is an imprint of the
Taylor & Francis Group, an **informa** business

CRC Press
Taylor & Francis Group
6000 Broken Sound Parkway NW, Suite 300
Boca Raton, FL 33487-2742

© 2000 by Taylor & Francis Group, LLC
CRC Press is an imprint of Taylor & Francis Group, an Informa business

Visit the Taylor & Francis Web site at
http://www.taylorandfrancis.com

and the CRC Press Web site at
http://www.crcpress.com

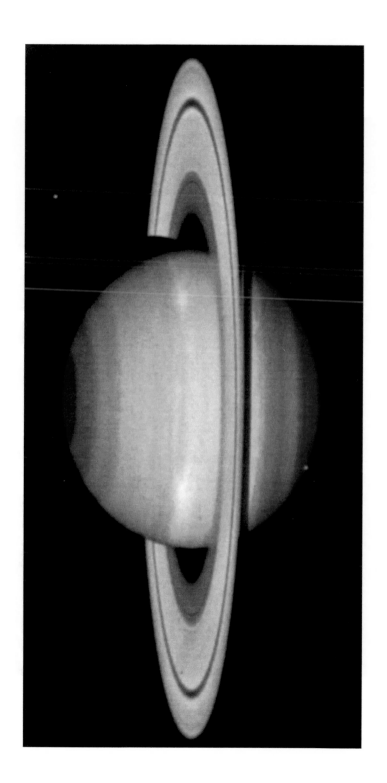

Contents

APPENDICES

Introduction

Since the time of Newton the basic structure of the solar system and the laws that govern the motions of the bodies within it have been well understood. One central body, the Sun, containing most of the mass of the system has a family of attendant planets in more-or-less circular orbits about it. In their turn some of the planets have accompanying satellites, including the Earth with its single satellite, the Moon. With improvements in telescope technology, and more recently through space research, knowledge of the solar system has grown apace. Since the time of Newton three planets have been discovered and also many additional satellites. A myriad of smaller bodies, asteroids and comets, has been discovered and a vast reservoir of comets, the Oort cloud, stretching out half way towards the nearest star has been inferred. Spacecraft reaching out into the solar system have revealed in great detail the structures of all the types of bodies it contains—the gas giants, terrestrial planets, comets, asteroids and satellites, both with and without atmospheres. At the same time observations of other stars have revealed the existence of planetary-mass companions for some of them. This suggests that theories must address the origin of planetary systems in general and not just the solar system. Observations of young stars have shown that many are accompanied by a dusty disk and it is tempting to associate these disks with planet formation.

In attempting to find a plausible theory the theorist has available not only all the observations to which previous reference has been made above but also a knowledge of the basic laws of physics, particularly those relating to conservation. It turns out that finding a theory consistent with both observation of the spins and orbits of solar system bodies and conservation of angular momentum is difficult, and has proved to be an unresolved problem for some current theories. In this respect it can be said that for some theories the post-Newtonian knowledge is irrelevant since an explanation of the origin of even the basic simple system, as known to Newton, has not been found.

This book describes the four major theories that have been under development during the last two or three decades: the Proto-planet Theory, the Capture Theory, the Modern Laplacian Theory and the Solar Nebula theory, and gives the main theoretical basis for each of them. Also discussed, but not so fully, is the Accretion Theory, an older model of solar-system formation with some positive features. These theories are examined in detail to determine the extent to which they provide a plausible mechanism for the origin of the solar system and their strengths and weaknesses are analysed. The only theory to essay a complete picture of the origin and evolution of the solar system is the Capture Theory developed by the author and colleagues since the early 1960s. This explains the basic structure of the solar system in terms of well-understood mechanisms that have a finite probability of having occurred. The way in which planets form, and the way that their orbits originate and evolve according to the Capture Theory, suggests the occurrence of a major catastrophic event in the early solar system. This event was a direct collision between two early planets, in terms of which virtually all other features of the solar system, many apparently disparate, can be explained. As new knowledge about the solar system has emerged so it has lent further support to this hypothesis.

There is a tendency in areas of science like cosmogony for a 'democratic principle' to operate whereby the theory that has the greatest effort devoted to it becomes accepted, without question and examination, by many people working in scientific areas peripheral to the subject. These individuals, highly respected in their own fields, swell the numbers of the apparently-expert adherents and, by a positive feedback mechanism, they enhance the credibility of the current paradigm—which is the Solar Nebula Theory in this case. Science writers and those producing radio and television programmes, accepting the verdict of the majority, produce verbal and visual descriptions of an evolving nebula that, if they were to illustrate any scientific principle at all, would be illustrating the *invalid* principle of the conservation of angular *velocity*. In scientific television programmes material is seen spiralling inwards to join a central condensation having jettisoned its angular momentum in some mysterious fashion on the way in. Computer graphics are not constrained by the petty requirements of science!

The 'democratic principle' is not necessarily a sound way to determine the plausibility of a scientific theory and there are many examples in the history of science that tell us so. The geocentric theory of the solar system, the phlogiston theory of burning and the concept of chemical alchemy were all ideas that persisted for long periods with the overwhelming support of the scientific community of the time.

The aim of this book has been to present the underlying science as simply as possible without trivializing or distorting it in any way. None of the important science is difficult—indeed most of it should be accessible to a final-year pupil at school. It is hoped that this book will enable those both inside and outside the community of cosmogonists to use their own judgement to assess the plausibility, or otherwise, of the theories described. For those wishing to delve more deeply into the subject many references are provided.

I must give special thanks to my friend and colleague, Dr John Dormand, for help and very useful discussions during the writing of this book. Gratitude is also due to Dr Robert Hutchison for providing illustrations of meteorites.

PART 1

THE GENERAL BACKGROUND

Chapter 1

The structure of the Solar System

1.1 Introduction

Before one can sensibly consider the origin of the Solar System it is first necessary to familiarize oneself with its present condition. Consequently this first chapter will provide an overview of the main features of the system of planets. The treatment will be particularly relevant to the study of solar-system cosmogony. Factors relating to the origin of stars and their evolution are left to the next chapter, as is a preliminary discussion of the structure of extra-solar planetary systems.

The salient features of the Solar System are split here into five sections, starting with its orbital structure. This exhibits many striking relationships that are still not fully understood but are now starting to yield to modern celestial mechanics. Secondly, the broad physical characteristics of the planets will be considered. The classification of planets into the major and terrestrial categories is a key feature here.

Most of the planets are themselves accompanied by satellites, thus comprising mini-systems reminiscent of the Solar System itself. The study of these smaller systems has been extremely important in the development of celestial mechanics and is greatly enhanced by spacecraft data from the outer Solar System. The fourth section will be concerned with the lesser bodies of the system, ranging from asteroids with radii up to some hundreds of kilometres down to microscopic particles that commonly cause meteor trails on entry into the atmosphere. The vast numbers of smaller bodies ensure frequent collisions with planets and the scars of their impacts are notable features of all solar-system bodies without an atmosphere.

The comets, responsible for some of the most spectacular celestial apparitions, will be the topic of the last section of this chapter. Inhabiting the furthest reaches of the Solar System the population of comets is, perhaps, the least well understood feature of the Solar System.

The conventional classification of solar-system objects is now challenged by recent discoveries of remote bodies inhabiting the region beyond Neptune. It is likely that these bodies have much physically in common with comets and so they are also included in the final section of this chapter.

1.2 Planetary orbits and solar spin

1.2.1 Two-body motion

The description of planetary orbits derives from the famous laws of orbital motion discovered by Johannes Kepler (1571–1630). These are:

(i)   Planets move in elliptical orbits with the Sun at one focus.
(ii) The line joining a planet to the Sun sweeps out equal areas in equal times.
(iii) The square of the orbital period is proportional to the cube of the average distance from the Sun (semi-major axis).

Kepler formulated these laws based on observations mainly of the planet Mars and he did not appreciate the dynamical aspects of planetary motion. This fundamental problem was solved by Isaac Newton (1642–1727) who analysed mathematically the motion of two gravitating bodies moving under an inverse square law of attraction. Kepler's laws are perfectly consistent with this solution.
    The equation of motion for the two-body problem can be written

$$\ddot{\boldsymbol{r}} = -\mu \frac{\boldsymbol{r}}{|\boldsymbol{r}|^3} \tag{1.1}$$

in which $\boldsymbol{r}$ is the position of one body relative to the other and $\mu = G(m_1 + m_2)$, G being the gravitational constant and m_1, m_2 the masses involved. It may be shown that $r = |\boldsymbol{r}|$ satisfies the equation of an ellipse (see figure 1.1) given by

$$r = \frac{p}{1 + e \cos \theta}, \quad p = a(1 - e^2), \tag{1.2}$$

where a is the semi-major axis of the ellipse of eccentricity e, and p is the semi-latus rectum. Other distances of interest in a heliocentric orbit are the perihelion and aphelion distances, q and Q respectively (figure 1.1), corresponding to the closest and furthest distances from the Sun. Another description of the ellipse is

$$r = a(1 - e \cos E),$$

where E, shown in figure 1.1, satisfies Kepler's equation

$$E - e \sin E = nt, \quad n = \sqrt{\frac{\mu}{a^3}}. \tag{1.3}$$

The quantities E, θ and n are termed *eccentric anomaly*, *true anomaly* and *mean angular motion* respectively. The mean angular motion is the average angular speed in the orbit.

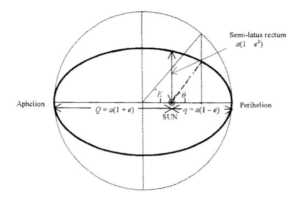

Figure 1.1. The characteristics of an elliptical orbit.

The second and third Kepler laws can be stated in these terms as

$$r^2\dot{\theta} = h,$$

$$P^2 = \frac{4\pi}{\mu}a^3,$$

where $P = 2\pi/n$ is the *orbital period* and $h = \sqrt{\mu p}$ is the *intrinsic angular momentum* or angular momentum per unit mass.

For a full specification of the orbit in space it is necessary to add to the two elliptical elements (a, e), which define the shape of the orbit, three orientation angles and a time fix. To define angles requires a coordinate system and, conventionally, the *ecliptic*, the plane of the Earth's orbit, is taken as the X–Y plane for a rectangular Cartesian system. The positive Z-axis is towards the north so all that is required to define the coordinate system completely is to define an X direction in the ecliptic. Relative to the Earth, during the year the Sun moves round in the ecliptic and twice a year, in spring and autumn, it crosses the Earth's equatorial plane. These are the times of the equinoxes, when all points on the Earth have day and night of equal duration. The equinox when the Sun passes from south of the equator to north is the *vernal* (spring) equinox. The direction of the vernal equinox, called the *First Point of Aires*, is taken as the positive X direction.

The first orientation angle for defining the orbit is the *inclination*, i, which is the angle made by the plane of the orbit with the ecliptic. However, this does not define the orbit completely since if the orbit is rotated about the normal to its plane a, e and i remain the same but the orientation changes. What does remain unchanged is the line of intersection of the orbital plane with the ecliptic. This line is called the *line of nodes*; the point on the line where the orbit crosses the ecliptic going from south to north is the *ascending node* and the *descending node* where it goes from north to south.

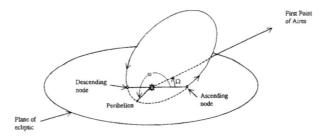

Figure 1.2. The longitude of the ascending node, Ω, and the argument of the perihelion, ω.

The other two angles that define the orbit in space are shown in figure 1.2. The first of these is the *longitude of the ascending node*, Ω, which is the angle between the ascending node and the first point of Aires. The second angle is the *argument of the perihelion*, ω, which is the angle between the ascending node and the perihelion in the direction of the orbiting body. Sometimes Ω and ω, which are not coplanar, are added together and referred to as the *longitude of the perihelion*.

To define the position of the body at any time also requires some time-dependent information and this is usually the *time of perihelion passage*, T_P, which is one of the times when the body is at perihelion. If all six quantities, a, e, i, Ω, ω and T_P, are given then the motion of the body is completely defined. Since the position, r, and velocity, v, together with a time also completely define the orbit it is clear that transformations between the two sets of quantities are possible.

1.2.2   Solar system orbits

The simple relationships listed so far are strictly true for an isolated two-body system. Clearly this is an idealized concept that cannot occur precisely in nature. The Solar System contains many bodies, not just two, but with the Sun being 1000 times more massive than Jupiter, the most massive planet, the motion of each planet is largely governed by the solar mass. The assumption of elliptical motion for each planet–Sun pair is useful and fairly accurate. Thus the equations of motion for the planets relative to the Sun may be written

$$\ddot{r}_i = -\mu_i \frac{r_i}{|r_i|^3} + F_i, \quad i = 1, 2, \ldots, 9, \tag{1.4}$$

in which the vectors F_i have small magnitudes and contain the perturbing effects on planet i of all the other planets and satellites and $\mu_i = (M_\odot + m_i)$. The symbol $\odot$ indicates quantities pertaining to the Sun. These perturbations cause the elliptic elements of the planetary orbits to vary but, as far as can be determined, only in a periodic fashion. As an example, the eccentricity of the Earth's orbit, currently 0.0167, varies in the range 0 to 0.06. At one extreme the distance of the Sun will

Table 1.1. The orbital characteristics of the planets.

Planet	a (AU)	e	i
Mercury	0.3871	0.2056	7°00'
Venus	0.7233	0.0068	3°24'
Earth	1.0000	0.0167	
Mars	1.5237	0.0934	1°51'
Jupiter	5.2026	0.0488	1°18'
Saturn	9.5549	0.0555	2°29'
Uranus	19.2184	0.0463	0°46'
Neptune	30.1104	0.0090	1°46'
Pluto	39.5447	0.2490	17°09'

1 AU (the mean Earth–Sun distance) = 1.496×10^{11} m.

vary by 12% during each year; this has important implications for the terrestrial climate. The present-day elliptic elements (a, e, i) of the nine planets are shown in table 1.1.

One of the most striking manifestations of order in the Solar System is in the regular spacing of the mean orbital radii. This was first noted in the 18th century, when the planets known were those out as far as Saturn, and it is easy to fit a rather simple formula to the semi-major axes of these planets. This formula is usually called the 'Titius-Bode (or just 'Bode's') law'. Many variants exist of this empirical rule, but the original and simplest version is

$$a_n = a_0 + 0.3 \times 2^{n-1}, \quad n = 1, 2, 3, \ldots \qquad (1.5)$$

where a_0 is the mean radius of Mercury's orbit in AU and $n = 1, 2$, represents Venus, the Earth and so on. Table 1.2 contains the values of orbital radii and the corresponding Titius-Bode values. The agreement is quite remarkable and belief in the law was reinforced by the discovery of Uranus by William Herschel in 1781. True, there was a gap between Mars and Jupiter but this was soon filled by Ceres, the largest asteroid, discovered by Giussepe Piazzi in 1801. The importance of this law seemed well established, but the discoveries of Neptune in 1846 (semi-major axis 30.1 AU, $a_8 = 38.8$) and Pluto in 1930 (semi-major axis 39.5 AU, $a_9 = 77.2$) have undermined its plausibility to some extent. Unlike Kepler's laws the Titius-Bode relationship does not emerge from any straightforward dynamical considerations.

The planetary system is now known to be stable over a period greater than its estimated age. This could not be the case in a system that permits close approaches between major bodies, as may occur in a system containing highly eccentric orbits.

The two extreme members of the system depart most strongly from circular orbits and from co-planarity with the remainder of the system. Pluto, in particular,

Table 1.2. The Titius-Bode relationship compared with the actual semi-major (s-m) axes for planets out to Uranus plus the asteroid Ceres.

				N				
	1	2	3	4	5	6	7
				Planet				
	Mercury	Venus	Earth	Mars	Ceres	Jupiter	Saturn	Uranus
s-m axis	0.4	0.7	1.0	1.5	2.8	5.2	9.6	19.2
a_n	0.4	0.7	1.0	1.6	2.8	5.2	10.0	19.6

has an orbit with a perihelion distance less than that of Neptune. In projection onto the plane of the ecliptic the orbits of these two planets would cross but because of the special relationship of the two orbits the planets never come closer together than 18 AU.

In recent years it has become technically feasible to study numerically the evolution of orbits of the Solar System over periods of time comparable with the age of the system. Computer simulations indicate that the planetary orbits may well have remained essentially the same over a period of 4.5×10^9 years. However, the injection of test particles into any of the perceived gaps always results in their ejection in a relatively short time. This implies that bodies, if they existed in such orbits, would relatively quickly be absorbed by collisions with planets or the Sun, or else be expelled from the inner Solar System following close encounters (Duncan and Quinn 1993).

1.2.3   Commensurable orbits

Another interesting feature of the planetary orbits is the existence of commensurabilities, that is pairs of bodies whose periods, and hence their mean motions, differ by a factor which is a simple fraction (Roy 1977). The most important of these is the Jupiter–Saturn or 'great' commensurability which satisfies the relation

$$5n_S - 2n_J = 0.007\,127 \text{ year}^{-1}.$$

With this near-perfect ratio of periods the mutual perturbations of the two planets are enhanced. The period associated with this is about 900 years, over which all mutual configurations will be repeated, as is implied by the discrepancy in their relative periods. The repetition increases the amplitude of the mutual perturbations but the two planets appear to be locked into this near resonance. All the planets exhibit rotation (precession) in their perihelion longitudes.

Another remarkable commensurability is that between Pluto and Neptune.

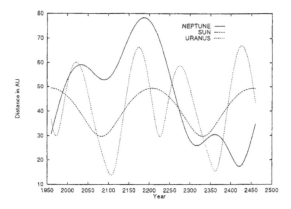

Figure 1.3. The distance from Pluto to the Sun, Neptune and Uranus over the 500 year period 1950–2450.

In this case the current elements give

$$2n_N - 3n_P = 0.000\,159\ \text{year}^{-1}.$$

Since the perihelion of Pluto is less than that of Neptune the orbits of these two planets approach each other quite closely, notwithstanding their different inclinations and the fact that their perihelion longitudes are currently nearly 180° apart. However, a close approach does not occur, even though the present discrepancy in the resonant frequency mode implies a period of about 40 000 years. It has been established that the angle Θ given by

$$\Theta = 3\lambda_P - 2\lambda_N - \varpi_P,$$

where λ is the mean longitude and ϖ_P is the longitude of the perihelion of Pluto, does not rotate but oscillates (librates) about 180° with amplitude 80° and period approximately 20 000 years (Williams and Benson 1971). In simple terms, conjunctions between these planets occur when Pluto is close to its aphelion. Computer simulations have demonstrated that this gravitational 'evasion' may persist for a period greater than the age of the Solar System. Interestingly, for Pluto the closest approaching planet is Uranus which can come as close as 11 AU. A graph of the separations of the three outer planets over a 500 year period is shown in figure 1.3. This special relationship is not unique since there are many commensurabilities which are observed between other solar-system bodies. In particular the ratio of the period of Neptune to that of Uranus, 1.962, is quite close to 2, although there are no 'evasion' processes going on between these two bodies. An explanation for commensurabilities and near-commensurabilities between planetary orbits is suggested in section 7.1.5.

1.2.4   Angular momentum distribution

A cosmogonically significant feature of the Solar System concerns the distribution of angular momentum within it. The Sun spins about an axis inclined at $6°$ to the vector representing the angular momentum for the whole of the system. The period of its outer layers varies from 25.4 days at the equator to 36 days near the poles. Internally the Sun appears to spin as a solid body with a period near 27 days. The spin angular momentum of the Sun has magnitude

$$H_\odot = \alpha_\odot M_\odot R_\odot^2 \omega_\odot = 2.5\alpha_\odot \times 10^{42} \text{ kg m}^2 \text{ s}^{-1},$$

where $M_\odot$, $R_\odot$ and $\omega_\odot$ are the solar mass, radius and angular speed and $\alpha_\odot$ is the *moment-of-inertia factor*. With a central density about 100 times the mean density $\alpha_\odot$ is about 0.055; for a uniform sphere α is 0.4 and becomes less as the central condensation in the body increases. The orbital angular momentum of a planet with semi-latus rectum, p_i, is

$$H_i = m_i \sqrt{\mu_i p_i}$$

and summing the contributions of the four major planets, Jupiter, Saturn, Uranus and Neptune, yields a total of 3.13×10^{43} kg m^2 s^{-1}, or more than 200 times that of the solar spin. Thus the Sun, containing 99.86% of the mass of the Solar System, contains less than 0.5% of its total angular momentum.

1.3   Planetary structure

1.3.1   The terrestrial planets

The basic characteristics of the planets are listed in table 1.3. With the exception of Pluto they are usually considered to be of two types. The inner group of four, of which the Earth is the largest member, are known as the terrestrial planets. The Moon is often included in any discussion of these planets. The terrestrials are all dense rocky bodies and almost certainly have cores, consisting of iron with a small proportion of nickel, overlaid by a silicate mantle. The interpretation of their densities is in terms of the relative size of the core to that of the whole body and also the total mass of the planet that will determine the degree of compression. The relative sizes of the five terrestrial bodies, together with an indication of their core sizes, are illustrated in figure 1.4.

Another common characteristic of the inner planets is that they all show signs of bombardment damage in the form of craters and large depressions. Mercury and the Moon show most damage superficially and these two bodies have a similar appearance. Crater sizes vary from the smallest capable of resolution up to the massive Caloris basin on Mercury, over 1000 km in diameter, which is almost matched by the lunar Orientale basin.

As a result of continuing geological processes, Venus and the Earth have generally less ancient surface features than the smaller planets. These processes are

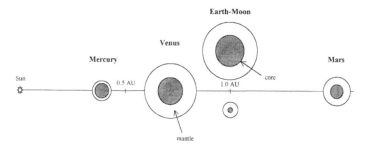

Figure 1.4. The relative orbital radii and sizes of the terrestrial planets. Planets are represented at 3000 times their natural linear dimensions relative to the depicted orbital radii.

Table 1.3. Characteristics of planetary bodies.

Planet	Mass (Earth units)	Diameter (km)	Density (10^3 kg m^{-3})
Mercury	0.0553	4 879	5.43
Venus	0.8150	12 104	5.24
Earth	1.0000	12 756	5.52
Mars	0.1074	6 794	3.94
Jupiter	317.8	142 984	1.33
Saturn	95.16	120 536	0.70
Uranus	14.5	51 118	1.30
Neptune	17.2	48 400	1.76
Pluto	0.0021	2 280	2.03

Mass of the Earth, $M_\oplus = 5.974 \times 10^{24}$ kg.

due to a greater retention of the original heat of formation and internal heating due to the decay of radioisotopes, mainly uranium (^{238}U), thorium (^{232}Th) and potassium (^{40}K). Conduction and convection in the mantle are responsible for tectonics and associated volcanism in which crustal material is being reformed from, and is reabsorbed by, the mantle. The process causes lateral movement in the crustal plates known as continental drift. Because of extensive cloud cover, large-scale observations of the surface of Venus are based only on radar, but these indicate that tectonic processes may have been important, thus implying an internal structure similar to that of the Earth. The atmosphere of Venus is very dense, mainly consisting of CO_2 with a surface pressure and density of 92 bar and 65 kg m^{-3}.

Being intermediate in mass, Mars shows surface features which might be interpolated from a study of the Earth and the Moon. Despite less internal heating from tides and radioactivity, Mars does exhibit ancient volcanic activity but this is now extinct. Like the Moon, Mars shows hemispherical asymmetry with heavily cratered uplands on one hemisphere and smoother 'filled' terrain on the other. On Mars the division is approximately north–south with the volcanoes in the north— in contrast to the Moon whose smooth hemisphere faces the Earth. Unlike the Moon the Martian surface has channel features which have almost certainly been caused by running water (Pollack *et al* 1990). The polar caps contain substantial permanent deposits of ice with the addition of solid CO_2 which comes and goes with the seasons. Since the orbit of Mars has an eccentricity which varies with time and may rise to 0.14 it is possible that Mars has had wet episodes in its existence. The present surface pressure is about 6 millibar (mb) and its atmosphere is 95% CO_2.

1.3.2   The major planets

The four major planets differ markedly in both structure and appearance from the terrestrials. Even a small telescope shows Jupiter as the most colourful and dynamic planet in the system. The banded appearance of its upper atmosphere, composed mainly of molecular hydrogen and helium, is due to the rapid rotation of the planet and has been studied for over three centuries. There is no visible solid surface and so no evidence of any collision history. However, the fact that Jupiter probably has absorbed many smaller bodies was well illustrated by the collisions of the broken-up Comet Shoemaker–Levy 9 in 1994. These collisions, by throwing up material from deep inside the planet, acted as probes for its internal composition.

The atmospheric bands parallel to the equator contain spots or ovals of various colours whose longevity seem to be size-dependent. The largest of these is the *Great Red Spot* (GRS) that has persisted for more than 300 years. This huge feature is roughly elliptical with axes some 25 000 by 13 000 km. Its colour is not constant but it is a notable feature even when its red colour fades. The ovals and spots are thought to be eddies formed between neighbouring bands moving with relative speeds of up to 150 m s^{-1}. This theory is a plausible one for application to small ovals with a lifetime of a few days but it seems not too successful in the case of the GRS (Ingersoll 1990).

In most respects Saturn is similar to Jupiter. The atmosphere has the same composition and the body of the planet has a banded appearance, although the differentiation of zones is far less prominent. With only about one-third of the mass of Jupiter, Saturn is less compressed and its rapid rotation makes it more oblate. Wind speeds in the upper atmosphere are greater even than those of Jupiter, reaching 500 m s^{-1}. The most remarkable feature of Saturn is, of course, its extensive ring system (figure 1.5). It is now known that all the major planets have one or more orbiting rings, but those of Jupiter, Uranus and Neptune are much less substantial than those of Saturn and more difficult to detect and observe. Uranus and Neptune also have hydrogen–helium atmospheres but have a much more uniform appearance than the two larger gas giants. Neptune does have a *Great Dark Spot*, a storm system similar to the GRS on Jupiter.

Figure 1.5. Saturn from the Hubble Space Telescope.

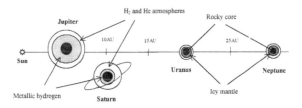

Figure 1.6. The relative orbital radii, sizes and internal structure of the major planets. Planets are represented at about 5000 times their natural linear dimensions relative to the depicted orbital radii.

The internal structures of the major planets are very different from those of the terrestrial planets, as illustrated in figure 1.6. Jupiter and Saturn, mostly hydrogen and helium, have compositions similar to that of the Sun, whereas Uranus and Neptune are formed from icy compounds such as water, methane and ammonia. It is probable that all the major planets possess rock-plus-metal cores but this type of information can only be inferred from theoretical studies (Jones 1984). Theory suggests that there is no sharp transition between gaseous and solid phases. At a depth of 20 000 km in Jupiter the atmosphere will resemble a hot liquid at 10^4 K; at greater depths the hydrogen enters a completely ionized metallic phase. Saturn also contains such a metallic hydrogen mantle but Uranus and Neptune, with less hydrogen and less compression, are unlikely to contain any of this exotic material.

The rock-plus-metal cores of Jupiter and Saturn, with perhaps ice as well, are variously estimated to have masses in the range 10–$20 M_\oplus$. The two outermost major planets might have only very small cores as it has been suggested that the higher central density could be entirely due to compression effects on the material forming the greater part of those planets.

1.3.3 Pluto

It is now clear that the outermost 'planet', Pluto, does not fit into either of the two main classes of planet. Estimates of the mass of Pluto have steadily declined

Figure 1.7. A Hubble Space Telescope view of Pluto and its satellite Charon.

since it was first discovered in 1930. Prior to its discovery it was postulated that a ninth planet should exist, of mass $6M_\oplus$, to explain the departures in the motions of Uranus and Neptune from those predicted. By 1978 this estimate had been lowered in several stages to $0.08M_\oplus$ but the discovery of a satellite of Pluto in 1979 (figure 1.7) gave the current estimate of $0.0021M_\oplus$. Since this is one-sixth of the mass of the Moon and gives a density less than one-half that of Mars it is obviously not similar to a terrestrial planet. It is reasonable to suppose that its origin might be ascribed to some process, or processes, different to that which produced the normal planets. Recent discoveries of trans-Neptunian objects (see section 1.7.3) make it logical to consider Pluto as a member of such a group.

1.4   Satellite systems, rings and planetary spins

1.4.1   Classification

Most of the planets are accompanied by smaller bodies, called satellites, in orbits around them. In the cases of the major planets these form regular systems similar to the planetary system itself. Several of the satellites are comparable in size to, or slightly larger than, the planet Mercury.

The only satellite known from ancient times is the Moon which, being so massive in relation to its primary, must be classified as irregular. The first satellites of another planet to be discovered were the four large Galilean satellites orbiting Jupiter, so named because of their discovery by Galileo Galilei in 1610. With telescope developments over the next three and a half centuries many smaller satellites were discovered and a further major boost to the known satellite population has been provided by spacecraft observation.

Many of the satellites of the major planets are relatively large and occupy near-circular orbits in the equatorial plane of the primary. These are termed *regular satellites* and they are linked to the plausible assumption that they originate as part of the process of planetary formation. Included in the *irregular* category of satellites there are some that are small but in regular orbits and one of the larger

Table 1.4. The satellite system of Jupiter. The spin period of Jupiter = 9 hr 55 min.

Satellite	Inclination of orbit to equator	Semi-major axis (10^3 km)	Eccentricity	Mass (10^{22} kg)	Average diameter (km)	Density (10^3 kg m^{-3})
Metis		128			40	
Adrastea		129			20	
Amalthea		181	0.003		190	
Thebe		222			100	
Io	0.0°	422	0.000	8.93	3630	3.5
Europa	0.5°	671	0.000	4.88	3138	3.0
Ganymede	0.2°	1 070	0.001	14.97	5262	1.9
Callisto	0.2°	1 880	0.007	10.68	4800	1.8
Group of four	25–29°	11 094– 11 737	0.102– 0.207		16–190	
Group of four	147–164°	21 200– 23 700	0.169– 0.410		30–50	

satellites, Triton, orbits Neptune in a close, circular but *retrograde* sense. Irregular satellites are usually interpreted in terms of some kind of capture event.

1.4.2   The Jovian system

The important orbital and physical properties of the satellites of Jupiter are listed in table 1.4. With periods measured in terrestrial days the orbital phenomena of the Galileans are particularly convenient for dynamical research and records of their motion cover many thousands of orbits. Thus it is confirmed that the mean motions of the three inner members of the quartet perfectly satisfy the relation

$$n_1 - 3n_2 + 2n_3 = 0,$$

where the suffices 1, 2 and 3 indicate Io, Europa and Ganymede. Furthermore the respective orbital longitudes satisfy

$$l_1 - 3l_2 + 2l_3 = 180,$$

indicating that the satellites cannot line-up on the same side of Jupiter. Allowed conjunctions and oppositions are shown in figure 1.8. The stability of this configuration was proved by Pierre Laplace (1746–1827) and so it is usually termed a *Laplacian triplet*.

The satellite systems exhibit regular orbital spacing in a similar way to the planetary orbits. In the Jovian system the expression

$$a_n = a_0 + 3 \times 2^{n-1}, \quad n = 1, 2, 3, 4,$$

gives a good fit to the Galilean orbital radii where a_0 is approximately the orbital radius of Amalthea and the unit of distance is the radius of Jupiter; this relationship should be compared to equation (1.5). The quality of this fit is shown in

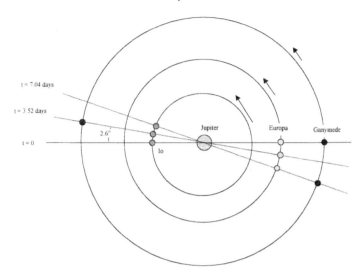

Figure 1.8. Three successive alignments of the three inner Galilean satellites which form a Laplacian triplet. Orbits and Jupiter are drawn to scale but angles are exaggerated.

Table 1.5. Titius-Bode type relationship compared with the actual semi-major axes for the Galilean satellites. Distances are in units of the Jupiter radius.

			n		
		1	2	3	4
			Satellite		
	Amalthea	Io	Europa	Ganymede	Callisto
a	2.5	5.9	9.4	15.0	26.4
a_n	3	6	9	15	27

table 1.5. Similar relationships may be constructed for the other major planet systems.

It is easy perhaps to overstate the similarities between the major satellite systems and the planetary system itself. One property of the planetary system that is not mirrored in the satellite systems is the ratio of angular momentum for the primary spin to that contained in the secondary orbits. As we have seen earlier the total magnitude of the orbital angular momentum of the major planets is more than 200 times that of the solar spin. By contrast, the spin of Jupiter has angular momentum 100 times that of the orbital value for the Galileans. Two major factors contribute to this difference.

(1) The major planets are 200 times more distant from the Sun in terms of primary radius. Thus the Galileans orbit Jupiter at 5.9–26.4 Jovian radii whereas the major planets orbit the Sun at between 1120 and 6470 solar radii.

(2) Jupiter rotates very rapidly compared to the Sun; its angular rate is 60 times greater.

The mass ratios in the two systems are more similar being 600:1 for the Sun–planets and 4000:1 for Jupiter–Galileans. The other satellite systems have similar ratios.

None of the other satellites of Jupiter is comparable to the Galileans, the next largest being Amalthea, an irregularly shaped body with dimensions 270 km × 170 km × 150 km. Two groups of outer satellites, one in direct and the other in retrograde orbits (inclinations greater than 90°—see table 1.4), are probably captured debris from some event in the vicinity of Jupiter.

The Galilean satellites are remarkable for the diversity of their surface structures. Io was the first body, other than the Earth, to show volcanic activity. Just before Voyager 1 reached the Jovian system it was predicted that volcanoes should be active on Io (Peale *et al* 1979) and, indeed, a volcanic plume was imaged by the approaching spacecraft. Altogether eight volcanoes have been observed on the satellite. The basis of the prediction was the 2:1 ratio of the periods of Io and Europa. Because of this the nearest approach, and hence the maximum perturbation, of Io by Europa is always at the same point of Io's orbit. Thus Io's orbit is not quite circular ($e \sim 0.0001$) and because of its proximity to Jupiter it undergoes a periodic tidal stress. Hysteresis converts some of the energy involved in this alternating stretching and compression into heat and it is estimated that the resultant energy generation in Io amounts to about 10^{13} W. It is this energy that drives the volcanism. The orange-yellow colour of much of Io's surface is due to sulphur and sulphur dioxide emission from the volcanoes. Since the surface is constantly being renewed there is no evidence of bombardment of the surface.

Europa, the next Galilean satellite, has a very different appearance. Its density is a little less than that of the Moon and it is the only Galilean satellite less massive than the Moon. The surface is extremely smooth, again showing little evidence of bombardment damage and indicating an active surface. It is covered with ice, which cannot be very thick if the density is taken into account. It is thought that water, in liquid form just below the surface of Europa, is occasionally released and then freezes, covering any underlying surface features. Although it is further from Jupiter than Io it may also have an input of tidal energy. This could contribute to the heat required to produce the liquid water. At the same time the tidal flexing could also give rise to the very distinctive cracked surface of Europa.

Ganymede, the next satellite outwards, is larger than Mercury and the largest and most massive satellite in the Solar System, having twice the mass of the Moon. It has an icy surface layer, perhaps 100 km thick, below which there is a much thicker layer of water or mushy ice. Older well-cratered regions are dark in colour, probably due to an undisturbed layer of dust from meteorites. There are also younger and brighter regions characterized by bundles of parallel grooves.

The final Galilean satellite, Callisto, has a thick icy crust that is dark and shows a large number of impact features. There is a very large 'bulls-eye' feature, Valhalla, in the form of a series of concentric rings. This is similar to the Orientale feature on the Moon and is certainly due to a very large impact.

Before Voyager I reached Jupiter in 1979 the ring system of Uranus had been detected from Earth observation and, with two known ring systems, there was interest in seeing if Jupiter also had a ring. A single thin ring was discovered— which then raised the possibility that rings were a universal feature of major planets and that Neptune too would have rings.

1.4.3   The Saturnian system

With 18 members identified so far Saturn has the most heavily populated satellite system (table 1.6). Only Titan, slightly larger than Mercury, matches the Galileans but four others—Tethys, Dione, Rhea and Iapetus—have diameters greater than 1000 km. The system has a number of striking commensurabilities (Roy 1977) with both Enceladus–Dione and Mimas–Tethys having mean motions in the ratio 2:1. In addition the 4:3 ratio for Titan–Hyperion ensures that this pair have conjunctions near the aposaturnium (furthest orbital point from Saturn) of Hyperion. The smallest separation of these two bodies is thus about 400 000 km rather than the 100 000 km implied by a simple consideration of the sizes of the two orbits

Spacecraft discoveries of smaller satellites show a number of 1:1 commensurabilities which are really examples of special solutions in the restricted three-body problem. It is well known that general solutions of the gravitational problem of n (≥ 3) bodies do not exist. However, Lagrange (1736–1813) showed that special configurations of three bodies do satisfy the equations of motion. These involve collinear and equilateral triangular arrangements of the bodies, as illustrated in figure 1.9, in which two of the bodies are placed at the points A and B and the third (C) can occupy one of the five points, L_1 to L_5, known as the Lagrange points. The whole system must rotate about the centre of mass. Generally these solutions are unstable and any small displacement will rapidly destroy the symmetry. Since no three-body system can properly be isolated from the perturbing effects of other bodies, this suggests that the Lagrange solutions cannot be achieved in practice. In certain restricted conditions the triangular solutions are stable; they require the third body, C, to be of negligible mass and for the ratio of the masses of A and B to exceed 25. In such cases small displacements of body C from L_4 and L_5 do not become unbounded and, of course, A and B execute two-body motion. The conditions are satisfied by Saturn–Tethys–Calypso, Saturn–Tethys–Telesto and also by Saturn–Dione–Dione B. Effectively Calypso and Telesto move in the same orbit as Tethys (hence the 1:1 commensurability) but maintain a position on average 60° in front and 60° behind Tethys in its orbit.

Saturn has a single very large satellite, Titan, which is very little below

Table 1.6. The satellite system of Saturn. The spin period of Saturn $= 10$ hr 14 min.

Satellite	Inclination of orbit to equator	Semi-major axis (10^3 km)	Eccentricity	Mass (10^{22} kg)	Average diameter (km)	Density (10^3 kg m^{-3})
Pan		134			20	
Atlas	0.3°	138	0.002		60	
Prometheus	0.0°	139	0.003		110	
Pandora	0.05°	142	0.004		90	
Epimetheus	0.1°	151	0.007		200	
Janus	0.3°	151	0.009		60	
Mimas	1.5°	186	0.020		390	1.2
Enceladus	0.0°	238	0.005		510	1.1
Tethys	1.9°	295	0.000	0.07	1060	1.0
Telesto		295			30	
Calypso		295			25	
Dione	0.0°	378	0.002	0.105	1120	1.4
Dione B		378			30	
Rhea	0.4°	527	0.001	0.250	1530	1.3
Titan	0.3°	1 222	0.029	14.22	5150	1.9
Hyperion	0.4°	1 483	0.104		280	1.9
Iapetus	14.7°	3 560	0.028	0.188	1440	1.2
Phoebe	159°	12 950	0.163		220	

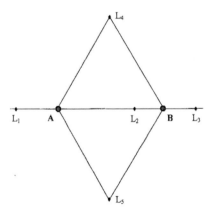

Figure 1.9. The Lagrange solutions for the three-body problem.

Ganymede both in mass and diameter. No features of the surface are visible because the satellite has a very thick atmosphere, 90% N_2 with most of the remainder Ar plus a little CH_4. The opaque clouds consist of hydrocarbon droplets. The surface pressure of Titan's atmosphere is 1.6 bar, greater than that of the Earth, but the *column mass*, the mass of atmosphere per unit area of surface, is ten times the Earth value.

The other satellites of Saturn have well-cratered icy surfaces and are obviously quite old. Mimas has a very large crater, Herschel, with a diameter almost one-third that of the satellite. It is clearly the scar of an impact that must have been close to destroying the satellite. Iapetus, in an extended orbit with semimajor axis more than 3.5×10^6 km, has a leading hemisphere that is much darker than the trailing hemisphere. The fractions of reflected light, the *albedoes*, for the two sides are 0.05 and 0.50 respectively. The reason for this difference has been the subject of much debate. Iapetus is certainly an icy satellite so the dark region must be due to something covering the ice, which is naturally white and bright. The most favoured explanation is that the dark material has come from the interior of Iapetus, although the nature of this material is very uncertain.

The outermost satellite of Saturn, Phoebe, is quite small and was not well observed by either Voyager 1 or Voyager 2. Its main claim to fame is its retrograde orbit that suggests that it is almost certainly a captured object.

The ring system of Saturn is one of the most striking and structurally interesting features of the Solar System. Seen from the Earth there are several prominent bands and divisions, notably the Cassini division, but imaged by Voyager 1 (figure 1.10(*a*)) the ring structure is seen to be very complex. The general structure, seen in figure 1.10(*b*), has divisions between various rings that correspond to orbits commensurate with the more massive inner satellites. The broad Cassini division corresponds to a period one-half that of Mimas, one-third that of Enceladus and one-quarter that of Tethys. The division between the B and C rings corresponds to one-third of the period of Mimas while the Encke division corresponds to three-fifths of Mimas' period. When the orbit of a particle is commensurate with one of the more massive inner satellites it tends to receive a perturbing kick at the same point or points in its orbit which reinforces the disturbance until the period changes to non-commensurability.

The F-ring has a peculiar braided structure that, at first, seemed inconsistent with the mechanics of a Keplerian orbit. However, the particles in this ring are influenced by the so-called *shepherd satellites*, Prometheus and Pandora, the positions of which bracket the ring. Not only do these satellites cause the non-Keplerian motions in the ring but they also lead to stability of the F-ring. A particle just inside the orbit of Pandora will overtake the satellite and be perturbed into an orbit just outside Pandora. It is then overtaken by Pandora and perturbed into an orbit inside Pandora and so on. Prometheus exerts a similar influence on the particles in its vicinity and so the satellites 'shepherd' the particles and keep them within the F-ring region.

1.4.4   Satellites of Uranus and Neptune

Uranus spins less rapidly than Jupiter and Saturn and its equator is inclined at 98° to its orbital plane, thus making its spin retrograde. The 15 known satellites, shown in table 1.7, all have orbits near the equatorial plane. The five outermost satellites, including four with diameters over 1000 km, were known from tele-

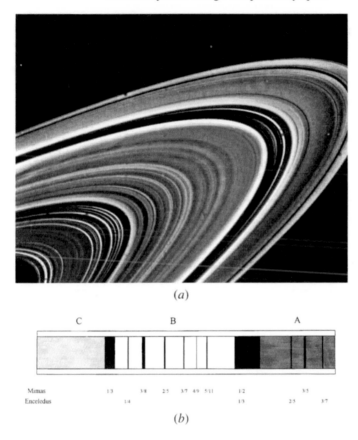

(*a*)

| C | B | A |

| Mimas | 1:3 | 3:8 | 2:5 | 3:7 4:9 5:11 | 1:2 | 3:5 |
| Enceledus | | 1:4 | | | 1:3 | 2:5 | 3:7 |

(*b*)

Figure 1.10. (*a*) A close-up view of Saturn's rings from Voyager 1. (*b*) A representation of the major divisions in Saturn's rings showing their commensurabilities with the periods of Mimas and Enceladus.

scope observation from Earth, the others being found from spacecraft. They are icy bodies with old cratered surfaces.

In 1977, during the observation of a stellar occultation by Uranus, several extinctions of light from the star were observed which were interpreted as being due to the existence of a system of five rings. Later occultation observations showed the presence of four further rings. These rings are very narrow, varying in width from a few kilometres to about 100 km.

Prior to observation by the two Voyager spacecraft only two satellites were known for Neptune but they were both rather remarkable. One of them, Titan, the seventh most massive satellite in the Solar System with a mass just under one-third that of the Moon, is in a close, perfectly circular but retrograde orbit. The other is Nereid in a direct but very extended orbit, which has the distinction of being the most eccentric in the Solar System for a satellite, with $e = 0.749$.

Table 1.7. The satellite system of Uranus. The spin period of Uranus = 17 hr 14 min.

Satellite	Inclination of orbit to equator	Semi-major axis (10^3 km)	Eccentricity	Mass (10^{22} kg)	Average diameter (km)	Density (10^3 kg m^{-3})
Nine small satellites		50–73			40–80	
Puck	0.0°	86	0.000		170	
Miranda	0.0°	130	0.000	0.0075	484	1.26
Ariel	0.0°	191	0.003	0.14	1160	1.65
Umbriel	0.0°	266	0.003	0.13	1190	1.44
Titania	0.0°	436	0.002	0.35	1610	1.59
Oberon	0.0°	583	0.007	0.29	1550	1.50

Table 1.8. The satellite system of Neptune. The spin period of Neptune = 16 hr 7 min.

Satellite	Inclination of orbit to equator	Semi-major axis (10^3 km)	Eccentricity	Mass (10^{22} kg)	Average diameter (km)	Density (10^3 kg m^{-3})
Niaid		48			60	
Thalassa		50			80	
Despoina		53			150	
Galatea		62			160	
Larissa		74			200	
Proteus		118			415	
Triton	160°	355	0.000	2.21	2705	2.07
Nereid	27.7°	5513	0.749	0.0021	340	

These two, plus another six found by spacecraft observation are listed in table 1.8.

Stellar occultation observations had indicated that Neptune should have a ring system and, indeed, these were seen and imaged by the Voyager spacecraft. The Earth-bound measurements had suggested that the rings were only partial but it turns out that they are complete but have a rather lumpy structure.

The presence of rings accompanying each of the major planets suggests that there is some common cause associated with their characteristics as large bodies with many satellite companions. The most likely origin for a ring system is that the orbit of a small orbiting satellite decayed to the extent that it strayed within the Roche limit (section 4.4.2). It would then have been tidally disrupted by the planet to give a vast number of small fragments that would have spread out to form the rings. Structure in the rings could then be produced by resonant perturbations by some of the inner satellites as described in section 1.4.3.

Table 1.9. The satellite systems of Mars and Pluto. The spin period of Mars = 24 hr 37 min; that of Pluto = 6.39 days.

Planet Satellite	Inclination of orbit to equator	Semi-major axis (10^3 km)	Eccentricity	Mass (10^{22} kg)	Average diameter (km)	Density (10^3 kg m^{-3})
Mars						
Phobos	1.1°	9.27	0.0210		23	
Deimos	1.8°	23.4	0.0028		12	
Pluto						
Charon	0.0°	19.6			1186	

1.4.5 Spins and satellites of Mercury, Venus, Mars and Pluto

Mercury and Venus have no satellites and they also happen to be the planets with the slowest spin rates in the Solar System. Mercury's spin rate, 58.64 days, is exactly two-thirds of its orbital period and this is consistent with its proximity to the Sun and the concomitant tidal forces. Since Mercury spins one and a half times every orbital period it presents the same face to the Sun every alternate perihelion passage. In the intervening perihelion passages it presents the opposite face to the Sun.

The rotation of Venus is retrograde with a period of 243 days which differs from the orbital period of 224.7 days. This combination of spin and orbital periods does have the curious result that Venus presents almost the same face to the Earth at each *inferior conjunction*, that is at closest approach of Venus and the Earth. This relationship is not an exact one and, since tidal effects between Venus and the Earth are negligible, must be regarded as purely fortuitous. The very slow spin of Venus marks it as a curiosity in the Solar System. No known evolutionary process would lead to this condition from a primitive fast spin such as that possessed by the Earth (McCue *et al* 1992).

The two satellites of Mars, Phobos and Deimos, are both small, of irregular shape and very close to the planet (table 1.9). Their appearance is similar to that of asteroids so they are usually regarded as captured bodies. However, their orbital inclinations and eccentricities are small, characteristics usually indicative of regular satellites.

Charon, the satellite of Pluto, has the distinction of being the largest and most massive satellite in relation to its primary. Its orbital and spin periods are both the same as the spin period of Pluto so the pair of bodies rotate about the centre of mass as a rigid system.

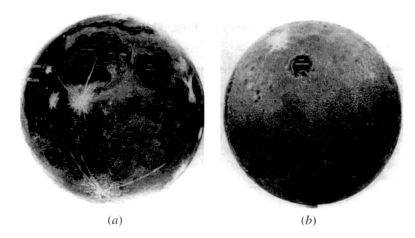

(a) (b)

Figure 1.11. A Moon-globe showing (a) the near-side (b) the far-side.

1.4.6   The Earth–Moon system

The Moon is the fifth most massive satellite in the Solar System. With a mass of 7.35×10^{22} kg and a diameter 3476 km it slots between Io and Europa, the innermost Galilean satellites, in both mass and density. However, while in its characteristics it is a normal large satellite, its association with a terrestrial planet clearly makes it anomalous and an explanation of the existence of the Earth–Moon system is a requirement of any well-developed cosmogonic theory.

1.4.6.1   Surface features of the Moon

The Moon has been examined in more detail than any body, other than the Earth, in the Solar System. It has been studied by telescopes from Earth for nearly 400 years, has been the subject of manned exploration, in the Apollo missions, and also exploration by automated vehicles designed to collect particular kinds of information.

The side of the Moon facing the Earth shows the full range of lunar features (figure 1.11(a)). There are two general types of terrain—the *highlands* and the *mare basins*. The highland regions consist of low-density heavily-cratered old crust. The mare basins are the result of large projectiles having struck the Moon and excavated large basins. These then were filled up from below by molten material by successive bouts of volcanism lasting over several hundred million years. Eventually the molten material retreated into the interior of the Moon and was no longer able to reach the surface. From radioactive dating of the mare basalts it appears that the main episodes of volcanism were between about 3.96×10^{9} and 3.16×10^{9} years ago. The mare regions show comparatively few craters, since they were excavated after the period of early bombardment by smaller bodies,

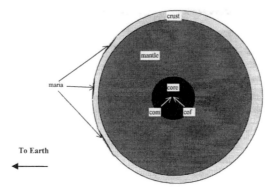

Figure 1.12. A schematic cross-section of the Moon showing the difference of crust thickness on the two sides and the centre-of-mass (com)–centre-of-figure (cof) offset (exaggerated).

which characterized the early Solar System, and all early damage was obliterated. Also present on the near side are rays, radial splashes of material thrown out of craters. The ejected material, when fresh, appears bright as is seen from the rays coming from the crater Copernicus. Due to the effect of the solar wind and covering by darker meteorite material the rays become less prominent as they age. There are also crack-like features known as *rills* probably due to a variety of causes but some of which may have been produced by flows of lava.

When the Soviet Lunik spacecraft photographed the far side of the Moon in 1959 it was found that it was quite different in appearance from the side facing the Earth. The face consisted almost completely of highland regions although there were some very small regions that could be designated as maria (figure 1.11(*b*)). This observation of the Moon's hemispherical asymmetry became an important Solar System problem. Altimetry measurements revealed that the cause of the hemispherical asymmetry was not due to asymmetric bombardment. The lunar far side showed several very large basins but these had not been filled by molten material from below. It is accepted from this, and some seismic evidence, that the crust on the far side of the Moon is between 25–40 km thicker than on the near side so that the molten material was much further from the surface when the basins were formed. This conclusion is also supported by the fact that the centre of mass of the Moon is displaced from the centre of figure by about 2.5 km towards the Earth due to the less thick, low density crust on the near side (figure 1.12).

1.4.6.2   *The mineralogy and composition of the Moon*

Highland rocks are all *igneous*, which is to say that they are formed by the crystallization of molten rock. The crystals in the rocks are large in size, so that the rocks are coarse-grained, which indicates that the highland rocks cooled slowly.

This contrasts with the maria basalts which cooled quickly and are fine-grained because the crystals had little time to grow. Like the maria material the highland rocks contain particulate iron and are also deficient in water and volatile elements but in terms of chemistry and mineral compositions the two types of material are very different. The main metallic components of the dark lava are iron, magnesium and titanium while the lighter-coloured highland material is rich in aluminium and calcium. More than 50% of highland rocks are *plagioclase*, a mixture of *albite* ($NaAlSi_3O_8$) and *anorthite* ($CaAl_2Si_2O_8$), with varying amounts of *pyroxene* $(Mg, Fe, Ca)SiO_3$, *olivine* $(Mg, Fe)_2SiO_4$ and some *spinel*, a metallic oxide. The lower-density crust material comes from differentiation of the bulk Moon as a result of large-scale melting of surface material early in the Moon's history.

The common minerals in the lunar basalt are clinopyroxene, a calcium-rich form of pyroxene and anorthite-rich plagioclase. There can also be up to 20% olivine but in most basalt it is absent.

The ages of the highland rocks, that is from the time they became closed systems, have been deduced from radioactive dating. They are usually in the range $4.0–4.2 \times 10^9$ years except for one Apollo 17 sample which has an age of 4.6×10^9 years, close to the accepted age of the Solar System and of the Moon itself. There seems to have been a 400 million year period when either rocks did not form on the surface or during which the rocks which had formed were being destroyed in some way. The ages of lunar basalts vary from 3.16 to 3.96×10^9 years, which shows that volcanism occurred on the Moon at least during that period. However, since older material gets covered by newer it is also possible that volcanism could have been earlier, even as far back as the origin of the Moon itself.

The surface of the Moon is covered by a thick blanket of pulverized material, described as lunar soil. A component of the lunar soil is called KREEP on account of its high component of potassium (K) rare-earth elements (REE) and phosphorus (P). It also contains more rubidium, thorium and uranium than is found in other lunar rocks. The majority of KREEP material is found in the vicinity of Mare Imbrium and could be material excavated from 25–50 km below the surface when the basin was formed.

A characteristic of the total surface is the general deficiency of volatile elements compared to the Earth. This is illustrated in figure 1.13 which shows the abundance of various elements relative to the Earth as a function of their condensation temperatures. It can be seen that there is a general trend for a lesser fractional abundance of the more volatile elements in the Moon with a balancing greater abundance of the more refractory materials. The discovery of small water-ice deposits in some well-sheltered parts of the Moon in 1998 goes against this trend. However, this ice was probably deposited by comet impacts long after the Moon's surface had become cool.

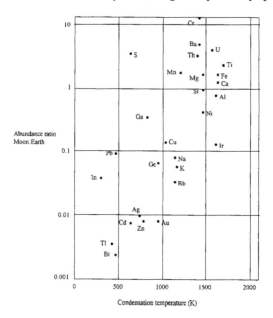

Figure 1.13. The relative abundance of elements on the Moon and Earth related to the elemental condensation temperatures.

1.4.6.3   Tides and the Earth–Moon system

The Moon, and its relationship to the Earth, has been studied for a very long time and the system provides a good illustration of the tidal mechanism which governs the behaviour of other bodies in the Solar System. It is the one system for which there exists direct evidence of its evolution.

Tides are produced in any extended body by a non-uniform gravitational field, such as might arise from a companion body. Thus the Moon produces tidal effects on the Earth and the Earth produces tidal effects on the Moon. In addition the Earth experiences significant tidal effects due to the Sun. However, the lunar tidal field exceeds that of the Sun by a factor greater than two because the much smaller mass of the Moon is more than compensated by its much greater proximity to the Earth. In crude terms the attractive force of the Moon at the sub-lunar point A (figure 1.14) is greater than at C, the centre of the Earth, which, in its turn, is greater than the lunar force on the opposite side of the Moon at B. The net effect is to produce a stretching force along the Earth–Moon direction AB. Perpendicular to this direction it is clear that the attractive forces of the Moon at D and E have inwards components towards C thus giving a compressive force perpendicular to AB.

Jeans (1929) gave a lucid description of the tidal phenomenon. The gravitational potential at the point $P(x, y, z)$ due to the two masses $M_\oplus$ and m is

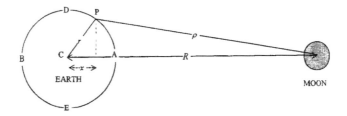

Figure 1.14. The geometry of the Earth–Moon system.

$-G(M_\oplus/r + m/\rho)$, if the bodies can be regarded as spherically symmetric. The tidal effect is due to the Moon, mass m, but part of its potential gives an acceleration of magnitude Gm/R^2 to the Earth. This arises from a force field with corresponding potential Gmx/R^2 where x is the coordinate relative to the centre of the Earth in the Earth–Moon direction. The effective tide-generating potential is thus

$$V = -Gm \left(\frac{1}{\rho} - \frac{x}{R^2} \right). \tag{1.6}$$

The components of the tidal acceleration at P are given by the partial derivatives

$$
\begin{aligned}
(\ddot{x}, \ddot{y}, \ddot{z}) &= \left(-\frac{\partial V}{\partial x}, -\frac{\partial V}{\partial y}, -\frac{\partial V}{\partial z} \right) \\
&= \left(Gm \left[\frac{R-x}{\rho^3} - \frac{1}{R^2} \right], -\frac{Gmy}{\rho^3}, -\frac{Gmz}{\rho^3} \right). \tag{1.7}
\end{aligned}
$$

Assuming that $R \gg r$ the accelerations at points A, B and D are

$$(\ddot{x}, \ddot{y}, \ddot{z})_A = \left(\frac{2Gmr}{R^3}, 0, 0 \right) \tag{1.8a}$$

$$(\ddot{x}, \ddot{y}, \ddot{z})_B = \left(-\frac{2Gmr}{R^3}, 0, 0 \right) \tag{1.8b}$$

$$(\ddot{x}, \ddot{y}, \ddot{z})_D = \left(0, -\frac{Gmr}{R^3}, 0 \right). \tag{1.8c}$$

The tidal forces over a polar section of the Earth are shown in figure 1.15.

The observed physical effect of this tidal force is the semi-diurnal rise and fall of sea level in the oceans. There are two high tides per day because the tidal forces cause bulges on opposite sides of the Earth rather than just on the side facing the Moon. The difference in sea levels in mid-ocean is around 1 m but this can be amplified considerably by coastal effects. The highest tides, *spring tides*, are experienced fortnightly when the Sun and the Moon are in line with the Earth but when the Earth–Sun and Earth–Moon directions are perpendicular the effects are subtracted giving the so-called *neap tides*.

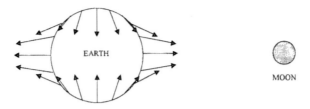

Figure 1.15. Tidal forces on the Earth due to the Moon.

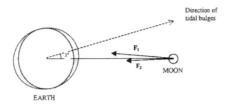

Figure 1.16. The net effect of the forces on the Moon due to the near and far tidal bulges, F_1 and F_2 respectively, increases the angular momentum of the Moon's orbit.

These short-period phenomena have been known for centuries but the tides have a slow secular effect on the Earth's spin. Since the Earth cannot react instantaneously to the tidal stress the tidal bulge is not a maximum at the sub-lunar point. With the Earth having a greater angular spin rate than the angular speed of the Moon in its orbit the maximum tide occurs approximately 3° ahead of the Moon (figure 1.16). This tidal bulge, which gives a departure from the spherical symmetry of the Earth, produces a force on the Moon with a tangential component in the direction of its motion that has the effect of increasing the orbital angular momentum of the Moon. This is partially offset by the gravitational effect of the advanced tide on the far side, which is less effective because it is further from the Moon. The net increase in the angular momentum of the Moon in its orbit is balanced by a reduction in angular momentum, due to a decrease in the rate of the Earth's spin. At the present time this effect gives an increase in the length of the day by 1.6 ms per century. Eventually the position will be reached when the lunar month and the day are of equal duration, about 50 present days. When that happens the high tide will be at the sub-lunar point, the force on the Moon due to the tidal bulges will be centrally directed and further dynamical evolution will cease. Actually, due to the effect of the Sun, that has been ignored in our argument, the situation will be rather more complicated.

There are several other effects on both the Earth and the Moon due to tidal interactions. The Earth raises a tide on the Moon and, if the Moon's spin had not been synchronous with the orbit, the gravitational attraction of the Earth on the Moon's tidal bulge would act in a sense to produce synchronization. This

Table 1.10. Characteristics of some important asteroids. The ones given are: a, the semi-major axis; e, the eccentricity; i, the inclination; and D, the diameter (U = unknown).

Name	Year of discovery	a (AU)	e	i (°)	D (km)
Ceres	1801	2.75	0.079	10.6	1003
Pallas	1802	2.77	0.237	34.9	608
Juno	1804	2.67	0.257	13.0	250
Vesta	1807	2.58	0.089	7.1	538
Hygeia	1849	3.15	0.100	3.8	450
Undina	1867	3.20	0.072	9.9	250
Eros	1898	1.46	0.223	10.8	U
Hildago	1920	5.81	0.657	42.5	15
Apollo	1932	1.47	0.566	6.4	U
Icarus	1949	1.08	0.827	22.9	~2
Chiron	1977	13.50	0.378	6.9	U

synchronization is found for all the satellites in the Solar System. Another effect is to cause a precession of the spin axis of the Earth. The spin of the Earth does not contribute to tides but it does create an equatorial bulge distorting it from spherical symmetry. The equatorial radius of the Earth is about 22 km greater than the polar radius. The spinning Earth acts like a gyroscope and the Moon exerts a torque on it due to the differential pull on the near and far regions of the equatorial bulge. It is this torque which gives a spin-axis precession period of about 26 000 years.

1.5   Asteroids

It was seen in table 1.2 that to fill a gap in Bode's law it was necessary to introduce a body between Mars and Jupiter. The small body Ceres, discovered in 1801, filled the gap admirably but the pattern was made more complicated by the discovery of many other bodies, all smaller than Ceres, in the same region of the Solar System. These bodies, called asteroids, raise many questions as to their origin but here we shall just consider their characteristics.

1.5.1   Characteristics of the major asteroids

In table 1.10 the data for a number of asteroids, chosen because they span the period from the first discovery to recent times and because they illustrate different orbital characteristics, are shown.

The sizes of asteroids are best measured by stellar occultation. By timing an occultation and knowing the orbit of the asteroid it is possible to obtain quite a precise estimate of the distance across the asteroid along the line of the star's motion across it.

Most asteroid orbits lie in the region between Mars and Jupiter. All known asteroid orbits are prograde, which is to say that they orbit the Sun in the same sense as do the planets, and most orbits have eccentricities less than 0.3 and inclinations less than 25°. Some notable exceptions to these general rules are shown in table 1.10. For example, Hildago has an inclination of 42.5° but a few asteroids have even larger inclinations up to 64°. The eccentricity of Icarus, the largest for any known asteroid, combined with its small semi-major axis, gives a perihelion distance of 0.19 AU, the closest approach of any asteroid to the Sun. It has an Earth-crossing orbit. Apollo was the first observed asteroid to have this characteristic. In the year of its discovery, 1932, Apollo came within three million kilometres of the Earth—just seven to eight times the distance of the Moon. Since then tens of other small asteroids, with diameters not more than a few kilometres, have been discovered with Earth-crossing orbits and these are known collectively as the *Apollo asteroids*. Another class of asteroids is the so-called *Aten* group with orbits that lie mostly within that of the Earth. Very few of these bodies are known and they are small but the known ones could be representatives of a much larger population which stay well within the Earth's orbit. Some Aten and Apollo asteroids have a theoretical possibility of striking the Earth and it is possible that in its long history the Earth has undergone collisions by asteroids from time to time. It has been postulated that an asteroid collision about 65 million years ago led to the demise of the dinosaurs, which became extinct within a short period having been the dominant living species on Earth for hundreds of millions of years.

A number of asteroids are Mars-crossing and have perihelia outside the Earth's orbit. The first such to be discovered was Eros, in 1898, but several more are now known. In a favourable conjunction Eros can get to within 23 million kilometres of the Earth and in such an approach in 1975 it was studied by radar and found to have a rough surface. As for most other small asteroids it is of irregular shape and it is somewhat elongated with a maximum dimension of about 25 km.

Two interesting groups of asteroids are the *Trojans* that move more-or-less in Jupiter's orbit, one group following Jupiter and 60° behind it and the other group leading Jupiter and 60° ahead of it. The dynamics of the Trojan asteroid configuration was discussed in section 1.4.3 in relation to Saturn's satellites.

It was originally thought that all asteroids were all confined to the region between Mars and Jupiter. The existence of the Apollo and Aten asteroid groups made it clear that this belief was not true. In 1977 the discovery of Chiron, which mostly moves in the region between Saturn and Uranus, raised the possibility that there were other families of asteroids in the outer Solar System that were too small to be observed from Earth. The diameter of Chiron is unknown but it must be at least 100 km to account for its observed brightness and may be much larger. At perihelion it crosses Saturn's orbit and in the 17th century it came within 16 million kilometres of that planet—not far beyond the orbit of Phoebe,

Saturn's outer retrograde satellite with an orbital radius of just under 13 million kilometres.

Several tens of other bodies have been detected, with estimated diameters in the range 150–360 km, which are close to, or further out than, the orbit of Neptune. These are *Kuiper-belt objects* and they are usually considered to be linked to comets rather than asteroids. There is uncertainty about the relationship between asteroids and comets—whether they are two manifestations of some common source of material or represent two different types of material with completely different origins. All the objects described as asteroids are in direct orbits and mostly have moderate inclinations and eccentricities. By contrast comets are frequently in retrograde orbits, have a wide range of inclinations and eccentricities and also have a considerable content of volatile material. They are also associated with regions well outside that occupied by the planets. Ideas about the origin of various types of object in the Solar System are heavily linked with ideas of the origin of the system itself.

1.5.2   The distribution of asteroid orbits: Kirkwood gaps

A diagram giving the frequency of asteroid periods, such as figure 1.17, indicates that the distribution has prominent gaps. These were first explained by the American astronomer, Daniel Kirkwood, in 1866. He pointed out that the two very prominent gaps, marked A and B, correspond to one-third and one-half the period of Jupiter, and that these gaps were a manifestation of some resonance phenomenon. For example, an asteroid with one-half the period of Jupiter will make two complete orbits while Jupiter is making one. Thus the two bodies will always be closest in the same region of the asteroid's orbit so that the perturbation by Jupiter at closest approach will always be modifying the asteroid orbit in the same direction. The asteroid's period will steadily change in one direction until the asteroid and Jupiter are sufficiently out of resonance for the nearest approaches, and hence maximum perturbations, to occur all round the asteroid's orbit with much diminished effect. Similarly, for the one-third resonance, the asteroid is perturbed at two points on opposite sides of its orbit. Other *Kirkwood gaps* at two-fifths and three-sevenths of Jupiter's period are also evident in figure 1.17. However, to illustrate the complexity of the resonance process there is a small *concentration* of asteroid orbits corresponding to two-thirds of Jupiter's period. The Kirkwood gap phenomenon is dynamically related to the formation of the gaps in Saturn's rings due to perturbation by the inner satellites Mimas and Enceladus.

1.5.3   The compositions of asteroids

Spacecraft observations of asteroids have given new information concerning their structure and composition. A near passage of the asteroid *Gaspra* by the Galileo spacecraft gave the very detailed photograph shown in figure 1.18. It has dimen-

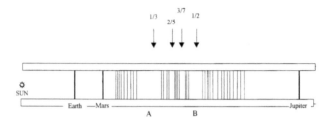

Figure 1.17. A schematic representation of asteroid orbits showing the gaps corresponding to commensurabilities with Jupiter's orbital period.

Figure 1.18. The asteroid Gaspra taken by the Galileo spacecraft on its way to Jupiter.

sions 11 km × 12 km × 10 km, is covered with craters and seems to be overlaid with rocky dust. In common with most other asteroids it is in a tumbling motion with a period of about 4 hr. It resembles the Martian satellites, Phobos and Deimos, that have long been thought to be captured asteroids. The pockmarked appearance of Gaspra, and of the Martian satellites, suggests that collisions involving asteroids take place and such collisions are almost certainly the source of most of the material that reaches the Earth in the form of meteorites.

Information about asteroid composition comes from visible and near-infrared spectroscopy. Reflectance spectra have been measured for many hundreds of asteroids in visible light and in the infrared range up to 1.07 μm and these spectra have been matched with those measured for meteorites in the laboratory. The main types of meteorite are *stones*, consisting mainly of various types of silicate, *irons* that are mostly iron with some nickel and *stony-irons* containing intimate mixtures of stone and iron regions. Within the stony classification is an important subclass, the *carbonaceous chondrites*, which are very dark in appearance and contain volatile materials. The match between spectra from various asteroids and meteorites clearly indicate the relationship between the two classes of object;

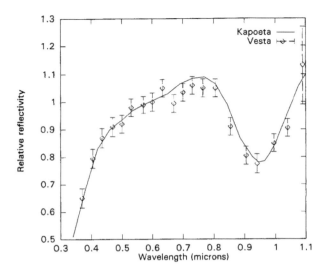

Figure 1.19. Comparison of the reflectivity of the calcium-rich achondrite meteorite Kapoeta and the asteroid Vesta.

an example of a match is shown in figure 1.19. On the basis of their spectral characteristics asteroids have been divided into six types. The two most common types, which together account for 80% of the spectrally observed asteroids, are designated as C, associated with carbonaceous chondrite material, and S, mostly associated with stony irons. These asteroid types occur at all distances within the main belt of asteroids between Mars and Jupiter but there is a distinct tendency for the C type asteroids to have larger orbital radii.

There seems little doubt on the basis of the observational evidence that the study of meteorites is also tantamount to the study of asteroids. The important question then is the way in which asteroids are related to planets because, if they are intimately related, the information from laboratory meteorite studies could be directly applied to the problem of the origin and evolution of the planets. There is no consensus on the form of this relationship. It was early thought that asteroids were the debris from a broken planet but this raised two difficulties. The first concerned the source of energy which could break up a planet and the second that of disposing of the planetary material since the total mass of known asteroids is much less than a lunar mass. A second theory, which assumes planetary formation by an accumulation of asteroid-sized objects, claims that Jupiter exerts considerable influence on objects in the asteroid-belt region and so prevents their accumulation by a continual stirring process. However, if there had been enough material in the asteroid region to produce a planet then this again raises the question of disposal. Certainly all observed solid bodies in the Solar System show signs of damage by large projectiles so many former asteroids can be accounted for in this way. Other projectiles could have been swept up by the major planets without leaving any visible evidence of their former existence.

Acceptance of the relationship between asteroids and meteorites throws up many interesting problems. Some meteorite material has come from bodies which were molten and in which segregation of material by density had taken place. An asteroid a few hundred kilometres in radius, formed by the accumulation of smaller bodies, would not release sufficient gravitational energy to melt silicates or metals. If the specific heat capacity of the material is c with latent heat of fusion, L, and it had to be raised by $\Delta\theta$ to bring it to melting point then the size of body for which gravitational energy would just produce complete melting is given by

$$\frac{3GM^2}{5R} = M(c\Delta\theta + L) \tag{1.9}$$

in which the left-hand side is the negative of the self-gravitational energy of a uniform sphere of mass M and radius R. Expressing mass in terms of density, ρ, and radius and rearranging one finds

$$R = \left\{ 1.25\frac{c\Delta\theta + L}{\pi G\rho} \right\}^{1/2}. \tag{1.10}$$

Inserting reasonable values, $\rho = 3400$ kg m^{-3}, $c = 1100$ J kg^{-1} K^{-1}, $L = 5 \times 10^5$ J kg^{-1} and $\Delta\theta = 1000$ K we find $R = 1675$ km.

For asteroids to have been melted some other source of heating must have been available. There is evidence in some meteorites for the one-time presence of a radioactive isotope of aluminium, ^{26}Al, which has a half-life of 720 000 years. If about two parts in 10^5 of the aluminium of the minerals in asteroids had been ^{26}Al then this would have been enough to melt asteroids with diameters as small as 10 km.

1.6 Meteorites

Somewhere between 100 and 1000 tonne of meteoritic material strikes the Earth per day. This amounts to one part in 10^7 of the Earth's mass over the lifetime of the Solar System and would cover the Earth uniformly with a 10 cm thick layer of material. The material is in a form ranging from fine dust to objects of kilometre size and must enter the atmosphere at more than 11 km s^{-1} (the escape speed from the Earth). Larger objects are decelerated by atmospheric friction but still strike the ground at high speed. Their passage through the atmosphere causes surface material to melt and a fusion crust to form. They may also fragment and form a shower of smaller objects. By contrast very tiny objects may survive the passage to Earth almost intact. Because of the large surface-to-volume ratio they radiate heat very efficiently, are quickly braked by the atmosphere and then gently drift down to Earth.

Table 1.11. Numbers of falls and finds up to the end of 1975.

	Falls	Finds	Totals
Stones	791	593	1384
Irons	46	610	656
Stony-irons	11	67	78

The largest meteorites found have masses up to about 30 tonne, most of them being of iron. The largest known stone meteorite that fell in Jilin, China, in 1976 had a mass of 1.76 tonne. Judging by what is observed on the Moon, much larger objects must have struck the Earth from time to time. As previously mentioned in section 1.5.1, a few-kilometre-size object may have fallen to Earth at the end of the Cretaceous period, some 65 million years ago. Marine clays deposited at that time have a high iridium content and iridium is a much more common element in meteorites than it is on Earth. More positive evidence for the fall of larger bodies can be seen in the craters that exist in various parts of the Earth. The largest of these is the Barringer crater in Arizona which is more than 1000 m in diameter and 170 m deep. Small amounts of meteoritic iron have been found in the vicinity and it is estimated that the crater was formed by the fall of an iron meteorite with a mass of approximately 50 000 tonne—that is with a diameter about 25 m if it was a sphere.

In 1908 there was a huge explosion in the Tunguska River region of central Siberia. The noise was heard at a distance of 1000 km and a fireball, brighter than the Sun, crossing the sky. The event was recorded on seismometers all over the world. In 1927 an expedition discovered a region of about 2000 km^2 of uprooted trees, with the direction of fall indicating that the explosion was at the centre of the region. However, neither a crater nor fragments were found that could be identified as of meteoritic origin. Fine fragments of meteoritic dust have been found embedded in local soils and the current belief is that the event was caused by the impact of a small comet with the explosion centre produced some 10 km above the surface.

1.6.1 Falls and finds

Recovered meteorites may be classed as either *falls* or *finds*. The former category consists of those objects that are seen to fall and are recovered shortly afterwards. Table 1.11 shows the proportions of falls and finds for the three major types of meteorite—stones, irons and stony-irons.

The distribution of the different kinds of meteorites differs for falls and finds and, in particular, the proportion of irons in the finds is much larger. The proportion of falls may be taken as representing the relative numbers of the different types of meteorites that fall to Earth. This just says that the chance of spotting a falling meteorite of a particular type is simply proportional to the number of that type that arrive on Earth. A 'find' depends on recognizing that the object, which may have fallen tens or hundreds of thousands of years previously, is actually a meteorite. A stone meteorite which lands in Europe, say, will be exposed to weathering that will erode its surface and soon make it indistinguishable from rocks of terrestrial origin. On the other hand we do not usually find large lumps of iron on the Earth's surface so that if we found a very dense iron object of blackish appearance then we could be certain that it was indeed a meteorite. Again an iron object would not weather in the same way as one of stone and would maintain its integrity for a much longer period.

Factors which assist in the recognition of stony meteorites are, first, that it should as far as possible maintain its original appearance and, second, that it should stand out in its environment. Two types of region where finds are often made are deserts and the Antarctic. In arid deserts weathering is a slow process so that meteorites retain their characteristic appearance for much longer. In the Antarctic, where the ice layer is kilometres thick, a silicate or iron object on or near the surface is bound to be a meteorite.

The ways in which meteorites are similar to, and different from, terrestrial materials are a very important source of information about the origin of the Earth and of the Solar System. A detailed description of meteorites and their properties is given in sections 11.2, 11.3 and 11.4. Here, we shall restrict the discussion to some salient features of meteorites.

1.6.2   Stony meteorites

There are two main types of stony meteorites—*chondrites* and *achondrites*—which differ from each other both chemically and physically. Most, but not all, chondrites contain *chondrules*. These are glassy millimetre-size spheroids, embedded in the fine-grain matrix that constitutes the bulk of the meteorite. The chondrules were certainly formed from molten silicate rock that was in the form of a fine spray. A section of a typical chondrite is shown in figure 1.20. An important type of chondrite is the *carbonaceous chondrite*. These meteorites contain carbon compounds, water and other volatile materials and tend to be rather dark in colour. They also contain lighter-coloured inclusions consisting of high-temperature condensates; these are known as CAI (calcium–aluminium-rich inclusions).

Achondrites contain no chondrules and virtually no metal or metallic sulphides, and are similar to terrestrial and lunar surface rocks in many ways. The ages of rocks, as determined by radiometric methods, gives the time from when the rocks became closed systems, retaining all the products of processes going on within them. For most meteorites the determined ages are about 4.5×10^9 years, which is the accepted age of the Solar System. However, there are a few achondrites, called SNC meteorites, that are much younger with ages around 10^9 years.

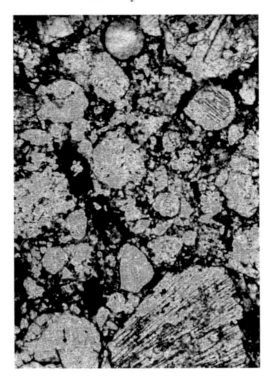

Figure 1.20. A section through a chondritic meteorite.

It has been suggested that these represent material ejected from the surface of Mars by a projectile. Mars could have been volcanically active in the sufficiently recent past to explain the SNC ages.

1.6.3 Stony-irons

Stony-irons consist of roughly equal proportions of stone and iron. A possible scenario for their formation is within an interface region of a cooling solid body in which there had been separation of dense metal and less dense stone. However, one type of stony-iron, *mesosiderites*, contain minerals only stable at pressures below 3 kbar suggesting that they are not directly derived from deep within a massive body.

1.6.4 Iron meteorites

Most iron meteorites consist of an iron–nickel mixture that was originally in a liquid state. There are, however, a few iron meteorites that look as though they have never been completely molten. Within the metal two iron–nickel alloys form—

Figure 1.21. An etched cross-section of an iron meteorite showing a Widmanstätten pattern.

taenite, which is nickel rich, and *kamacite*, which is nickel poor. When the metal has cooled sufficiently to become solid, but is still hot enough for the atoms within it to be mobile, then it separates into taenite and kamacite regions. A cut etched surface of an iron meteorite shows a characteristic *Widmänstatten* pattern (figure 1.21) consisting of dark taenite rims around plates of kamacite. From the scale of these patterns it is possible to estimate the time between when the material became solid and when it became so cold that the atoms ceased to be mobile. The cooling rates so deduced can be in the range 1–10 K per million years, indicating that cooling was either in the interior of an asteroid-size body or close to the surface of a much larger body.

1.6.5 Isotopic anomalies in meteorites

Most elements have more than one stable isotope and all elements have a multiplicity of isotopes, many of which are radioactive. Thus oxygen has three stable isotopes ^{16}O, ^{17}O and ^{18}O and the characteristic ratios for these on Earth are 0.9527:0.0071:0.0401, a composition referred to as SMOW (Standard Mean

Ocean Water). There are departures from SMOW on Earth but these are due to mass-dependent fractionation. In any chemical or physical process dependent on isotope mass—for example, diffusion in a thermal gradient—the difference in the diffusion rate between ^{18}O and ^{16}O will be twice that between ^{17}O and ^{16}O. This will be seen by the fractional change in the proportion of ^{18}O being twice that of ^{17}O. Thus, although terrestrial oxygen samples may have different isotopic ratios, they can be recognized as coming from a common source.

Samples from CAI material in carbonaceous chondrites and from some ordinary chondrites give oxygen isotope ratios that can be interpreted as having been produced by adding pure ^{16}O to some terrestrial standard mixture of isotopes. This has given rise to much speculation about possible sources for the pure ^{16}O.

Another observation from the CAI regions involves the stable isotopes of magnesium, ^{24}Mg, ^{25}Mg and ^{26}Mg which occur in the ratios 0.790:0.100:0.110. In some CAI, for different grains in the meteorite, there is an excess of ^{26}Mg that is proportional to the amount of aluminium present. This is due to the decay of ^{26}Al that formed a small proportion of the original aluminium (for which the only stable isotope is ^{27}Al). The half-life of ^{26}Al is 720 000 years which suggests that the interval between some radio-synthetic event that preceded the formation of the Solar System and the CAIs becoming closed systems was a few times the half-life, say, less than 10 million years.

Normal neon has three stable isotopes, ^{20}Ne, ^{21}Ne and ^{22}Ne, in the proportions 0.9051:0.0027:0.0922. Gases trapped in a meteorite can be driven out by heating and there are some stony meteorites which are found to contain pure, or nearly pure, ^{22}Ne—the so-called neon E. The most probable source of this is the radioactive sodium isotope ^{22}Na (only stable isotope ^{23}Na). There is reluctance to accept ^{22}Na as the source of neon E because the half-life of ^{22}Na is only 2.6 years and this would imply that the meteorite rock had become a closed system within, say, 20 years of some radio-synthetic event. Alternative origins by particle irradiation of the meteorite after it formed or by neon E having come from some unknown source outside the Solar System have also been proposed.

A number of interesting isotopic anomalies have been found in grains of silicon carbide, SiC, which occur in some chondrites. Silicon has three stable isotopes, ^{28}Si, ^{29}Si and ^{30}Si, but systematic variations in the ratios are found that cannot be related to mass-dependent fractionation or, as in the case of oxygen, to the addition of various amounts of the dominant isotope, ^{28}Si. The carbon in these grains also shows a very variable ratio of $n(^{12}C)/n(^{13}C)$, sometimes below 20, whereas the terrestrial value is 89.9. This is termed 'heavy' carbon. There is also 'light' nitrogen for which the ratio of $n(^{14}N)/n(^{15}N)$ is much higher than the terrestrial figure and some 'heavy' nitrogen as well. Another anomaly found in SiC samples is 'heavy' neon in which both ^{21}Ne and ^{22}Ne are enhanced and a linear relationship exists between $n(^{20}Ne)/n(^{21}Ne)$ and $n(^{21}Ne)/n(^{22}Ne)$ from different grains.

There are other isotopic anomalies but those described here are important ones and they illustrate the range and complexity of those that occur. Clearly these anomalies contain a message about conditions in the early Solar System. In section 11.6 isotopic anomalies will be described in more detail together with the ideas which have been put forward to explain them.

1.7   Comets

The general appearance of a comet is well known—a luminous ball with a long, often double, tail (figure 1.22). It was Edmund Halley who, with knowledge of Newton's analysis of planetary orbits, first recognized that comets were bodies in orbits around the Sun. He postulated that the comet seen in 1682 was identical to the comets in 1607, 1531 and possibly that of 1456, that had similar orbits and he predicted the comet's return in 1758 although he did not live to see his prediction confirmed. From Newton's work Halley deduced that the semi-major axis of the orbit had to be about 18 AU. That figure, combined with its close approach to the Sun, implied a high eccentricity, and this was the first deduction that there were bodies moving around the Sun in very eccentric orbits.

1.7.1   Types of comet orbit

The periods of cometary orbits show a very wide variation and it is customary to divide comets into two categories: *short-period* for those with periods less than 200 years and which therefore stay mainly in the region occupied by the planets, and *long-period* otherwise. About 100 short-period comets have been observed. Those with periods of more than about 20 years have more-or-less random inclinations; Halley's comet, with a period of 76 years, has a retrograde orbit with an inclination of 162°. Comets are significantly perturbed by planets, especially by the major planets, and the period of Halley's comet can vary between 74 and 78 years due to this cause.

There are about 70 short-period comets with periods, mostly between three and ten years, which have direct orbits and have fairly small inclinations, less than about 30°, and modest eccentricities, mostly in the range 0.5–0.7. Their aphelia are all about 5 AU and they form the *Jupiter family* of comets. They are presumed to have originally been long-period comets which interacted with Jupiter either in a series of small perturbations on occasional incursions into the inner Solar System or, possibly, in one massive perturbation. This is more likely to happen for comets with small inclinations and in direct orbits, for then their speeds relative to Jupiter during the interaction will be smaller and there will be more time to generate a strong perturbation

Another extreme class of orbits is where the periods extend from tens of thousands to millions of years. Such comets have very large major axes and since comets can only be observed when they have small perihelia, usually less than 3 AU, this implies that the orbital eccentricity must be close to unity. It is difficult to measure the orbital characteristics of such comets well enough to distinguish an extreme elliptical orbit from a marginally hyperbolic one. The characteristic

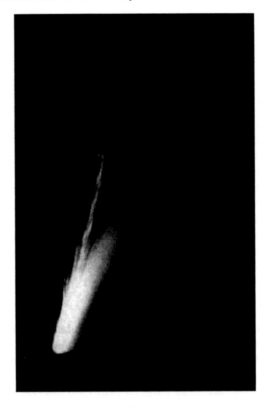

Figure 1.22. The comet Mrkos, photographed in 1957, showing the long plasma tail and the shorter, but thicker, dust tail (Mount Palomar Observatory).

of interest is the orbit before it approached the inner Solar System making it necessary to correct for planetary perturbation. A comet approaching the inner Solar System with a marginally hyperbolic orbit would have had almost zero velocity relative to the Sun at a large distance. This is extremely improbable and, taking the possible errors of measurement into account, it is safe to assume that all such comets actually approach the inner Solar System in extreme elliptical orbits. Comets with such orbits are called *new comets*, implying that the comet has never been so close to the Sun on a previous occasion and will never do so again on an orbit with such extreme characteristics. It is implied from these observations that there is a cloud of comets, the Oort cloud, surrounding the Solar System at distances of tens of thousands of AU. This will be described in greater detail in section 11.8.1.

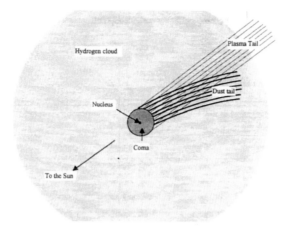

Figure 1.23. The structure of a comet in the near perihelion part of its orbit.

1.7.2 The physical structure of comets

Comets are difficult to see when they are more than about 5 AU from the Sun. At that distance they are solid inert objects, usually with low albedo and diameters of a few kilometres. As they approach the Sun their appearance changes dramatically. Vaporized material, plus some dust, escapes from the solid *nucleus*, and forms a large, approximately-spherical gaseous *coma* some 10^5 to 10^6 km in radius. The coma becomes visible due to the action of sunlight on its constituents. Outside the coma, with ten times its dimension, there is a large cloud of hydrogen that emits no visible light but that can be detected by suitable instruments. Finally the comet develops a tail, or often two tails, one of plasma and one of dust, which are acted on by the stream of particles from the Sun, called the *solar wind*, so that the tails point approximately in an anti-solar direction. These features are shown in a schematic form in figure 1.23.

Current belief about the structure of the nucleus is that it is, as Fred Whipple once described it, a 'dirty snowball'. The best model is of an intimate mixture of silicate rocks and ices—perhaps similar to the composition of frozen swampy ground on the Earth although in the comet's nucleus the ices would not all be water ice. The outer parts of comets that have made several perihelion passages would be relatively deficient in volatile material and would probably be in the form of a frangible rocky crust. As the nucleus approaches the Sun so it will absorb solar radiation and heat energy will eventually penetrate into the interior, causing sublimation of the volatile material. The pressures so produced will eventually fracture the crust in weaker regions and jets of vapour will escape to form the coma. The molecules in the coma will *fluoresce*, due to excitation by ultraviolet radiation from the Sun, and from the fluorescent spectrum the composition of the coma can be found.

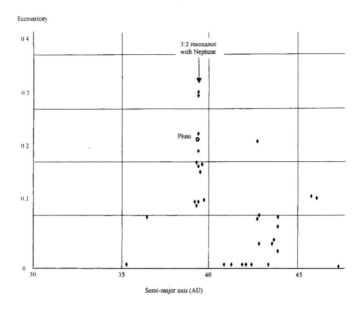

Figure 1.24. Characteristics of the Kuiper-belt objects.

The polarization of the reflected light from silicate grains in the dust tail suggests that they are typically of 1 μm dimension. The gravitational attraction of the Sun on the dust particles tend to send it into Keplerian orbits while the action of the solar wind tends to move it radially outwards from the Sun. The net effect is a tail that is almost antisolar but is usually quite distinct from the plasma tail. However, dust tails are generally much shorter than plasma tails and rarely exceed 0.1 AU in extent. The two tails show up particularly well in figure 1.22, a photograph of comet Mrkos taken in 1957.

Although the apparition of Halley's comet of 1986 was a rather poor one for Earthbound observers it was well observed from space. The European Space Agency's spacecraft *Giotto* passed within 600 km of the nucleus. Photographs were taken of the nucleus and measurements were made of the charged-particle density, magnetic fields and the compositions of the dust particles. Many of the particles were clearly silicates, similar in composition to carbonaceous chondrites, but others were rich in H, C, N and O and were presumably grains containing organic material. The rate of loss of icy material is of the order of 50 tonne s^{-1} during the perihelion passage and such a rate implies that the lifetime of Halley, and of other comets, as vapour-emitting bodies must be limited. The expected lifetime of comets is estimated to be in the range of hundreds to thousands of orbits, after which they will be small, dark, inert objects very difficult to detect. The fact that short-period comets have such a short life compared with the age of the Solar System indicates that somewhere there must be a reservoir of comets that constantly replenishes their number.

1.7.3   The Kuiper belt

In 1951 Gerard Kuiper suggested that outside the orbit of Neptune there would exist a region in which comet-style bodies orbited close to the mean plane of the Solar System. His main argument was that it seemed unlikely that the material in the Solar System abruptly ended beyond the orbits of Neptune and Pluto. In addition, in the outer reaches of the Solar System, material in the form of small bodies would be too widely separated ever to have aggregated into a larger body.

In 1992 the first such body was found, 1992 QB_1, with an estimated diameter of 200 km at about 40 AU from the Sun. By 1997 more than 30 bodies had been located in what is now known as the *Kuiper belt*. All the bodies have orbits close to the ecliptic and have semi-major axes larger than that of Neptune, although the eccentricities of some bring them within Neptune's orbit. In figure 1.24 the values of a and e for the first 32 of these discovered objects, plus Pluto, are shown; it will be seen that there is a family of 12 objects which share Pluto's 3:2 orbital resonance with Neptune.

It has been suggested that the Kuiper belt could contain more than 35 000 bodies with a diameter greater than 100 km. Perturbations in the Kuiper-belt region are sufficiently small to enable most objects there to have survived for the lifetime of the Solar System but perturbation by Neptune is sufficient to cause a small transfer of bodies inwards. Once they have penetrated the inner Solar System they can be further perturbed either to become short-period comets or to be thrown outwards to regions well beyond the Kuiper belt.

Chapter 2

Observations and theories of star formation

2.1 Stars and stellar evolution

2.1.1 Brightness and distance

Even in the most powerful telescopes stars are normally seen just as point sources of light. Sometime in the second century BC the Greek astronomer Hipparchus produced a star catalogue in which stars were categorized numerically according to brightness as having magnitudes from one to six, with one indicating the brightest stars. In the second century AD this information was included by Ptolemy in his *Almagest*, the 13 books which contained all the astronomical knowledge of the period. Certainly, by the 18th century, astronomers had refined the Hipparchus magnitudes and by carefully comparing pairs of stars they were able to refine brightness estimates, referring to a magnitude of, say, 4.4. In the 19th century instruments became available for quantitative measurements of the brightness of stars. It was found that the Hipparchus range from one to six (five units of increment) corresponded to a factor of about 100 in brightness, defined as the energy received at the Earth per unit area normal to the star's direction. A scientific scale was established where each unit step in magnitude corresponds to a brightness factor of $100^{1/5}$ ($\sim$2.51) and the scale was extended well outside the original range. Thus with modern telescopes equipped with CCD (charge-coupled device) detectors, faint objects with magnitudes of 28 or even fainter can be picked up. At the other end of the scale the magnitude of the Sun, the brightest object seen from Earth, is -26.74. The relationship between brightness, b, and magnitude, m, for two sources so established is

$$\frac{b_1}{b_2} = \frac{10^{-2m_1/5}}{10^{-2m_2/5}} = 10^{2(m_2-m_1)/5}. \tag{2.1}$$

The brightness of a body seen from Earth is dependent both on its intrinsic brightness, measured by its *luminosity* or total power output, and on its distance, so the determination of distance is important if the intrinsic brightness or any

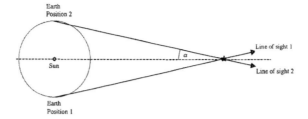

Figure 2.1. Parallax measurements on a star from opposite sides of Earth's orbit.

other properties of the body are to be deduced. In terrestrial surveying the relative positions of ground features are found by the process of triangulation. A feature, F, is observed from two ends of a baseline AB; with a theodolite the angles subtended by F and B at A and by F and A at B are found. In this way the position of F, and hence its distance from either A or B, is found from simple geometry. To measure distances even to the nearest stars a baseline is required that is longer than any that could be established on Earth, but points on Earth's orbit around the Sun separated by six months provides a suitable baseline of 2 AU. To measure the angle subtended by this baseline at the star, what is actually observed is the motion of the star relative to the background of very distant stars (figure 2.1). This observed relative motion due to motion of the observer is known as parallax and the parallax angle, α, shown in figure 2.1, is used to define a unit of measurement suitable for stellar distances. If α is $1''$ of arc, so that 1 AU subtends $1''$ at the star's distance, then the distance of the star is 1 parsec (pc) and, in general, the distance in parsecs is given by

$$D = \frac{1}{\alpha} \tag{2.2}$$

where α is the parallax in arc-seconds. The parsec equals 3.26 light-years, 3.08×10^{16} m or 206 625 AU. Ground-based telescopes can measure to about $0.005''$ so estimates of distances out to approximately 100 pc are possible, although with very poor accuracy at the upper end of the range. In 1989 the astrometric satellite *Hipparcos* was launched by the European Space Agency and it is capable of measurements down to about $0.001''$. This extends the range of direct distance measurement by a factor of five or so and the total number of stars within range of direct measurement to something of the order of one million.

For a star at a known distance it is possible to derive the *absolute magnitude*, M, and *absolute brightness*, B, the apparent magnitude and brightness the star would have if it was at a standard distance of 10 pc. Since brightness varies as the inverse-square of distance it can be found from equation (2.1) that

$$M = m + 5 - 5 \log D \tag{2.3}$$

where m is the measured magnitude and D the distance of the star in parsecs.

Distance measurements are greatly extended by observations of Cepheid variables. In 1912 Henrietta Leavitt, observing Cepheid variables in two nearby Galaxies, the Magellanic Clouds, found that their brightness and periods were correlated. This relationship means that from the period of the variable, the absolute brightness or magnitude can be found and hence, by measuring the apparent magnitude, the distance can be deduced from equation (2.3). If a Cepheid variable occurs in a cluster, or can be seen in a distant galaxy, then the distance of the cluster or galaxy can be found and hence the various intrinsic characteristics of other observable stars in the same stellar association.

2.1.2 Luminosity, temperature and spectral class

Once the distance of a star is known it is possible to estimate its luminosity. If the brightness of the star is measured as b (W m^{-2}) then its luminosity, L, the total rate of energy production is given by

$$L = 4\pi D^2 b \qquad (2.4)$$

for which, in this case, D is in metres.

Apart from the brightness of stars, another characteristic that could be seen by early naked-eye observers was their colour. With modern instruments the spectral output of stars can be measured over a wavelength range much greater than just the optical region and their intensity versus wavelength curves are found to match theoretical Planck radiation curves reasonably well. These curves indicate the equivalent black-body temperature of the source—the Sun, for example, has a surface temperature of ∼5800 K. A selection of radiation curves for different temperatures is shown in figure 2.2 and from these it will be seen that a good estimate of the temperature can be obtained just from the relative intensities at two well-separated wavelengths. The magnitude of a star as seen from light passing through a blue filter, B, compared to that from light passing through a yellow filter, V (V from *Visible*), gives the *colour index*, $B - V$, from which the temperature may be assessed. The colour index of a star is independent of its distance from the observer since changing distance changes B and V by the same amount. A problem with very distant stars is that the light may be slightly reddened by passage through the *interstellar medium* (ISM). The effect of this can be eliminated by using a third magnitude measurement, U, of the light passing through an ultraviolet filter.

From the surface temperature, T, and luminosity, L, of a star its radius, R, can be estimated. These quantities are linked by

$$L = 4\pi R^2 \sigma T^4 \quad \text{or} \quad R = \frac{1}{T^2}\left(\frac{L}{4\pi\sigma}\right)^{1/2} \qquad (2.5)$$

in which Stefan's constant, $\sigma = 5.67 \times 10^{-8}$ W m^{-2} K^{-4}.

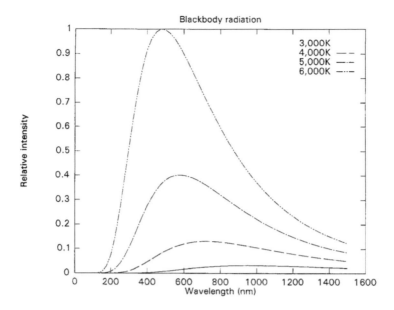

Figure 2.2. Black-body radiation curves for a range of possible stellar temperatures.

When the spectra of stars are examined with a high-resolution spectrometer they are found to consist of a continuous background with absorption lines corresponding to electron transitions in atoms and ions. The strengths of individual absorption lines vary greatly from one star to another, and may even be absent, and if the spectra of many stars are compared they can be arranged in a sequence in which each varies only a little from that of its neighbours. According to which spectral lines are prominent, stars are assigned to one of the *spectral classes* O, B, A, F, G, K or M; each class has ten subdivisions indicated by B0, B1, B2, ..., B9, for example. The main characteristic of a star that determines its spectral class is its temperature where O-type stars are the hottest and M-type stars the coolest. We can understand how temperature affects the strength of an absorption line by taking as an example the Balmer-series hydrogen line, Hγ, at 434.0 nm as seen in figure 2.3. Absorption of radiation at this wavelength corresponds to an electron transition in the atom from the energy level $n = 2$, the first excited state, to energy level $n = 5$. The $n = 2$ excited state, that is 10.2 eV above the ground state, requires a sufficiently high environmental temperature for collision by electrons to be able to produce such an excitation. The temperature of an M-type star, about 3500 K, is insufficient to excite the electrons of many hydrogen atoms to the state $n = 2$, the necessary ground state to give the Hγ absorption line. Moving up the spectra in figure 2.3, for spectral class G0, corresponding to about 6000 K, a sufficient proportion of the hydrogen is excited to give an observable absorption line and the line gets stronger with increasing temperature up to class

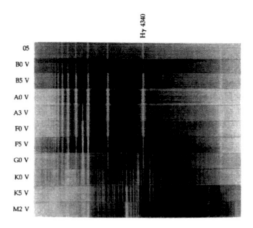

Figure 2.3. A sequence of stellar spectra with the Hγ absorption line identified.

A0 stars, at about 11 000 K, and thereafter declines. The reason for the decline is that at very high temperatures hydrogen atoms are excited to levels beyond $n = 2$ and even completely ionized so that fewer suitable ground-state atoms for the Hγ absorption become available.

2.1.3   The motions of stars relative to the Sun

To find the velocity of a star relative to the Sun, it first has to be measured relative to Earth and then transformed from knowledge of Earth's motion relative to the Sun. Methods are available that are able to give separately the radial and transverse components of the velocity. When the parallax method of estimating the distances of stars was described it was implicitly assumed that the star had not moved between the two measurements six months apart. The transverse motion of the star can be corrected for by making a third observation one year after the first. The apparent movement of the star relative to the background of very distant stars in one year enables a correction to the six-month observation to be made and hence the correct distance to be found. With the distance known, together with the angular shift of the star in the one-year period, the transverse component of velocity, v_t, can then be calculated.

Estimating the radial velocity component involves measuring the Doppler shifts of spectral lines in the light from the star. If the laboratory wavelength of a spectral line is λ and that measured in the spectrum of the star is $\lambda + \delta\lambda$ then the radial velocity, v_r, comes from

$$\frac{v_r}{c} = \frac{\delta\lambda}{\lambda} \tag{2.6}$$

in which c is the speed of light. It is found that the relative velocities of stars in

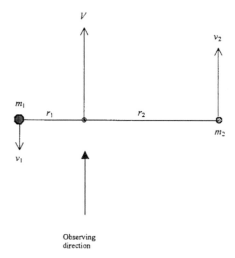

Figure 2.4. Characteristics of a binary system.

the solar neighbourhood have magnitudes mostly in the range 20–30 km s^{-1} with more or less random directions.

2.1.4 The masses of stars

About one-half of all the visible stellar light sources are not single stars but binary systems—which means that more individual stars belong to binary systems than those which do not. These binaries have a wide range of characteristics from *visible binaries*, where the individual stars can be resolved by a telescope, to *spectroscopic binaries* that cannot be resolved and can only be inferred as binaries by spectroscopic measurements. In the following discussion, for simplicity, it will be assumed that the sight line to the binary is in the plane of its orbit. We shall also assume circular orbits, which is certainly true for closer binaries where tidal dissipation effects impose near-circular motions.

In figure 2.4 the two stars, moving around their centre-of-mass, are moving directly towards and directly away from the observer. If their speeds relative to the centre-of-mass are v_1 and v_2 and the radial speed of the centre-of-mass is V then, as shown, the star radial-velocity components relative to the observer are $v_{1,1} = V - v_1$ and $v_{1,2} = V + v_2$. One-half period later the radial velocity components are $v_{2,1} = V + v_1$ and $v_{2,2} = V - v_2$. The measurements of $v_{1,1}$, $v_{1,2}$, $v_{2,1}$ and $v_{2,2}$ can be made by direct Doppler-shift measurements on the individual stars for a visual binary and from the splitting of spectral lines for a spectroscopic binary. The four speed measurements (actually only three are needed) enable V, v_1 and v_2 to be found. From the period, P, v_1 and v_2 it is then possible to infer the masses and physical dimensions of the binary system. From P one finds the angular speed, $\omega = 2\pi/P$. This then gives the distances of the stars from the centre-of-mass, r_1 and r_2, from

$$\omega = \frac{v_1}{r_1} = \frac{v_2}{r_2}. \tag{2.7}$$

The sum of the stellar masses then comes from

$$P = 2\pi \left\{ \frac{(r_1 + r)^3}{G(m_1 + m_2)} \right\}^{1/2} \tag{2.8}$$

and the ratio of the masses from

$$\frac{m_1}{m_2} = \frac{r_2}{r_1}. \tag{2.9}$$

From (2.8) and (2.9) the individual masses can be found.

For the majority of stars the mass is quite well correlated with the spectral class so that mass estimates of many stars are available—even if the stars are not members of a binary system. The general pattern observed is that very massive stars are rare and that the frequency of occurrence increases as the mass decreases. The mass probability density is found over a range of masses to be of the form

$$f(m) \propto m^{-\mu} \tag{2.10}$$

in which $f(m)$ is the proportion of stars per unit mass range and $-\mu$ is the *mass index*. Observation suggests a mass index somewhere in the range -2.3 to -2.6 for stars between one-tenth to ten times the solar mass. Actually $f(m)$ dips below the expected value for low-mass stars but this is probably an observational effect since such stars have larger magnitudes and hence are more difficult to detect.

2.1.5   The Hertzsprung–Russell diagram and main-sequence stars

At the beginning of the 20th century Ejnar Hertzsprung and Henry Norris Russell independently investigated the relationship between the luminosity of a star and its temperature and the culmination of their work is what is now known as the *Hertzsprung–Russell (H–R) diagram*. A typical H–R diagram is shown in figure 2.5 where the horizontal axis is labelled as spectral class, which is an alternative to temperature, and the vertical axis is labelled as absolute magnitude, which is an alternative to luminosity. Most stars are seen to fall in a band running from top left to bottom right; this band contains *main-sequence* stars in which the source of energy generation is the conversion of hydrogen to helium within the stellar core. The Sun will spend about 10^{10} years as a main-sequence star and is at present about halfway through this stage of its life. However, the duration of the main-sequence stage is very heavily dependent on the mass of the star. From measurements of mass, M_*, and luminosity, L_*, of main-sequence stars it is found that L_* is approximately proportional to $M_*^{7/2}$. Luminosity is a measure of the rate of hydrogen-to-helium transformation and mass is a measure of

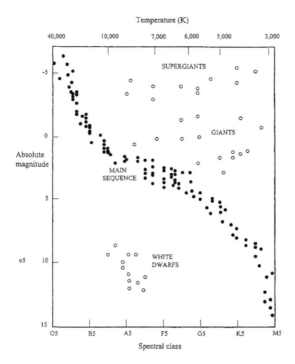

Figure 2.5. A Hertzsprung–Russell diagram for stars close to the Sun.

the amount of hydrogen available. Based on this, simple analysis shows that the lifetime of a star on the main sequence will be proportional to $M_*^{-5/2}$. Thus a star with 10 times the mass of the Sun will have a main-sequence lifetime of 3×10^7 years while for one with one-tenth of the mass of the Sun the lifetime would be 3×10^{12} years.

Not all stars are on the main sequence and their temperatures and radii may be inferred from their position on the diagram. There is a prominent group of stars in the top right-hand side of the diagram corresponding to temperatures generally less than that of the Sun but with luminosity anything from 10^2 to 10^5 greater than the solar value. From equation (2.5) it is found that they must have very large radii and on this account they are referred to as *red giants* or, in extreme cases, *supergiants*. Another prominent group of stars is seen in figure 2.5 below the main sequence. These have quite high temperatures but low luminosities between 10^{-2} and 10^{-5} that of the Sun and hence must be very small stars—smaller than the Earth in some cases. These stars are known as *white dwarfs* and represent one possible final stage in the development of stars.

Returning to main-sequence stars it is found that their physical characteristics are strongly related to their spectral class and these relationships are illustrated in table 2.1. These relationships are not precise because they are affected

Table 2.1. Approximate properties of main-sequence stars related to their spectral class.

Spectral class	Approximate temperature (K)	Mass (Solar units)	Radius (Solar units)	Luminosity (Solar units)
O5	40 000	40	18	3×10^5
B5	15 000	7.1	4.0	700
A5	8 500	2.2	1.8	20
F5	6 600	1.4	1.2	2.5
G5	5 500	0.9	0.9	0.8
K5	4 100	0.7	0.7	0.2
M5	2 800	0.2	0.3	0.008

by differences in the composition of stars and also by how long they have spent on the main sequence. For example over the Sun's lifetime on the main sequence its luminosity will change by a factor of five or six, from less than one-half of the present luminosity when it entered the main sequence to more than twice its present luminosity when it leaves. The values given in the table are representative of those given by different published sources, which are not entirely consistent with each other.

2.1.6   The spin rates of stars

A characteristic of main-sequence stars that cannot be inferred from an H–R di-agram is the rate at which they are spinning. This can be estimated by observing the thickness of spectral lines. If the spin axis of the star is perpendicular to the line of sight then the equatorial material at opposite edges of the star are moving towards and away from the observer, relative to the motion of the star as a whole. There will be an average Doppler shift of a spectral line due to the whole motion of the star and a broadening of the line due to additional Doppler shifts in oppo-site directions from light coming from opposite sides of the stellar image. The information given directly is the equatorial speed associated with spin and this is found to correlate with spectral class. The mean equatorial speed as a function of spectral class is given in figure 2.6; it will be seen that *late-type stars*—those with mass less than about $1.4 M_\odot$—have low equatorial speeds and that the maximum average speeds are for stars of spectral class B5 or thereabouts.

2.1.7   Evolution of stars away from the main sequence

The most massive stars get through the hydrogen-to-helium conversion stage in the core quite quickly and thereafter they evolve away from the main sequence. How they do so can be found observationally by looking at H–R diagrams for stellar clusters. Many stars occur in clusters that are of two main types. The

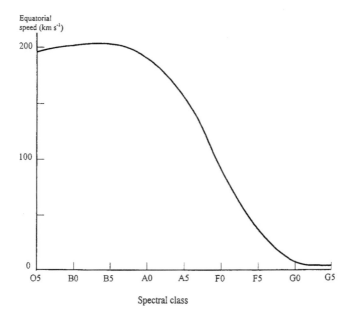

Figure 2.6. The variation of average stellar equatorial speed with spectral class.

first is the *galactic* or *open cluster*, that is an association of typically a few tens to a few hundred stars. These clusters are 'open' in the sense that they occupy regions of diameter from about 2 to 20 pc so that individual stars can be seen and 'galactic' in that they are exclusively found within the galactic disc (figure 2.7(a)). The other type of association is the *globular cluster* containing from 10^4 to 10^6 stars and with diameters from 10 to 30 pc so that the stars cannot be individually resolved in the centre of the image (figure 2.7(b)).

In figure 2.8 an H–R diagram for a typical globular cluster, M5, is shown. Stars have left the main sequence above the *turn-off point* and it is clear that the evolutionary tracks for these stars take them towards the red-giant region. Since main-sequence lifetime depends on the spectral class of the star the turn-off point gives a measure of the age of the cluster.

The detailed evolutionary development of a star depends on its mass but plausible scenarios have been developed and here we describe what happens to a star of one solar mass. An important feature of the development process is that from time to time the material in the inner core becomes *degenerate*. Degeneracy in this context implies that the properties of the material are dominated by electrons which, at suitable combinations of density and pressure, resist being pushed together and obey Fermi–Dirac (F–D), rather than Maxwell–Boltzmann, statistics. Degeneracy is favoured by high density and low temperature but at almost any temperature degeneracy can set in if the density is high enough. The mean energy

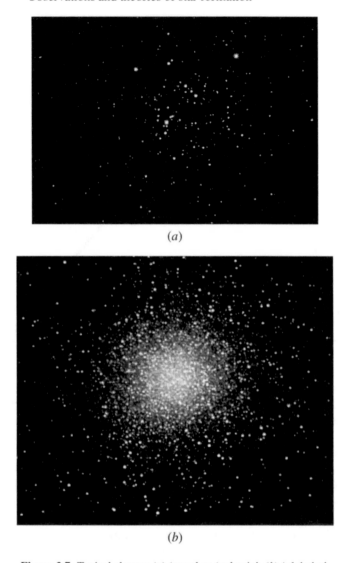

(a)

(b)

Figure 2.7. Typical clusters (*a*) 'open' or 'galactic', (*b*) 'globular'.

of electrons obeying F–D statistics varies very little with temperature and, since pressure is just a measure of energy-density, this means that pressure is virtually independent of temperature for a degenerate gas. The evolutionary pattern can be followed in figure 2.9 and can be described thus:

(1) After its formation and arrival on the main sequence the star spends 10^{10} years on the main sequence converting hydrogen to helium, at first through the proton–proton chain reaction but later, when the core heats up, increas-

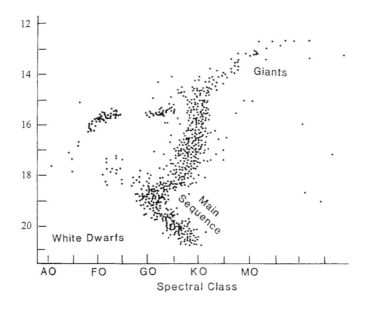

Figure 2.8. A Hertzsprung–Russell diagram for the globular cluster M5.

ingly through the C–N–O cycle.

(2) The core burning is reduced due to a lack of hydrogen but hydrogen-to-helium conversion goes on in a shell surrounding the core. This applies a pressure both inwards, compressing the core and heating it up, and outwards, expanding the star and cooling it down. The star becomes a red giant.

(3) Compressing the core has made the material degenerate and it heats up while its pressure remains constant. Eventually it reaches a temperature at which helium is converted to carbon (the triple-α reaction). Since the material is initially degenerate the heat produced does not lead to pressure-induced expansion and cooling until the temperature reaches a value where the degeneracy is removed. While the degeneracy persists the helium burning proceeds in an accelerating runaway fashion giving the 'helium-flash' stage.

(4) With degeneracy removed the main energy production is by helium burning in the core and the conditions for equilibrium are rather similar to those of the initial core-hydrogen-burning stage. Consequently the star moves towards the main sequence in the H–R diagram.

(5) With helium becoming exhausted in the core the helium burning takes place in a shell and conditions similar to stage 2 lead to a red-giant form again.

(6) The core shrinks and becomes degenerate while outer material is completely removed from the star in the form of a massive solar wind. Nuclear reactions cease and the star becomes a degenerate white dwarf. A white dwarf of mass $1M_\odot$ has a diameter similar to that of the Earth.

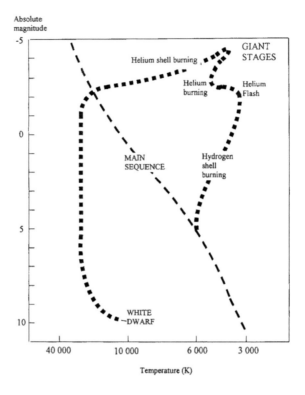

Figure 2.9. The evolution of a solar-mass star away from the main sequence.

In a more massive star events take place more quickly and higher temperatures are involved. Consequently other stages of nuclear reactions involving carbon can take place. If the final mass of the star when nuclear reactions have ceased somewhat exceeds the *Chandrasekhar limit* (Appendix I), ~1.44$M_\odot$, then the degeneracy pressure of the electrons is unable to resist pressure due to gravity. Instead of a white dwarf, the final outcome is a *neutron star*, where electrons and protons have combined to form neutrons. For more massive stars this is preceded by a *supernova* when the star explodes expelling a great deal of matter into interstellar space and leaving behind the neutron-star core. A neutron star has the density of a nucleon and with mass 2$M_\odot$ would have a diameter of about 10 km.

If the mass of the final stellar residue is greater than a limit somewhere in the range 3–6$M_\odot$ then even a neutron star would be crushed by gravitational pressure and it is believed that in this case a *black hole* would be the outcome (Appendix I).

2.2   The formation of dense interstellar clouds

2.2.1   Dense interstellar clouds

All the condensed objects in our galaxy are formed from the material in the interstellar medium (ISM). This material has density 10^{-21}–10^{-22} kg m^{-3} and a kinetic temperature (measuring the random component of its kinetic energy) up to 10^4 K or even higher. Most of it is gas consisting of hydrogen ($\sim$75%), helium ($\sim$23%), with the remainder being heavier elements. A few percent of the total mass of the ISM is in the form of dust or grains that account for its silicate and metallic component. The ISM also has an energy density unrelated to the matter within it, that comes from:

(i)   the galactic magnetic field ($\sim$$10^{-10}$ T) giving an energy density of 3×10^{-14} J m^{-3};

(ii) cosmic rays with an energy density of 8×10^{-14} J m^{-3}; and

(iii) star light with an *average* energy density of 8×10^{-14} J m^{-3} although the local density depends greatly on the proximity or otherwise of stars.

It is this very unpromising material which is somehow transformed into stars and ultimately into objects with the densities of neutron stars and black holes.

Observations indicate that the first stage in the condensation process is the formation of interstellar clouds, regions of much higher density and much lower temperature than that of average ISM material. The densest type are *dark molecular clouds* (DMC) which may be a factor 10^3–10^6 as dense as the mean ISM but with typical temperatures about 10–20 K. They contain H_2 and other molecules and radicals, such as CO and OH, and also a few per cent by mass of grains that may be of icy materials, silicates or metal or some combinations of the different materials. Their masses vary between a few hundred to a few thousand solar masses with a typical radius of 2 pc. The way in which such clouds form is clearly of interest and there may be more than one process at work. A common assumption is that a portion of the interstellar medium may be subjected to shock waves or mass flows from some violent source such as a supernova and that this triggers the collapse. Here we describe another possible mechanism that depends on heating and cooling processes that affect diffuse galactic material.

2.2.2   Heating and cooling in the ISM

It has been indicated that the ISM is pervaded by energy sources. Cosmic rays and star radiation will be absorbed by the ISM and so act as a source of heating. Cosmic rays are the more penetrating form of radiation and if the medium becomes opaque to visible light then it may be the dominant source. Heating by cosmic rays is uncertain but probably within two orders of magnitude of 5×10^{-6} W kg^{-1} and is not heavily dependent of the density and temperature of the ISM. However, heating by stellar radiation could be anything from negligible to dominant, depending on the proximity or otherwise of stars and on the state of the medium.

The main effect of the galactic magnetic field is that it may influence the way in which the ISM moves, in particular if the ISM is ionized significantly and becomes coupled to the field. Any motion of the medium, such as collapse to a higher density, which changes the density of field lines and hence the energy associated with the field, will be inhibited—although not necessarily prevented.

If the ISM is to be in a state of equilibrium then the heating must be balanced by cooling processes. There are several of these that will be discussed individually.

(1) Grain cooling

This form of cooling was described by Hayashi (1966) as part of a study of the way in which proto-stars evolve. The grains maintain a temperature, θ_g, of about 15 K in equilibrium with the various sources of radiation in the galaxy and this temperature is, in general, different from the kinetic temperature, θ, of the ISM or of DMCs. Gas molecules which strike a grain with average kinetic energy appropriate to a temperature θ leave with kinetic energy appropriate to temperature θ_g; if $\theta > \theta_g$ then the ISM or DMC will be cooled. The equation for the rate of cooling in W kg^{-1} given by Hayashi (1966), as modified by Woolfson (1979), is

$$Q_g = \frac{kn_g r_g^2}{\rho} \left\{ \frac{n_1}{(\gamma_1 - 1)\sqrt{m_1}} + \frac{n_2}{(\gamma_2 - 1)\sqrt{m_2}} \right\} (8\pi k\theta)^{1/2}(\theta - \theta_g) \quad (2.11)$$

in which n_g and r_g are the number density and mean radius of the grains, ρ the density of the gas, n_1, m_1 and γ_1 the number density, mass and ratio of specific heats for hydrogen atoms, with subscript 2 indicating the same quantities for hydrogen molecules. Gausted (1963) and Hayashi (1966) gave the following typical values: $r_g = 0.2\ \mu$m and $n_g = 10^{-13}\rho/m_1$.

(2) Ionic and atomic cooling

Interstellar material is partially ionized by the action of cosmic rays and so contains energetic electrons which, because of their small mass, move with high speed and frequently interact with the ions. Seaton (1955) showed that the excitations of C^+, Si^+ and Fe^+ were particularly effective in this respect. The basic mechanism is that the colliding electron loses energy by exciting one of the ion electrons into a higher energy state. When the excited electron falls back to its previous state it emits a photon which escapes from the local system thus removing energy from it. The equation given by Seaton, which includes an assumed electron density, is

$$Q_i = 1.79 \times 10^{14} \rho \theta^{-1/2} \left\{ 0.58 \exp\left(-\frac{92}{\theta}\right) + 5.0 \exp\left(-\frac{413}{\theta}\right) \right.$$
$$\left. + 1.7 \exp\left(-\frac{554}{\theta}\right) + 2.2 \exp\left(-\frac{961}{\theta}\right) \right\} \text{ W kg}^{-1}. \quad (2.12)$$

For lower temperatures the first term within the main bracket, related to C^+ excitation, is the dominant one. The second term relates to Si^+ and the last two terms to excitation to two different levels of Fe^+. The numerical coefficients in (2.12) depend on assumptions about the composition of the ISM and different coefficients have been given by Field *et al* (1968). McNally (1971) has expressed the view that (2.12) may not be valid at low temperatures because the elements giving these ions may then be locked up in solid grains.

Atomic oxygen is also an effective coolant where, in this case, the excitation is by the collision of hydrogen atoms, an interaction with a very high cross section. Field *et al* have given detailed expressions for oxygen cooling as a function of density and temperature based on calculations by Smith (1966) for two different excitation modes. Disney *et al* (1969) gave an empirical formula for an oxygen-to-hydrogen abundance ratio of 5.4×10^{-4}, fitted to Smith's analysis in the form

$$\log_{10}(Q_{\text{oxy}}) = 10.12 + \log_{10} \rho + 5.7 \log_{10} \theta - 1.55(\log_{10} \theta)^2 + 0.15(\log_{10} \theta)^3 \tag{2.13}$$

where the units of Q_{oxy} are W kg^{-1}.

(3) Cooling by molecular hydrogen

Molecular hydrogen is a dumb-bell-shaped molecule which can go into quantized rotational modes and can be excited by collision with hydrogen atoms. The square of the total angular momentum for a particular mode is of the form

$$L^2 = l(l+1)\hbar^2 \tag{2.14}$$

and the associated energy is

$$E_1 = \frac{L^2}{2I} \tag{2.15}$$

where I is the moment of inertia of the molecule which, for hydrogen, is about 8×10^{-48} kg m^2. The lowest possible energy for a tumbling mode, with $l = 1$, is thus 1.4×10^{-21} J which corresponds to a temperature of about 100 K. This means that this cooling process cannot set in much below that temperature. Another consequence is that at temperatures well below 100 K the only degrees of freedom for a hydrogen molecule are those of translational motion so the ratio of specific heats, γ, will then be the same for atomic and molecular hydrogen. This leads to a slight simplification of equation (2.11). Field *et al* (1968) gave implicit expressions for molecular hydrogen cooling due to a series of allowed transitions. However, a simplified cooling-rate equation for molecular hydrogen was given by Hayashi (1966) in the form

$$Q_{\text{H}_2} = \frac{1.08 \times 10^{-4} \rho \exp(-512/\theta)}{\rho + 6.69 \times 10^{-19} \theta^{-1/2}}. \tag{2.16}$$

The form of the cooling rates per unit mass of the ISM are given for the various mechanisms as functions of temperature in figure 2.10. These are only

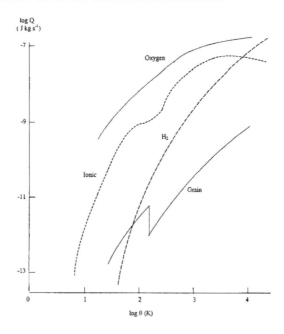

Figure 2.10. The cooling rates per unit mass for various mechanisms in the ISM.

of limited accuracy and contain assumptions about composition of the interstellar medium, including the degree of ionization within it and the relative abundance of molecular to atomic hydrogen. Although they cannot be relied on to within a factor of two, or even more, it turns out that the results obtained by modelling with these formulae are not very sensitive to factors of that size. The main point is that there is an abundance of coolant and the behaviour of the ISM is greatly influenced by this.

2.2.3 The pressure-density relationship for thermal equilibrium

To a good approximation the heating rate of the ISM, or a DMC formed within it, is approximately constant but the cooling rates for various processes are both density and temperature dependent. For any particular heating rate it is possible to find combinations (ρ, θ) which gives a cooling rate equal to the heating rate and thus a condition for thermal equilibrium. Since the low-density material behaves like a perfect gas the pair of quantities (ρ, θ) can be transformed to (ρ, p) where p is the pressure. The general form of the relationship between $\log(p)$ and $\log(\rho)$ for thermal equilibrium is shown in figure 2.11. It will be seen that it has a sinuous form and there is a range of pressures for which there are three possible densities giving equilibrium, all corresponding to the same pressure. However, the point B corresponds to a state of unstable equilibrium since for a slight *increase* in density the pressure would have to *decrease* to maintain the thermal equilibrium. Points A and C represent conditions of stable equilibrium and show that, for a particular pressure, two states of the medium are possible—one corresponding to low density and high temperature and the other to high density and low temperature. It is thus possible to have a DMC embedded in the interstellar medium which is both in thermal equilibrium and in pressure equilibrium with the ISM. This then raises the question of how the ISM can be triggered into forming the DMC in the first place.

2.2.4 Jeans' stability criterion

In all the preceding discussion no account has been taken of gravitation and how this will affect the stability or otherwise of a DMC. The conditions for the stability of an *isolated* gaseous cloud were first investigated by Jeans in 1902. There are several approaches to deriving the stability equation he gave, which give results differing by small numerical factors. Here we shall use the powerful Virial Theorem (Appendix II) which is a special form of a theorem originally given by Poincaré (1911). The general theorem says that for a system of particles for which the translational kinetic energy is E and the potential energy is V then

$$2E + V = \frac{1}{2}\frac{d^2 I}{dt^2}.$$ (2.17)

The quantity I is the *geometrical moment of inertia* given by

$$I = \sum_{i=1}^{N} m_i r_i^2$$ (2.18)

where there are N particles and the ith particle has mass m_i and is distant r_i from the centre of mass of the system. The Virial Theorem has zero on the right-hand side of (2.17) for the case when I, a measure of the total spread of the system, does not change with time.

We now consider a uniform gaseous sphere of radius R and density ρ for which the mean molecular mass is μ and the temperature is θ. The mass of the sphere is

$$M = \tfrac{4}{3}\pi\rho R^3$$ (2.19)

and hence the total energy associated with translational motion from thermal energy is

$$E = \frac{M}{\mu} \times \frac{3}{2}k\theta = \frac{2\pi\rho k\theta R^3}{\mu}.$$ (2.20)

The gravitational potential energy of the sphere is

$$V = -\frac{3}{5}G\frac{M^2}{R} = -\frac{16}{15}\pi^2\rho^2 R^5.$$ (2.21)

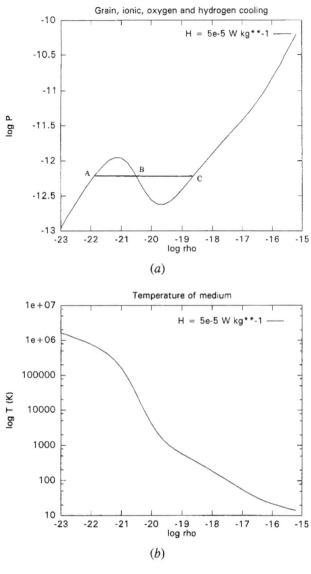

Figure 2.11. (*a*) The $\log P$ versus $\log \rho$ curve for a heating rate of 5×10^{-5} J kg^{-1} s^{-1}. (*b*) The temperature versus $\log \rho$ curve corresponding to (*a*).

Inserting (2.20) and (2.21) into the Virial Theorem and using (2.19) to substitute for R in terms of M we find Jeans' critical mass, M_J as

$$M_J = \left(\frac{375 k^3 \theta^3}{4\pi G^3 \mu^3 \rho} \right)^{1/2}.$$ (2.22)

The meaning of this equation is that if the temperature and density are fixed then any mass greater than M_J will collapse while any mass less than M_J will expand. Corresponding to M_J there is a Jeans radius, R_J, which can be found from (2.19).

A DMC buried in the ISM is not isolated but if it is sufficiently massive it will begin to collapse. The time for free-fall collapse to high density, just under gravity and without taking into account pressure-gradient forces that will slow down the collapse, is

$$t_{ff} = \sqrt{\frac{3\pi}{32G\rho}}. \tag{2.23}$$

From a static start, free-fall collapse is initially very slow and only accelerates appreciably after a considerable fraction of t_{ff} has passed. Hence (2.23) may be taken as a collapse time-scale although the collapse may not be completely free-fall and may be slowed down by pressure-gradient forces in its later stages.

It is interesting to see what (2.22) gives as a critical mass under various circumstances. For the ISM with a density of 10^{-21} kg m^{-3} and temperature 10^4 K for $\mu = 2 \times 10^{-27}$ kg, somewhat higher than for atomic hydrogen, $M_J = 1.8 \times 10^{38}$ kg or about $10^8 M_\odot$. For a DMC with $\rho = 10^{-18}$ kg m^{-3} and $\theta = 20$ K the critical mass is about $500 M_\odot$, which, as has been previously indicated, is of the same order as the mass of a galactic cluster.

2.2.5 Mechanisms for forming cool dense clouds

It has been shown that a cool dense cloud can co-exist with the interstellar medium in thermal equilibrium with the galaxy and also in, or close to, pressure equilibrium with the ISM. With a sufficiently high mass it may then collapse to form a stellar cluster.

One possible mechanism which has been suggested for forming dense clouds is through compression of the ISM either by shock waves, from supernovae or novae, or by the ram pressure of streams of matter leaving stars at some stages of their evolution. The shock-wave mechanism has been explored by Grzedzielski (1966) albeit on a galactic scale. The fragmentation of a pre-galaxy is described in terms of the effects of random shock waves that compress material to densities at which it will spontaneously collapse.

Another possible triggering mechanism due to a supernova is the injection into the local ISM of extra coolant material in the form of grains and heavier atoms giving rise to augmented atomic and ionic cooling. The affected ISM region will cool and the pressure within it will fall; this is illustrated in figure 2.12 as a change from state A to state B. Compression of the region by the external unaffected ISM will follow, leading to an increase in both pressure and density and the material will move from state B to C—not precisely on the original pressure–density curve because the extra coolant corresponds to a different pressure–density relationship. This description, first given by Dormand and Woolfson (1989), is somewhat idealized and assumes that during the initial cooling from A to B the density will not have changed. However, the mechanism has since been modelled numerically by Golanski (1999) and his model and results will now be described.

The model has been based on the use of *smoothed-particle hydrodynamics* (SPH) first implemented by Lucy (1977) and Gingold and Monaghan (1977). A short description of SPH is given in Appendix III. Briefly, it is a Lagrangian system in which a fluid is represented by a discrete distribution of points, each of which is endowed with properties (mass, internal energy, distribution in space) so that forces due to gravity, pressure gradients and viscosity may be simulated and changes of internal energy estimated. In the present model 8441 SPH particles, with a total mass $1000M_\odot$, are placed within a sphere of radius 25.3 pc, corresponding to a density of 10^{-21} kg m^{-3}, taken as the ISM density. The initial ISM temperature was 10^4 K. This region has coolants enhanced by a factor of two and is surrounded by normal ISM material, the gravitational and pressure effects of which were simulated by an envelope of SPH particles. Cooling is very slow at first because the material is so diffuse. As the density of the gas increases so the cooling becomes more efficient and the evolution speeds up. Eventually the system starts to collapse quite quickly, the density going from 3.1×10^{-20} kg m^{-3} to 5×10^{-19} kg m^{-3}. Several condensed regions, *proto-clouds*, develop with a variety of characteristics. The highest density increases to 10^{-18} kg m^{-3} and stays at about that level unless a collision between proto-clouds occurs. Such a collision compresses the material in the collision interface to such an extent that the density can increase up to 10^{-15} kg m^{-3}.

Figures 2.13, 2.14 and 2.15 show sliced density plots of the cooled regions at various times. Within each slice the velocity field, the temperature and the density are indicated. The velocity field is represented by vectors. The scale is given by the maximum velocity, V, the value of which is given in the figure legends. Similarly each slice has a maximum density which, again, is given in the figure legends. Density contours correspond to increments of a factor of one-half an order of magnitude ($10^{1/2}$). Temperature variation is represented by shading such that the darker the shading the higher is the temperature.

Two main features can be seen in the figures. First, proto-clouds form separately from each other and second, some of them collide to give further enhancement of density. A proto-cloud such as C2 (figures 2.13(*c*), 2.14(*c*) and 2.15(*d*)) is an example of an isolated cloud giving rise to a DMC. Once it forms it stays at about the same temperature (17 K) and density (2×10^{-18} kg m^{-3}) but grows by slow accretion of material from the hot diffuse ISM.

Proto-clouds C1 (figure 2.13(*b*)) and C3 (figure 2.13(*d*)) have densities about 10^{-19} kg m^{-3} when they form. The velocity field shows that they are moving towards each other and they are also accreting material. In figure 2.14(*c*) C1 and C3 start colliding. The density at the interface has gone up to 10^{-18} kg m^{-3} and the temperature is about 14 K. The Mach number of the collision is about 5.4 and, according to the model presented by Woolfson (1979), such a combination of Mach number and density should lead to star formation. Figure 2.15(*c*) shows the collision region when the density has reached about 10^{-15} kg m^{-3}. The com-

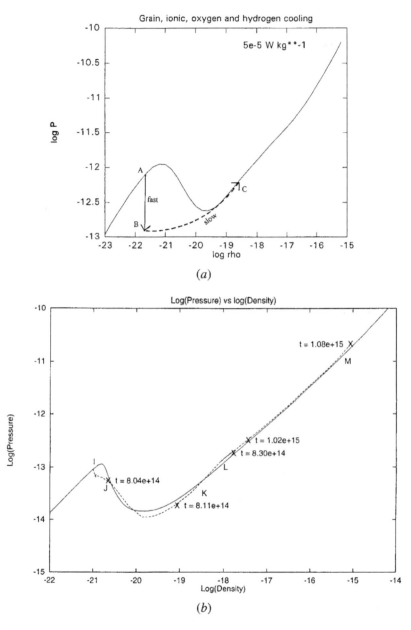

Figure 2.12. (*a*) The idealized path of material from low density, high temperature to high density, low temperature after the addition of coolant material. (*b*) The path from low density, high temperature to high density, low temperature as computed by Golanski (1999). Times are given from the beginning of the simulation. The final stage, from L to M, represents collapse under gravity.

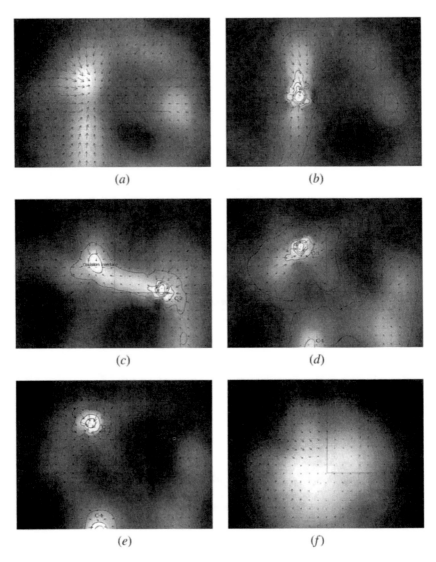

Figure 2.13. Sections for various values of z through the collapsing region at time 8.65×10^{14} s.

	z (m)	ρ_{max} (kg m^{-3})	V_{max} (km s^{-1})	θ_{min} (K)	θ_{max} (K)
(a)	3.03×10^{17}	2.22×10^{-22}	7.50×10^{2}	3.27×10^{2}	1.17×10^{4}
(b)	1.90×10^{17}	5.60×10^{-19}	2.15×10^{2}	2.01×10^{1}	1.07×10^{4}
(c)	0	1.35×10^{-18}	2.93×10^{3}	1.96×10^{1}	1.02×10^{4}
(d)	-1.38×10^{17}	4.99×10^{-19}	3.19×10^{3}	2.05×10^{1}	1.04×10^{4}
(e)	-2.07×10^{17}	5.18×10^{-20}	2.01×10^{3}	4.29×10^{1}	1.07×10^{4}
(f)	-6.17×10^{17}	7.72×10^{-22}	3.76×10^{2}	9.63×10^{3}	1.25×10^{4}

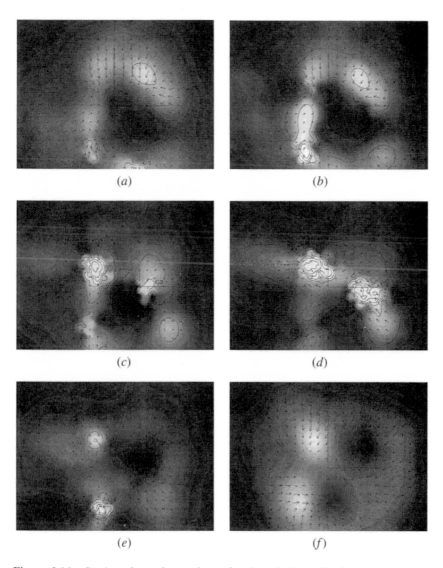

Figure 2.14. Sections for various values of z through the collapsing region at time 1.02×10^{15} s.

	z (m)	ρ_{max} (kg m^{-3})	V_{max} (km s^{-1})	θ_{min} (K)	θ_{max} (K)
(a)	1.38×10^{17}	1.15×10^{-19}	1.60×10^{3}	4.03×10^{1}	1.18×10^{4}
(b)	9.57×10^{16}	5.25×10^{-19}	2.15×10^{3}	2.05×10^{1}	1.17×10^{4}
(c)	3.48×10^{16}	3.29×10^{-18}	4.82×10^{3}	1.44×10^{1}	1.44×10^{4}
(d)	-2.52×10^{16}	1.28×10^{-18}	3.24×10^{3}	1.77×10^{1}	1.12×10^{4}
(e)	-1.12×10^{17}	1.28×10^{-18}	2.10×10^{3}	2.31×10^{1}	1.12×10^{4}
(f)	-2.77×10^{17}	1.90×10^{-21}	7.11×10^{2}	3.35×10^{3}	1.21×10^{4}

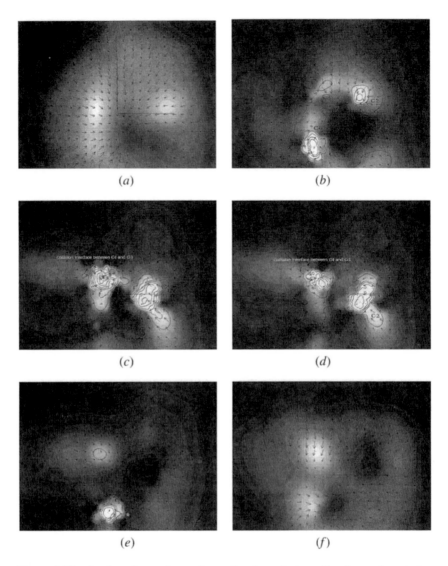

Figure 2.15. Sections for various values of z through the collapsing region at time 1.08×10^{15} s.

	z (m)	ρ_{max} (kg m^{-3})	V_{max} (km s^{-1})	θ_{min} (K)	θ_{max} (K)
(a)	2.69×10^{17}	1.33×10^{-21}	6.31×10^{2}	5.12×10^{3}	1.22×10^{4}
(b)	8.70×10^{16}	5.09×10^{-19}	2.64×10^{3}	2.05×10^{1}	1.15×10^{4}
(c)	0	1.31×10^{-15}	3.44×10^{3}	4.80×10^{0}	1.12×10^{4}
(d)	-2.50×10^{16}	1.11×10^{-18}	4.82×10^{3}	1.72×10^{1}	1.11×10^{4}
(e)	-1.21×10^{17}	7.32×10^{-19}	2.12×10^{3}	1.91×10^{1}	1.12×10^{4}
(f)	-2.25×10^{17}	1.80×10^{-21}	8.93×10^{2}	3.06×10^{3}	1.17×10^{4}

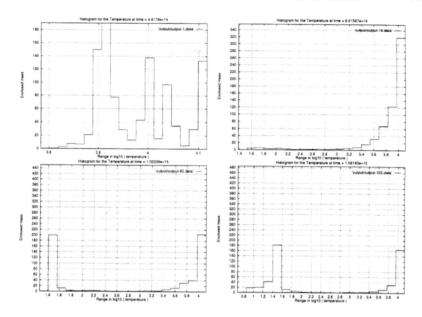

Figure 2.16. The temperature histogram for the ISM collapse simulation at four times.

bination of the two proto-clouds has produced a DMC of mass 2.15×10^{32} kg or about $108 M_\odot$. At this stage the DMC C2 has, by accretion, reached about $55 M_\odot$. In addition the velocity field shows that clouds C1/3 and C2 are heading for a collision at a relative speed of about 3 km s^{-1} corresponding to a Mach number of about 6.2. This would result in a very high density cloud of total mass $163 M_\odot$. The resultant DMC is extremely turbulent and the Jeans critical mass within it will be of the order of a solar mass. Collisions between gas streams within the cloud can then lead to form either single stars (Woolfson 1979) or perhaps binary or other multiple-star systems (Turner *et al* 1995). The latter workers started with much more massive colliding clouds, each a few hundred $M_\odot$, and with densities between 10^{-18} to 10^{-12} kg m^{-3}. Their collision occurred with a Mach number of 5.6—very similar to that found here. Proto-clouds such as C4 (figures 2.13(*e*), 2.14(*e*) and 2.15(*e*)) and C6 (figures 2.14(*b*) and 2.15(*b*)) will probably join the existing DMC, augmenting both its mass and its density. The final mass of the DMC would then be about $300 M_\odot$, more in keeping with what was suggested by Turner *et al* (1995).

The progression of $\log P$ versus $\log \rho$ for the highest density regions, as found by Golanski, is shown in figure 2.12(*b*), with times from the beginning of the simulation, and differs markedly from that assumed by Dormand and Woolfson (1989). Figures 2.16 and 2.17 show temperature and density histograms for the SPH points at four times. These enable the development of the high-density, low-temperature regions to be followed in more detail.

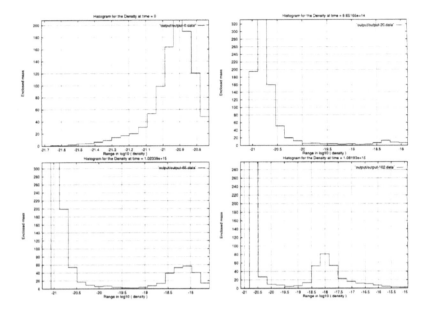

Figure 2.17. The density histogram for the ISM collapse simulation at four times.

The impregnation of the ISM by supernova material over the collapsing volume of the ISM would take of the order of several hundred to one thousand years, judging by the size and age of the Crab Nebula.

2.3 The evolution of proto-stars

2.3.1 The Hayashi model

Hayashi (1961) calculated the evolution of a star of $1M_\odot$ from the state of being a diffuse proto-star to the time of joining the main sequence and the evolutionary path is shown on a H–R diagram in figure 2.18. The initial proto-star state is at point A where the temperature and density are approximately 25 K and 4.5×10^{-11} kg m^{-3} and where its luminosity is little more than one-tenth that of the Sun. It slowly collapses towards B but the gravitational energy released by the collapse does not appreciably heat up the proto-star because it is diffuse and transparent to infrared radiation so that most of the energy is radiated away. As the collapse progresses so it becomes faster, as happens in free-fall collapse, but at the same time the proto-star becomes more opaque. At point B the opacity increases to the point where released gravitational energy is retained, heating up the proto-star and slowing down the collapse as pressure gradients build up within it. The collapse continues to point C where a bounce occurs as the proto-star moves through an equilibrium position and back again. This bounce, which lasts about

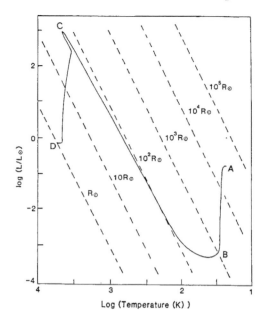

Figure 2.18. The path of a proto-star on to the main sequence. The broken lines give the radius in solar units.

100 days, is accompanied by an increase in luminosity; in 1936 Herbig observed an increase in luminosity of the young evolving star FU-Orionis by a factor of over 200 in a period of less than one year followed by a slow decline in luminosity. This could be related to the evolution at the bounce stage predicted by Hayashi's analysis.

The final descent of the star to point D, on the main sequence, is the slow Kelvin–Helmholtz (K–H) stage of contraction which is when the star is in a state of quasi-equilibrium. As the star radiates energy so it slowly contracts to restore equilibrium and in the process it increases in temperature. Eventually, at D, the internal temperature rises to the level where thermonuclear reactions involving hydrogen set in and a state of very slow evolution begins which maintains the star on the main sequence for about 10^{10} years.

The lifetime on the main sequence of a star of mass M_* was shown in section 2.1.5 to be proportional to $M_*^{-2.5}$. Another time-scale of interest for a newly forming star is the time taken for the K–H contraction and its dependence on mass is given in table 2.2.

The K–H stage of development has also been independently investigated by Ezer and Cameron (1965) and by Iben (1965). The evolutionary tracks calculated by Iben are shown in figure 2.19 for various stellar masses. They differ in detail

Table 2.2. Times spent by model stars on the Kelvin–Helmholtz stage of evolution.

Mass (solar units)	Time (10^6 years)
15	0.062
9	0.15
5	0.58
3	2.5
2.25	5.9
1.5	18
1.25	29
1.0	50
0.5	150

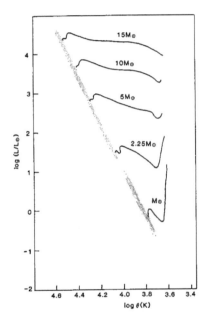

Figure 2.19. The Kelvin–Helmholtz contraction stage for stars of different mass.

from the results of Hayashi (1961) and Ezer and Cameron (1965) but the general form of the tracks is similar.

It has been shown in section 2.2.5 how a DMC could form and be massive enough to go into gravity-induced collapse and the work of Hayashi and others shows clearly how a proto-star evolves towards the main sequence. To fill the gap in the process of star formation from ISM material it is now necessary to find the process by which proto-stars could be produced within a collapsing cloud.

Before considering this matter further we shall see what can be observed about the processes of star formation.

2.4 Observations of star formation

2.4.1 Infrared observations

When stars are in the early stages of their evolution they are at temperatures for which the main part of the electromagnetic-radiation output is in the infrared. It is possible to observe from the surface of the Earth in the near infrared, at about 1 μm or so, as is done by the *United Kingdom Infrared Telescope* (UKIRT) situated at the Mauna Kea observatory in Hawaii. However, due to absorption by the atmosphere, to get information at much longer wavelengths requires observation from space and NASA's *Infrared Astronomical Satellite* (IRAS) and, later, ESA's *Infrared Space Observatory* (ISO) and the *Hubble Space Telescope* (HST) have enabled very detailed infrared maps of the sky to be produced. In the context of star formation well-defined infrared sources, emitting radiation in the 30–100 μm region, have been located within nebulae—for example, in the Orion nebula which is a rich source of newly-forming stars.

2.4.2 Radio-wave observations

The structure of the galaxy has been mainly explored by the radio emission at 1421 MHz due to hyperfine transitions in atomic hydrogen. Although it is an intrinsically weak source there is so much hydrogen in the galaxy that it is easily recorded by radio telescopes. Other fine-structure transitions giving radio frequencies occur for other atomic species but cannot be picked up because the amount of radiating material is too small.

A property of DMCs, as their name implies, is that they contain molecules such as H_2O, CO and CO_2 and also free radicals, such as CH and OH which readily react but which have a long lifetime in very diffuse material. Molecules have vibrational modes and rotational modes with quantized energies and transitions between different vibrational modes and between different rotational modes give electromagnetic radiation at discrete frequencies. The frequencies associated with transitions between vibrational states lie in the infrared but those associated with rotational state transitions may have radiofrequencies. The energies associated with these latter states are of the form

$$E_{\mathrm{J}} = \frac{J(J+1)\hbar^2}{2I} \tag{2.24}$$

where J is an integer and I the moment of inertia of the molecule about an appropriate axis. For heavier diatomic molecules (e.g. CO) or poly-atomic molecules, with larger moments of inertia, transitions between lower rotational states correspond to energies small enough to fall in the microwave or radio region of the

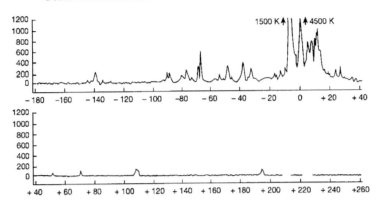

Figure 2.20. The water spectrum of the maser source W49 (Sullivan © 1971 The University of Chicago). The aerial temperature (K), proportional to intensity, is plotted against Doppler shift as a velocity (km s^{-1}).

spectrum. However, the rotational or vibrational state may be split into two or more levels by internal interactions and transitions between these levels for less massive molecules, such as OH, may also fall into the microwave or radio part of the spectrum.

Strong radio-emission has been detected, particularly at wavelengths corresponding to emission by OH and H_2O and radiation from several tens of other kinds of molecules has also been picked up. It is clear from the character of the radiation—its state of polarization and the width of the spectral lines—that it has not been produced by *spontaneous emission* from a gas in thermal equilibrium but rather that it is *stimulated emission*, similar to that occurring in a laser. The radiation is probably produced in narrow beams and we observe just what happens to be coming in Earth's direction. The exact mechanism that produces these *masers* (lasers for microwave radiation) is not precisely understood but for our purpose it is enough to know that these maser sources come from active regions where there is an abundance of molecules—which is consistent with what would be expected in star-forming regions.

The radial velocity of the source can be found from Doppler shifts of the maser frequencies; for hydroxyl radiation at 1650 MHz a velocity of 1 km s^{-1} corresponds to a frequency shift of 5.5 kHz, which is easily measured. Some emitting regions show a velocity spread of 500 km s^{-1} and consist of several discrete sources each with its own Doppler shift. In figure 2.20 the water spectrum from the source region W49 is shown as found by Sullivan (1971) and it is clear that it consists of many individual sources moving at different radial speeds relative to Earth. The structure of the moving sources changes with time on a fairly short time-scale, as is illustrated in figure 2.21 by the water spectrum for the source region W3.

By the use of two or more radio dishes in interferometric mode the size of

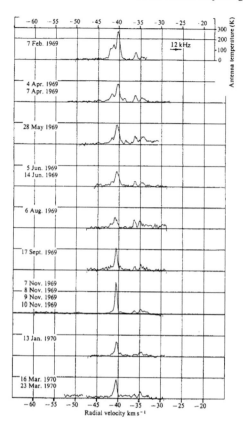

Figure 2.21. Changes over a 13-month period of the water source W3. (Sullivan © 1971 The University of Chicago).

the maser source regions can be found. The overall dimensions range typically from 10^{11} km to several times 10^{12} km while the discrete sources within them have diameters between 10^8 and 10^{10} km.

A detailed account of maser sources has been given by Cook (1977). He suggests that the maser radiation comes from star-forming regions. An obvious interpretation of the results is that individual stars, the discrete sources of diameter 10^8–10^{10} m, are being formed within a region of overall size $\sim 10^{12}$ km, within which there are turbulent motions with characteristic speeds 20 km s^{-1}.

2.5   Observation of young stars

2.5.1   Identifying young stellar clusters

From sections 2.1.7 and 2.3.1 it is clear that stars evolve towards the main sequence from the right-hand side of the H–R diagram and also evolve away from the main sequence in the same direction. If an isolated star was observed at some general position in the H–R diagram, fairly close to the main-sequence line but to its right, then there would be no way of knowing whether it was approaching the main sequence or moving away from it. However, if it were a member of a cluster, particularly a young galactic cluster, then the uncertainty would be removed. If a cluster contains O and B stars *on* the main sequence then one can be absolutely certain that the cluster is young since, as stated in section 2.1.5, the lifetime of an O-type star on the main sequence is only $\sim 10^7$ years. Thus, for such a cluster, if a star is located off the main sequence, but not far from, say, the G spectral class region of the main sequence, then it is certain that it is evolving towards the main sequence and not away from it.

2.5.2   Age–mass relationships in young clusters

The evolutionary tracks of stars during the K–H stage of their evolution are shown in figure 2.19. An important feature of these theoretically derived tracks is that those corresponding to different masses do not cross each other. That means that any particular position on the H–R diagram for a star in this part of its evolution can be identified with a unique stellar mass and a unique age, where age is the time from the beginning of the K–H stage. Of course the actual estimate of mass and age will depend on the particular model being used and the results of Hayashi (1961), Ezer and Cameron (1965) and Iben (1965) are similar in form but different in detail. As it turns out the conclusions that are drawn by analysis of young stellar clusters in terms of the different models are very similar.

The K–H evolutionary curves have been used by Iben and Talbot (1966) and Williams and Cremin (1969) to find the masses and ages of stars in several young stellar clusters. Here the results of the latter workers will be examined. They examined four clusters, NGC2264, NGC6530, IC2602 and IC5146 but here we shall give their results just for NGC2264, which are quite typical.

In figure 2.22(*a*) the mass–age relationship is shown. With the exception of the group of 12 stars older than 10^7 years, the pattern is that star formation began about 8×10^6 years ago with stars that were of somewhat more than $1 M_\odot$. Thereafter stars of lesser mass were produced but 5×10^6 years ago another stream of development began where the stars became more massive with the progress of time. This pattern is indicated in figure 2.22(*a*) by the shaded bands which take in most of the stars although, clearly, there is some scatter outside the bands.

From figure 2.22(*a*) the way in which the rate of star formation changed with time can be found as given in figure 2.22(*b*). The pattern is that star formation was slow at first but then accelerated to the present time in almost exponential fashion. The fluctuations in the rate-of-formation curve are not believed to be significant.

The final relationship which can be extracted from figure 2.22(*a*) is the mass distribution function. From equation (2.10) it will be seen that, for stars in general, plotting $\log\{f(M)\}$ against $\log(M)$ should give a straight line of slope $-\mu$. Figure 2.22(*c*) shows such a plot, where the masses are expressed in solar-mass

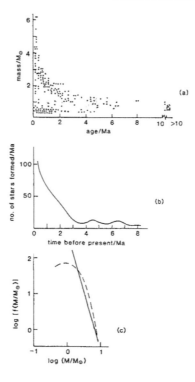

Figure 2.22. Observational data for the young stellar cluster NGC 2264. (*a*) Stellar mass against age. (*b*) Variation of rate of star formation. (*c*) The mass distribution function.

units. A line of slope -2.7 is drawn on the plot and is in reasonable agreement with the observed curve except for smaller masses where there are fewer stars than expected—probably due to observational limitations.

These investigations by Iben and Talbot (1966) and Williams and Cremin (1969) provide a model with which the results from any theoretical investigation of star formation in a galactic cluster may be compared.

2.6   Theories of star formation

2.6.1   Stars and stellar clusters

It is estimated that our galaxy contains about 10^{11} stars and we observe that some of these exist in clusters. More than 130 globular clusters have been observed, each containing anything from 50 000 to 50 million stars. From their turn-off ages (section 2.1.7) they are found to be the oldest entities in the galaxy and the material within them is representative of the original product of the 'big bang'. Because they are so old they have little to contribute observationally to knowledge about the processes of star formation. On the other hand there are more than 1000 known galactic clusters, each containing typically several hundred stars, and some of these are very young with stars still evolving towards the main sequence. They consist of material containing dust and heavier atoms that are the product of one or more cycles of reprocessing within stars, similar to the material of the Sun.

Only a tiny fraction of the stars of the galaxy are known to occur in clusters yet it is obvious from the existence of so many of them that there must be something conducive to star formation in clusters. A star formed in a cluster will not stay within it indefinitely. The stars within a cluster are constantly interacting and exchanging energy and occasionally a star near the edge of the system will be moving outwards with enough energy to escape. This evaporation process will continue until the system is left in some stable configuration—a binary system or some other simple system containing a few stars. The characteristic time-scale for the dispersal of a cluster, which can be defined as the time for which a fraction $1/e$ of the original population of stars will remain, is of the order 10^8–10^9 years for a galactic cluster and 10^{10} years for a globular cluster.

This raises the question of what proportion of stars originate in clusters and it is a question that cannot be confidently answered. It is difficult to envisage that an individual star is produced from original ISM material just because the Jeans mass for the material is so high. Some degree of pre-condensation of the ISM is necessary to produce stars and condensations must be much more massive than that required for just a single star. Globular clusters consist of Population II stars, which is to say stars formed from the primordial material resulting from the big bang. It is, therefore, not possible for the Sun to have originated in a globular cluster. Large-scale regions of Population I star formation, that is of stars like the Sun consisting of processed material, are giant molecular clouds. These are dense clouds hundreds to thousands of times as dense as the ISM with temperatures ~ 10 K, diameters which can be up to 100 pc and with total mass up to $10^6 M_\odot$. An example of such a star-forming region is the Orion nebula. However, we also have evidence from the existence of young galactic clusters such as NGC 2264 that star formation is actually going on within isolated regions with total mass $1000 M_\odot$ or less. While it is not certain, it is probable that many, if not most, field stars similar to the Sun began their existence as members of a galactic cluster from which they eventually escaped.

2.6.2 A general theory of star formation in a galactic cluster

Up to the late 1970s most work on star formation (e.g. Larson 1969, Black and Bodenheimer 1976) had been concerned with the evolution of isolated spherical proto-stars without regard to the way in which such entities might be produced in the first place. Work by Hunter (1962, 1964) had shown that any small density perturbation in a cloud would grow and later work by Disney *et al* (1969), who numerically studied the linear wave flow collapse of an interstellar cloud, supported that conclusion. The idea was thus established that a cloud would be unstable to

small perturbations and would spontaneously fragment into Jeans mass condensations so that it was only necessary to follow the evolution of the condensations.
This model of proto-star origin could only be sustained if the cloud was non-
turbulent. Spontaneous fragmentation would be a slow process with a free-fall
time-scale and unless the cloud was very static the incipient fragment would be
stirred back into the cloud long before it took on a stable separate existence. From
maser and other observations it is clear that star formation takes place in a highly
turbulent environment and any theoretical approach to star formation must take
account of this.

2.6.2.1   *The turbulent-cloud model*

The first approach to following the evolution of a collapsing cloud with turbulence, including a mechanism for producing proto-stars, was described by Woolfson (1979). While there were a number of simplifications in the model, all the
important features in the evolution of a cloud were included. The spherical cloud
was taken as having some initial turbulence and it was assumed that the cloud
remained of uniform density during the collapse. The Virial Theorem, in the form
(2.17), was applied to the collapsing cloud in which there were three sources of
translational energy:

(i)   The thermal energy of the material

$$E_\theta = \frac{3Mk\theta}{2m} \tag{2.25}$$

in which M is the mass of the cloud, θ its temperature and m the average
molecular mass for the cloud material.

(ii) The energy of linear-wave flow, for which each element of the cloud moves
radially in such a way as to give a homologous collapse. This is of the form

$$E_1 = \tfrac{3}{10}M(\dot{R})^2 \tag{2.26}$$

in which R is the radius of the cloud and $\dot{R}$ the speed of the boundary material.

(iii) The turbulent energy of the cloud, due to randomized motions of cloud elements superimposed on the linear-wave flow. This is written as

$$E_t = \tfrac{1}{2}M\varepsilon \tag{2.27}$$

where $\varepsilon = \langle u^2 \rangle$ is the mean-square turbulent speed.

For a uniform sphere the geometrical moment of inertia is

$$I = \tfrac{3}{5}MR^2 \tag{2.28}$$

and the gravitational potential energy is

$$V = -\frac{3}{5}\frac{GM^2}{R}. \tag{2.29}$$

Inserting (2.25)–(2.29) into the general theorem (2.17) gives an equation for the acceleration of the boundary

$$\ddot{R} = 5\frac{k\theta}{mR} + \frac{5\varepsilon}{3R} - \frac{GM}{R^2}. \tag{2.30}$$

The model used for turbulence in the cloud was to take it as consisting of roughly spherical elements, each of mass equal to a Jeans critical mass, moving with a combination of linear-wave flow and random motion. The linear-wave motion was radial and at a speed, $v(r)$, required to give homologous collapse, i.e.

$$v(r) = \frac{r}{R}\dot{R} \tag{2.31}$$

in which r is the distance of the centre of the element from the centre of the cloud. The turbulent speed of each element was made the same, $\varepsilon^{1/2}$, but in a random direction. A proper description of turbulence would be that the motion of neighbouring elements would be correlated but beyond a certain distance, the *coherence length*, l_c; the correlation would break down and relative motions of material separated by such distances would be random. From general considerations the coherence length was taken as a Jeans diameter. There follows from this a *coherence time*, t_c, which is the time required to traverse one coherence length at the turbulent speed, u. The coherence time is an expression of how long it takes for a complete redistribution of matter within the cloud. Velocities of corresponding regions of the cloud at the beginning and end of an interval much less than a coherence time will be correlated; if the interval is much more than a correlation time they will be uncorrelated.

The gravitational energy released by the cloud collapse is transformed into other forms of energy. It leads to enhancement of E_l and E_t and some of it goes into heating the cloud. Another form of heating, other than by cosmic rays and external starlight, is the radiation from stars forming within the cloud. An equation was developed giving the rate of change of temperature within the cloud. Because of the effectiveness of the cooling mechanisms, which were just taken as grain cooling according to (2.11) and ionic cooling according to (2.12), it was found that the temperature of the cloud changed comparatively little during the cloud collapse. In addition it turned out that the outcome of the model was very little affected by the cloud temperature, which was varied between 8 and 30 K in various simulations. For these reasons the temperature was kept constant during the collapse.

For most of the simulation period the collisions between turbulent elements were supersonic and the elements were compressed by collisions quickly and almost adiabatically to a density ρ_2, then cooled due to the action of cooling agents and subsequently expanded more slowly and isothermally to the original density ρ_1. For the collision of a pair of elements colliding head-on, each moving at speed u, the loss of thermal energy per unit mass in the collision plus re-expansion is

$$E_{\text{coll}} = \frac{k\theta}{m} \left\{ \frac{1}{2\gamma} \left[\left(\frac{v}{c}\right)^2 + \frac{2uv}{c^2} \right] - \ln\left(\frac{\rho_2}{\rho_1}\right) \right\} \tag{2.32}$$

where

$$\frac{v}{c} = -\left(\frac{3-\gamma}{4}\right)\frac{u}{c} + \left[\left(\frac{\gamma+1}{4}\right)^2 \frac{u^2}{c^2} + \gamma \right]^{1/2}, \tag{2.33}$$

$$\frac{\rho_2}{\rho_1} = \frac{v+u}{v} \tag{2.34}$$

and c is the speed of sound in the uncompressed material. The time between collisions was taken as the coherence time, t_c, giving a rate of loss of thermal energy per unit mass as E_{coll}/t_c. If all turbulent elements collided head-on in pairs at the suggested rate then this would be the rate of loss of turbulent energy but there would actually be oblique collisions and multiple collisions taking place. To allow for uncertainties in time-scale and the pairing-off assumption the rate of loss of turbulent energy was written as

$$Q_t = \beta E_{\text{coll}}/t_c \tag{2.35}$$

in which β, a variable parameter of order unity, was taken as unity in most simulations.

A further part of the released gravitational energy goes into compressing the cloud material and, for homologous collapse, the rate of doing work on the gas per unit mass is found to be

$$Q_1 = -\frac{3k\theta}{mR}\dot{R}. \tag{2.36}$$

The equation for energy conservation can be written in the form

$$\dot{V} + \dot{E}_1 + \dot{E}_t + M(Q_1 + Q_t) = 0$$

which reduces to

$$\dot{\varepsilon} = -\frac{2\varepsilon}{R}\dot{R} - 2Q_t. \tag{2.37}$$

Simply interpreted, the first term on the right-hand side of (2.37) is that part of the released gravitational energy that feeds the turbulence; as long as some turbulence exists in the first place, then it can grow. The second term represents the reduction in the turbulence due to interactions of the turbulent material.

2.6.2.2   Conditions for star formation

When two turbulent elements collide they form a compressed region of high density. Since each of the elements just satisfies the Jeans criterion it might be thought that just merging two of them, however gently, would produce an aggregation of material able to collapse to form a star but this is not so. For a star to be produced it would be necessary for the density to be enhanced to the point where the free-fall time (2.23) is appreciably less than the coherence time—otherwise the material of the combined elements is re-stirred into the cloud before it can actually produce a star. This sets a minimum compression and hence a minimum velocity for a head-on collision. On the other hand if the compression is too large then the combined elements will take on a thin disc, or pancake, form which is unfavourable to condensation. When these two conditions are considered in the light of the free-fall time, the coherence time for the cloud and the Jeans stability condition, it was found that, to produce a star for a head-on collision required the compression factor, ϕ, to satisfy the condition

$$3.5 < \phi < 4.0. \tag{2.38}$$

The upper limit of ϕ was found to be dependent on the geometry of the collision. In figure 2.23(a) an oblique collision is depicted of two turbulent elements where the total amount of compressed material, shown shaded, is a fraction η of the whole. The upper limit of (2.38) is less than 4.0 for $\eta < 1$ and for $\eta = 0.86$ the upper limit equals the lower limit, 3.5. The range of values of ϕ which can give rise to star formation is shown in figure 2.23(b).

2.6.2.3   The rate of star formation

If two spherical turbulent elements are considered with turbulent velocities u_1 and u_2 of equal magnitude then the approach velocity of their centres can be calculated as can η, the proportion of the material of the elements which will be compressed. If $\eta < 0.86$ then no star can form and if $\eta > 0.86$ then a star will form only if $3.5 < \phi < \phi_{max}$ where ϕ_{max} is as indicated in figure 2.23(b). Thus for any pair of directions of u_1 and u_2 the probability of a star forming can be found. By integrating over all combinations of directions the overall probability that any pair of colliding elements will give a star will be found. From (2.33) and (2.34) it is clear that this probability will depend on the Mach number of the turbulent velocity and can be written in the form $P(u/c)$, which is shown for $\gamma = 5/3$ in figure 2.24. The time-scale associated with each collision of turbulent elements is t_c so the rate of star formation can be written as

$$\frac{dS}{dt} = \frac{N\beta' P(u/c)}{2t_c} \tag{2.39}$$

in which N is the number of turbulent elements and β' is a factor, similar to β in (2.35), which allows for departures from the assumption of paired-off elements.

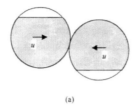

(a)

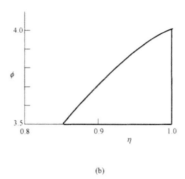

(b)

Figure 2.23. (*a*) The oblique collision of two turbulent elements each of mass M_c. The shaded material, of mass $2\eta M_c$, is compressed. (*b*) The values of ϕ and η that can give star formation are within the thick-line enclosed region.

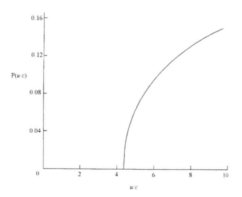

Figure 2.24. The probability function, $P(u/c)$, for star formation for $\gamma = 5/3$, $\phi_{min} = 3.5$ and $\phi_{max} = 4.0$.

2.6.2.4 *Numerical calculations*

The solution of the differential equations (2.30), (2.37) and (2.39) enables the evolution of the cloud and the rate of formation of stars with their masses to be followed. The initial radius is chosen to give the mean density about 10^{-20} kg m^{-3} and the initial dR/dt as that which makes $E_1 = E_t$. The pattern of events is not critically dependent on the starting conditions as long as the Mach number of the initial turbulence is not too low; in the simulations reported by Woolfson (1979) the initial values of u/c were between 0.66 and 1.24. In all the cases reported the mass of the cloud was $1500 M_\odot$. For the first 47 million years of the simulation no stars were produced because turbulent speeds were not sufficiently high (see figure 2.24). Stars were then formed at an ever-increasing rate with the average mass of the stars being formed decreasing with time. The simulation was terminated when no stars were produced with masses greater than $0.07 M_\odot$, the observable minimum mass, or when the total number exceeded some preset limit in the range 400–1000 although this limit was hardly ever invoked. For a real cloud the formation of stars would be terminated by dispersion of the cloud material by radiation from the already-formed stars (Herbig 1962).

Since the results of many different simulations were all very similar in character just one of them is shown here in figure 2.25 corresponding to the starting conditions in table 2.3. Figure 2.22(a) shows the build-up of u/c and the density and the fall in radius as the collapse proceeds. Because of the steady increase in turbulent energy the collapse is controlled and does not display the runaway feature of a free-fall collapse. After about 47 million years u/c has become large enough for star formation to begin at which stage the mean density of the cloud has become greater than 10^{-18} kg m^{-3}.

In figure 2.25(b) the number of stars and their mass range for intervals of 2.5×10^5 years backwards from the present time is shown. The general pattern is seen that the mass of the stars diminishes with the passage of time while the rate of star formation increases almost exponentially. Finally, the mass distribution is shown in figure 2.25(c) and the straight line indicates a mass index of -2.6, close to the observed value for stars in general.

Many features of the Williams and Cremin observations of young stellar clusters are reproduced by these results with the notable exception of the development stream of larger mass stars seen in figure 2.22(a). What is seen is the lower stream, which is referred to as the *primary stream*, which starts at a mass of $1.4 M_\odot$ about 4.5×10^6 years ago down to the lower limit of $0.07 M_\odot$ at the present time.

2.6.2.5   *Massive stars by accretion*

When a proto-star is first produced it has density $\phi \rho_1$, where ρ_1 is the background density and ϕ the compression produced by the turbulent collision. The proto-star, collapsing in approximately free-fall fashion at first, will become progressively denser relative to the cloud density since the cloud collapse is inhibited by the turbulence. While the star is within the cloud it can accrete material and there are two types of accretion process which can operate. The first of these was described by Eddington (1926) and assumes that all the cloud material that falls onto the star

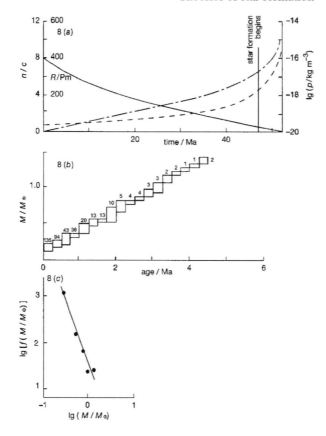

Figure 2.25. The collapse of a cloud and star formation. The characteristics of the cloud are: mass 3.0×10^{33} kg; initial density 1.2×10^{-20} kg m^{-3}; initial $\dot{R} = -492$ m s^{-1}; initial temperature 20 K; initial $u/c = 0.71$. (*a*) Variation with time of radius (R) ——, turbulence (u/c) – – –, density $\log(\rho/\text{kg m}^{-3})$ — · —. (*b*) Number of stars formed per 2.5×10^5 years with the mass range of stars formed. (*c*) Frequency of star formation against mass (slope of line -2.7) (Woolfson 1979).

is accreted. If the star has a radius r then its accretion cross-section is greater than πr^2 because of deflection of material by the star's gravitational field. If the star moves relative to the cloud material at a speed V then the rate of gain of mass is found to be

$$\frac{\mathrm{d}M}{\mathrm{d}t} = \pi r(r + 2GM/V^2)^{1/2}V\rho. \tag{2.40}$$

This corresponds to an accretion radius of

$$r_{\mathrm{a}} = \{r(r + 2GM/V^2)\}^{1/2}. \tag{2.41}$$

The second mechanism additionally takes as accreted material that which

Table 2.3. Initial parameters for the cloud collapse and star formation results displayed in figure 2.25.

Temperature (K)	20
γ	5/3
Mass of cloud (kg)	3×10^{33}
m (kg)	2.5×10^{-27}
$\phi_{\min} : \phi_{\max}$	3.5:4.0
$\beta : \beta'$	1.0:1.0
Initial density (kg m^{-3})	1.2×10^{-20}
Initial dR/dt (m s^{-1})	-492
Initial u/c	0.71

interacts along the line of motion of the star relative to the medium and forms an accretion column along the downstream axis. Bondi (1952) suggested for this type of accretion a modification of an expression given by Bondi and Hoyle (1944, Appendix IV). This gives

$$\frac{dM}{dt} = \frac{2\pi GM^2 \rho_1}{(V^2 + c^2)^{3/2}} \tag{2.42}$$

in which c is the speed of sound in the gas. The accretion radius in this case is

$$r_b = GM \left(\frac{2}{V}\right)^{1/2} (V^2 + c^2)^{-3/4}. \tag{2.43}$$

In any particular situation the larger of r_a and r_b should be chosen as the accretion radius.

It might be thought that when proto-stars were newly formed then, because of their large size, they would accrete material very rapidly but this is not so. Material striking the star would do so at more than the escape speed from the surface of the star, in general at a speed $(V^2 + V_e^2)^{1/2}$ where V_e is the escape speed. Initially V_e is small compared with V so material striking the star shares its energy with proto-star surface material some of which will escape. When the star is very diffuse the 'accretion' mechanism is actually an 'abrasion' mechanism and the star will lose mass. The abrasion loss becomes negligible when the radius of the proto-star is such that $V_e \approx V$ and thereafter accretion will commence. The imposition of this condition for accretion as against abrasion made $r_b > r_a$ in all the cases considered.

Another factor that affects the rate of accretion is turbulence in the cloud. Equations (2.40) and (2.42) have the built-in assumption that the star is moving through a quiescent cloud. An estimate can be made of the distance from the star over which the motion of the cloud material is correlated with that of the star,

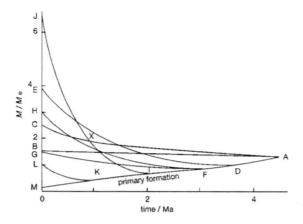

Figure 2.26. Accretion lines for stars starting with those on the primary stream. The conditions for the accretion lines are given in table 2.4 (Woolfson 1979).

which involves an arbitrary numerical parameter, α, and the accretion radius is taken as some fraction, g, of that distance. The accretion radius, taking turbulence into account, then becomes

$$r_c = g \frac{R_J V^2}{4u^2} \left\{ 1 + \left(1 + \frac{16 G M u^2}{\alpha R_J V^4} \right)^{1/2} \right\} \qquad (2.44)$$

in which R_J, the Jeans radius, comes in as a measure of the coherence length in the cloud. In numerical work α and g were taken as 0.1 and 0.6 respectively but results were not sensitive to this choice of parameters. When turbulence is low then (2.44) can give $r_c > r_b$, in which case the accretion radius used is r_c.

One more factor must be taken into account in considering accretion and that is the fact that a more-realistic non-homologous model of cloud collapse could give a density in the central region of the cloud up to one hundred times the average density (Disney *et al* 1969). An accreting star would be moving through cloud material of ever-changing density and without a knowledge of the actual motion there is no way of calculating the overall effect of this on accretion. The procedure adopted by Woolfson was to assume that accretion took place at a constant factor, f, of density enhancement over the average density throughout the accretion period. The values of f taken for illustration of possible accretion processes were 1, $10^{1/2}$, 10 and $10^{3/2}$ together with five values of $V/u - 3^{-1}, 3^{-1/2}$, 1, $3^{1/2}$ and 3. It was found in practice that only the smaller values of V/u gave appreciable accretion.

Various accretion lines are shown in figure 2.26 corresponding to the entries in table 2.4. Under favourable conditions the final mass can be more than $6M_\odot$ corresponding to the largest mass stars in figure 2.22(*a*).

Table 2.4. Accretion lines for various initial times and parameters f and V/u. M_0 is the initial mass and M_f is the final mass.

Line	$M_0/M_\odot$	$M_f/M_\odot$	f	V/u
AB	1.35	1.65	1.00	$3^{-1/2}$
AC	1.35	2.50	$10^{1/2}$	$3^{-1/2}$
DE	0.97	3.89	10	$3^{-1/2}$
FG	0.80	1.50	10	3^{-1}
FH	0.80	3.00	10	$3^{-1/2}$
IJ	0.71	6.64	$10^{3/2}$	$3^{-1/2}$
KL	0.42	1.01	$10^{3/2}$	$3^{-1/2}$

In a real cloud the accretion pattern for an individual star on the mass/time diagram might be quite complex. For example, a star that had accreted along AX and suddenly entered a dense region could then continue along XJ.

While it is evident that the upper development stream in figure 2.22(a) can be accounted for, this cannot be done quantitatively so the effect on, say, the mass distribution is unpredictable. Another factor that complicates the comparison between the model and observation is the estimation of the age of a star that has accreted. The path of such a star on the H–R diagram is unknown although von Sengbusch and Temesvary (1966), considering rapid accretion, concluded that Hayashi's treatment of the evolution of a proto-star would need to be considerably modified.

Despite its limitations, the theory described by Woolfson does indicate the general way in which a cluster of stars can be formed in a DMC and reasonable agreement with the observations is found.

2.6.2.6 Angular momentum

In figure 2.6 the mean equatorial speed as a function of spectral class is shown. For stars with masses less than about $1.4M_\odot$ (spectral class F5) equatorial speeds are low and, on the whole, this is the mass range corresponding to the primary stream of development. Most stars in this mass range would have acquired their mass by the collision of two turbulent streams of material without much further accretion. We shall now see why this mode of formation should lead to a star with little angular momentum.

The head-on collision of two streams of gas with equal densities and speeds is shown in figure 2.27(a). If the speed of the gas is fairly uniform across the streams then the compressed region, which is to form the star, will have comparatively little angular momentum. However, the star will have only several times the density of the background material and will be strongly coupled to that material through abrasion and, until it reaches a density such that interchange of material

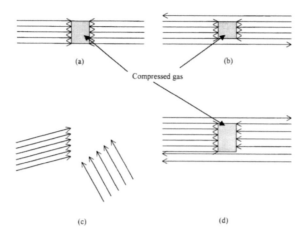

Figure 2.27. (*a*) Head-on collision of two gas streams with similar densities and speeds. (*b*) An offset collision. All angular momentum is associated with the uncompressed material. (*c*) Two streams with non-parallel motions. (*d*) The situation in (*c*) referred to the centre of mass of the gas streams.

with the medium is negligible, it will not collapse with conservation of angular momentum. At the time that the star becomes decoupled from the environment, abrasion and accretion would be in balance and the expected angular momentum of accreted material would then be that of material in the outer parts of the star. With an estimate of the velocity gradient in the cloud material as u/R_J, the angular momentum per unit mass of accreted material is found to be

$$\frac{dH}{dM} = \frac{GM^2 u}{8R_J V^4}. \tag{2.45}$$

If this is the intrinsic angular momentum of stellar equatorial material which applies when the star has collapsed to the main sequence radius r_* then the final equatorial speed will be

$$V_{eq} = \frac{G^2 M^2 u}{8R_J V^4 r_*}. \tag{2.46}$$

For the model which gave figure 2.25, when $M = M_\odot$ then $u = 1.65 \text{ km s}^{-1}$ and $R_J = 8 \times 10^{12}$ km. With $r_* = R_\odot$ and $V = 4 \text{ km s}^{-1}$ the value of V_{eq} is 2.6 km s^{-1} which is not much above the value for the Sun. However, it will be seen from (2.46) that V_{eq} depends very sensitively on V and for $V = 2 \text{ km s}^{-1}$ the value of V_{eq} increases to 41 km s^{-1}.

Various mechanisms have been suggested from time to time to explain the slow rotation of the Sun or, more precisely, how late-type stars lose angular momentum during their collapse. A review of suggested mechanisms is given in chapter 6 but for now we may take it that *modest* reductions in angular momentum are possible and that the head-on collision of two streams can lead to the small equatorial velocities which are observed for late-type stars.

The previous discussion has been in relation to the head-on collision, centre to centre, of two streams of gas of similar characteristics. If the streams are moving in anti-parallel directions but with an offset, as shown in figure 2.27(*b*) then the combined streams will have net angular momentum but none of it will be associated with the compressed material, which may form a star if it satisfies the necessary conditions. An oblique collision, as shown in figure 2.27(*c*), is redepicted relative to the centre of mass in figure 2.27(*d*) and is seen to be a head-on collision with some offset and perhaps with different speeds and densities for the two streams but the conclusion is still that the compressed material will have little angular momentum.

Stars that have had an appreciable gain of mass by accretion would have crossed many turbulent regions during the mass-gain process and in each region the gain of angular momentum, considered vectorially, will be in a different direction. The calculation of the total gain of angular momentum takes the form of a random-walk problem and the expected magnitude of the final angular momentum is given by

$$H = \left\{ \int_{t_0}^{t_f} t_r \left(\frac{dH}{dM} \right)^2 \left(\frac{dM}{dt} \right)^2 dt \right\}^{1/2} \tag{2.47}$$

where t_r is the mean time spent within a turbulent region, taken as $4R_J/3V$, dH/dM is given by (2.45) and dM/dt is the rate of gain of mass as described in section 2.6.2.5. For an accreting star this integral can be evaluated numerically.

The known characteristics of a real star are its mass, M_*, radius, R_* and equatorial speed, V_{eq*}. To use these observed quantities to estimate angular momentum requires a knowledge of the moment-of-inertia factor, α_*, which then gives the angular momentum as

$$H_* = \alpha_* M_* R_* V_{eq*}. \tag{2.48}$$

Chandrasekhar (1939) gave for main-sequence stars with different masses the proportion of the radius, ξ_*, within which 90% of the mass of the star is contained. Woolfson (1979) showed that from this information alone it is possible to estimate α_* within 10% or so and for the Sun a reasonable estimate is 0.055.

The estimated values of $\log(H_*)$ from a combination of observation and Chandrasekhar's theoretical results are shown for various M_* in figure 2.28 together with values calculated for a number of model stars from (2.47). The quantitative agreement is quite good—better than might be expected in view of the variable parameters in the star-forming model—but the general form of relationship including the fall-off at lower mass, is of more significance. It turns out that when (2.47) is solved numerically most of the mass, and therefore most of the angular momentum, is gained at the end of the accretion period. If it is assumed

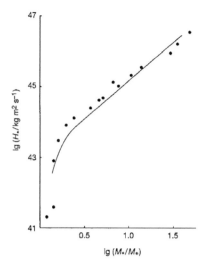

Figure 2.28. The observed relationship between the angular momentum of a star H^* and its mass M^* (full line). The points are derived from the star-forming model (Woolfson 1979).

that the gain is all in one turbulent element then the vector-addition feature of (2.47) can be ignored and an analytical solution for the angular momentum can be found. This solution shows that for large final star masses $H \propto M_f^2$ which agrees with the calculated results; the results from observation and α_* suggest the slightly different relationship $H \propto M^{2.1}$.

2.6.2.7 *A critical review of Woolfson's star-formation theory*

The assumption of a homologous collapse of a cloud is quite a severe one and the development of a stellar cluster is likely to depend critically on the cloud's inhomogeneity. This was introduced artificially when considering mass accretion by appealing to the results of Disney *et al* (1969)—but then their results did not specifically include the effect of turbulence, which is an important factor in the overall evolution of the cloud.

The results of the theory with respect to the mass index for stars produced in the primary stream, which agreed with observation, would be disturbed in an unpredictable way by the accretion process. However, the angular momentum calculations are reasonably robust and do give good agreement.

As given the theory produces individual stars, whereas observation suggests that most stars will be in the form of binaries. Woolfson referred to work by Aarseth (1968) who studied interactions between stars in a cluster and concluded that these could lead to binary formation. However, it seems unlikely that as many as two-thirds of stars would finish in binary systems due to this type of mechanism.

The conditions predicted by the model agree quite well with observations of maser sources (section 2.4.2) with respect to turbulent speeds, overall size of source region (cloud size) and size of individual sources (proto-stars). The theory is the only one at present that has explored in detail the development of a cloud through all stages up to producing proto-stars together with estimates of the characteristics of the stars produced. The work of Golansky (1999) shows that a cloud will collapse and break up in a more complicated way than the original theory suggests. However, the 1979 model can be applied to the individual clumps shown in figures 2.13, 2.14 and 2.15. Overall the general theory has plausibility in terms of its main features but it also has a number of deficiencies that would require much more detailed modelling to remove.

2.6.2.8   *The formation of binary and multiple star systems*

In a series of papers, originating in the Department of Physics and Astronomy at Cardiff, the supersonic collisions of gas clouds has been modelled using SPH (Appendix III). To some extent this work supplements the Woolfson star-forming model but the range of conditions explored by the Cardiff group is often outside that considered by Woolfson.

Pongracic *et al* (1991) modelled the collision of two identical sub-clouds with masses $75M_\odot$ and radii 1 pc colliding at 1.62 km s^{-1} (Mach 4). The material of the sub-clouds was taken as molecular hydrogen and cooling was roughly incorporated into the model by taking the temperature as 100 K for number densities of H$_2$, n, less than 3×10^8 m^{-3}, as 10 K for $n > 10^{10}$ m^{-3} and as $100(n/3 \times 10^8)^{-2/3}$ K at intermediate densities. Where the sub-clouds collide a dense shocked layer is formed and the model shows a proto-star condensing out of this layer. If the collision speed is increased then a bound binary system is formed, further increase of collision speed gives two unbound stars and at very large impact speeds no stars are produced. This last conclusion agrees with that of Woolfson and corresponds to $\phi > \phi_{max}$ in figure 2.23(*b*). In another simulation 20 sub-clouds, each of mass $5M_\odot$ and radius 0.1 pc, with a velocity dispersion of 1 km s^{-1} interacted to give an outcome in which stars of masses $20M_\odot$ and $6M_\odot$ formed a close binary system of diameter 800 AU while a third star of mass $2M_\odot$ orbited the close binary system at a distance of 3500 AU. The SPH particle positions projected on the x–y plane for the initial sub-clouds and the final three-star system are shown in figure 2.29. Other simulations of a similar kind were reported by Chapman *et al* (1992); one simulation gave two close binary systems orbiting each other.

Another mechanism for binary star formation, numerically modelled by Whitworth *et al* (1995), is that in which a proto-stellar disc, produced by the collision of sub-clouds, accretes material and its specific angular momentum increases with time. This has some relationship to the angular momentum increase with ac-

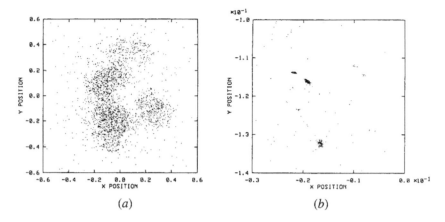

Figure 2.29. An SPH simulation of a collision of 20 clouds each of mass $5M_\odot$. (*a*) The initial configuration of SPH points. (*b*) The final configuration showing three stars in a hierarchical binary configuration (Pongracic *et al* 1991).

cretion described in section 2.6.2.6. If the angular momentum of the proto-stellar disc increases rapidly then the disc becomes unstable in a bar mode and breaks up into two components similar in size. Further accretion by these fragments takes up angular momentum into orbital motion rather than spin and a binary system is formed in this way. For a slower increase in angular momentum of the disc it develops a system of spiral arms which, if they are massive enough, become detached and form secondary condensations. These sometimes recombine with the disc but may also form aggregations to give massive separate condensations. Again the final outcome of this scenario includes binary systems.

These numerical simulations, and also analytical approaches by Whitworth *et al* (1994a, b), all lead to stellar masses of $2M_\odot$ or more and have been suggested as appropriate to the formation of OB associations as in the Orion cloud. However, they do reveal that the assumption by Woolfson that a collision of a pair of turbulent elements under the conditions of his model will necessarily give an individual star is probably invalid and points to the need for detailed modelling. If it turned out that binary systems could form from a pair of turbulent elements, each of mass $1M_\odot$ or less, then a major deficiency in the Woolfson model will have been addressed.

2.7 Planets around other stars

In 1992 two planets, with masses similar to that of the Earth and with periods 66.6 and 98.2 days were detected in orbit around the pulsar PSR1257+12, with the possibility of a third planet being present. Their detection depended on the extreme regularity of the radio pulses from pulsars. The pulsar moves around the centre of mass of the pulsar–planet system; when it is in that part of its orbit which causes it to move away from Earth relative to the centre of mass then there will be a cumulative delay in the arrival of the pulses as they have further to travel. The opposite effect occurs when it moves towards the Earth relative to the centre of mass.

Although it is an intriguing observation it may not be relevant to the formation of planetary systems around normal stars. A pulsar is a neutron star that is the outcome of a supernova. There is no way of knowing whether the planetary bodies were part of a pre-existing system that somehow survived the supernova event, had been captured subsequently as previously-formed bodies or had condensed from the debris of the supernova.

An unambiguous observation of a planet around a normal star, 51 Pegasus, was made in 1995 by Mayor and Queloz. The technique they used depended on monitoring the Doppler shift of spectral lines over a long period of time. For a star of mass M_* with a companion planet of mass M_P at distance D, the centre of mass will be at distance $D \times M_P/(M_* + M_P)$ from the star. Assuming a circular orbit the angular velocity of the bodies about their centre of mass is

$$\omega = \left\{ \frac{G(M_* + M_P)}{D^3} \right\}^{1/2} = \frac{2\pi}{P} \qquad (2.49)$$

where P is the period of the orbit, which can be measured. The other quantity which can be estimated from the Doppler shifts is the orbital speed of the star in its orbit—or, more correctly, the radial component of that speed. If the normal to the orbital plane makes an angle i with the line of sight then the radial component of the orbital speed, $_iV_*$, is related to the orbital speed, V_*, by

$$V_* = \frac{_iV_*}{\sin i}. \qquad (2.50)$$

The orbital speed of the star is

$$V_* = D \frac{M_P}{M_* + M_P} \frac{2\pi}{P}. \qquad (2.51)$$

Assuming that $M_P \ll M_*$ equations (2.49), (2.50) and (2.51) can be rearranged to give

$$M_P = \frac{_iV_*}{\sin i} \left(\frac{PM_*^2}{2\pi G} \right)^{1/3}. \qquad (2.52)$$

With the exception of the inclination angle all quantities on the right-hand side of (2.52) can be estimated so a minimum planetary mass can be calculated, where the actual planetary mass will be larger by an unknown factor $1/\sin i$.

A planet around 47 UMa (Ursa Major) was detected by Butler and Marcy (1996). Initially, Doppler velocities could be measured to a precision of 10 m s^{-1} and 34 observations of 47 UMa were made over a period of 8.7 years starting in

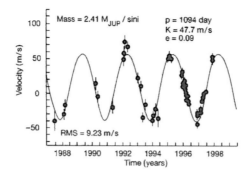

Figure 2.30. Measured Doppler shifts from the star 47 UMa (Butler and Marcy).

the middle of 1987. Observations after November 1994 were made with an upgraded spectrometer that reduced the error of velocity measurements to 3 m s^{-1}. The Doppler shifts are shown in figure 2.30. The period is 3.00 years and the rms variation of the velocities is 35 m s^{-1}. The star is of spectral class G0 and was taken as having a mass $1.05 M_\odot$. Based on the observations it was deduced that the mass of the orbiting body is $2.41 M_J / \sin i$, where M_J is the mass of Jupiter and that the orbital radius is 2.11 AU. The unknown inclination of the orbit gives the slight possibility that what is being observed is not a planet but a *brown dwarf*, a body not supporting nuclear reactions that is intermediate in characteristics between a star and a planet and with mass in the range 13–$75 M_J$. The probability of this is very low as it would imply a very small, and statistically unlikely, inclination.

The Doppler-shift method favours the detection of massive planets. Again, if the orbital radius is small then the period is shorter and more orbits can be followed for a given period of observation. In addition a small orbital radius also gives larger stellar orbital speeds and hence less difficulty in measuring those speeds. In principle planets with a tenth or so of Jupiter's mass could be detected if they were in extended orbits but the penalty would be that observations would have to made over a very long period.

Table 2.5 gives the characteristics of the first few planets that have been detected around normal stars. Now that the technique for their detection has been established there is a steady rate of increase in the number reported.

There has been some speculation about the way in which large planets could end up so close to the parent star—for example, the orbital radius of 51 Peg is only about one-seventh of that of Mercury. A popular scenario is that the planets were formed much further out but then spiralled in due to the presence of the material surrounding the star from some of which the planet had previously formed. Lin *et al* (1996) have suggested two possible mechanisms which would stop the planet being absorbed by the parent star. The first is that when the planet is close enough it raises a tide on the star. If the star is spinning with a shorter period than the

Table 2.5. Characteristics of planets detected around normal stars.

Star	Minimum planet mass (M_J)	Orbital radius (AU)
47 Uma	2.41	2.10
51 Peg	0.45–0.7	0.05
55 Cancer	0.80	0.21
70 Vir	6.60	0.66
τ Boolis	3.8	0.05
υ Andromedae	0.68	0.05
16 Cygnus B	1.7	1.7
HD 114762	9.00	0.60

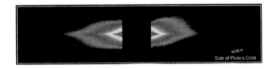

Figure 2.31. False-colour image of a disc around β-Pictoris (Hubble Space Telescope).

planet's orbital period then the tidal bulge will be dragged ahead of the radius vector. This bulge is then in a position to pull on the planet in the direction of its motion thus tending to increase its energy that will counteract the decay due to the resisting medium. The second mechanism invokes the stellar magnetic field to clear away the gas and dust close to the star. Once the planet entered this region then any tendency to further spiral inward will be halted.

Since planets of mass much less than Jupiter cannot be detected at present it is not possible to estimate the proportion of stars with planetary companions. Large numbers of stars are being monitored at the Lick Observatory and Butler and Marcey (1996) have suggested that 3% of stars may have planetary companions of mass greater than $2M_J$. The establishment of this proportion clearly has implications for possible theories of the origin of planetary systems.

2.8 Circumstellar discs

Some theories of the origin of planetary systems require planets to be formed from a disc of gas and dust around a newly-formed star. In the 1980s evidence began to appear that many YSOs (Young Stellar Objects—either young stars or proto-stars) are accompanied by circumstellar discs and this gave support to the theories which required such discs.

The most convincing visual evidence for discs come from direct imaging either in the infrared or visible parts of the spectrum. An image of a disc around β-Pictoris is shown in figure 2.31 and similar images of discs around other stars, e.g. Vega, have been produced. These stars are not YSOs and the discs have a very low mass, $\sim 10^{-7} M_\odot$, and are hundreds, or even thousands, of AU in extent; they are interpreted as the residues of much more massive discs that surrounded the stars when they were younger. Imaging of YSOs in the near infrared (Grasdalen *et al* 1984, Beck and Beckwith 1984, Strom *et al* 1985) has shown discs for HL Tauri, R Monocerotis and the infrared source L1551/IRS 5. The radiation from which the image is produced is that scattered by micrometre and sub-micrometre dust particles within the disc. The image of HL Tauri, a star less than 10^5 years old and at a distance of 150 pc, shows gas emission in a disc-like form with an additional fast jet moving outwards along the disc axis. The total mass of the disc is estimated as $0.1 M_\odot$ of which the solid component would be 300 or so Earth masses, more than enough to provide the solid components of planets in the Solar System.

For the most part the evidence for discs around YSOs comes from the spectral energy distribution in the light coming from the source. The energy distribution can be interpreted as the sum of that coming from two separate types of source. The first source is the central star giving a typical black-body emission corresponding to its temperature (figure 2.2) and the second the disc, a low temperature but very extended source emitting mainly in the infrared. The disc is thus detected from the *infrared excess* in the light emission from the star. For some stars, e.g. T Tauri, there is a very vigorous solar wind. Light from the central star is scattered by solid particles which are moving outwards and thus will be red-shifted on the far side of the star and blue-shifted on the near side. Observations with [O I] and [S II] spectral lines nearly always show only the blue-shifted component. This is interpreted as being due to a disc sufficiently opaque to block out radiation from the far side.

Observations of young stars of spectral classes A, F, G and K with ages less than about 3×10^6 years indicate that about 50% of them have dusty environments with extents from 50 to several thousand AU and with masses typically in the range 0.01–$0.1 M_\odot$. The evolutionary time-scale for these discs is of order 3×10^6–10^7 years but this merely indicates the time for which the material exists in the form of very fine grains and hence scatters radiation effectively. If the material had organized itself into large objects, even of metre size, then it would be quite invisible. The implications of this evolutionary time- scale for planetary formation will be discussed at greater length in section 5.4.

Chapter 3

What should a theory explain?

3.1 The nature of scientific theories

3.1.1 What is a good theory?

A scientific theory is a systematic explanation in terms of basic principles of some body of information. To say that the sky is blue is not the statement of a theory, since it is a verifiable observation about which there can be no dispute, but it is possible to give a theory for *why* it is blue. Although at any time there are many scientific theories which are generally accepted it is important to understand that there is no such thing as a *correct* theory. Any theory that is currently accepted and used must be regarded as only plausible, or an approximation, until either it is found to disagree with some observation, old or new, with some experiment or until some internal inconsistency is found theoretically. Newton's law of gravitation, which could not explain the precession of the orbit of Mercury, was eventually replaced by Einstein's theory of gravity, which includes Newtonian gravity as an approximation. From Einstein's theory a prediction was made about the bending of light passing a massive object, which was confirmed by observations during the solar eclipse of 1919. Another example is the Bohr theory for the structure of the hydrogen atom that explained the hydrogen spectrum but failed for many-electron atoms and was eventually replaced by models based on quantum mechanics.

Older theories that are replaced by better theories may still sometimes be useful. Nobody these days would teach the Bohr model, except as an historical illustration, but Newtonian gravity is still a useful tool because it closely approximates what is given by relativity theory so that for most practical purposes there is no discernible difference. In line with what has been stated about the status of current theories both quantum mechanics and general-relativity theory can themselves be only regarded just as plausible. In the light of new knowledge they may eventually be found wanting and then eventually be replaced by more comprehensive theories to which they are just approximations.

A good theory, in the sense of being plausible and worthy of serious consideration, should have the following characteristics:

(i) It should explain what is known.
(ii) Ideally it should have predictive power, that is to say that it should predict some result or observation which can then be tested.
(iii) Related to (ii) a theory should be sufficiently detailed to be subjected to experimental or theoretical test and so be *vulnerable*. Theories that are so vaguely formulated that they are not vulnerable are also of little value.

3.1.2 The acceptance of new theories

To be truly plausible a theory must be consistent with *all* the observations to which it is relevant. Even if it explains a multitude of observations but fails to explain or, even more importantly, actually contradicts one well-established observation then the theory must be presumed to be wrong. The theory might still have a role to play because it is useful and either there is no replacement for it or the replacement is too complicated to use in practical applications. Here we are echoing the situation relating Newtonian gravity to general relativity. It might be possible that the theory is not completely wrong but just has one aspect that can be revised to fit in with the new information without disturbing its agreement with previous observations. However, there is always the other possibility—that the theory is completely wrong and that its replacement will involve very different basic principles.

A good illustration of the replacement of one theory by another is the supplanting of the Ptolemeic Earth-centred theory of the structure of the Solar System by the Copernican heliocentric theory. The reason that Copernicus proposed a Sun-centred system was not that Ptolemy's theory did not explain the observations. Within the limitations of the observations which existed in the early part of the 16th century the arbitrary system of deferents and epicycles proposed by Ptolemy explained the motions of the planets quite adequately. Actually, since Copernicus wanted all his planetary orbits to be circles he had to take the centres of the orbits offset from the Sun and also introduce small epicycles to obtain agreement with observations, so his model also had its arbitrary features. Copernicus preferred the Sun-centred model because it explained the observations equally well and it was somewhat less complicated. He was also influenced by the systematic progression of orbital radii revealed by the heliocentric model that gave the impression of order in the system. Consciously or not, Copernicus was applying a principle first enunciated by the English philosopher William of Occam (1285–1349) and known generally as *Occam's razor*. Loosely translated from the Latin (*Entia non sunt multiplicanda praeter necessitatem*) this is taken to imply that 'if alternative theories are available that explain the observations equally well then the simpler is to be preferred'.

In 1610 Galileo observed the phases of Venus with his telescope, which was inconsistent with the Ptolemeic model but consistent with the heliocentric model. Now, apart from any question of simplicity, the theories did *not* explain the observations equally well and, following the precepts of Occam's razor, it might be thought that the Copernican view would have been quickly accepted—but this was not so. The mechanisms by which failing theories are replaced by new theories have been analysed in detail by the American philosopher Thomas S Kuhn (1970). He shows by examples throughout the ages that there is a great resistance to change within the scientific community. If a new observation conflicts with the prevailing dominant theory the first reaction to the crisis is to attempt to patch up the theory in some way to accommodate the new information. This may be a very extended process and attempts to patch up the old theory may persist for long after reason may indicate that it is untenable—even if a new theory already exists or becomes available which explains the new observation as well as all the previous ones. For almost a century after Galileo's critical observations of Venus, the Earth-centred model still had its strong adherents.

3.1.3   Particular problems associated with the Solar System

In section 2.7 an account was given of the observations of planets around other stars so it is clear that a general theory is required for the formation of planets in general and not just the Solar System in particular. Nevertheless there are still some doubts about whether or not the Solar System is a typical example of a large number of systems of a similar kind. All that has yet been observed around other stars are up to three but usually single very massive planets some of which are in orbit very close to the parent star. Since this is the combination of characteristics which is easiest to observe, because it gives the greatest Doppler shift of light from the star and the shortest period, there is no way of knowing whether or not these solitary planets are members of planetary *systems*. Although it is by no means certain that the Solar System is typical of many systems of the same kind, it seems reasonable, on balance, to think in terms of theories that would sometimes give planetary systems with several planets.

Chapter 4 gives descriptions of the many different kinds of theory that have been advanced to explain the origin of the Solar System and all of them can explain at least some features of the system. If one or other of these theories explains *nearly* all features of the Solar System but fails to explain one or more important features, and seems incapable of being amended to provide an explanation, then the theory must be regarded as almost certainly wrong—or at least highly suspect. In this case to think of it as an *approximate* theory, in the sense that Newtonian gravitation is an approximation to Einstein's gravitation, is not really sensible. There may be some other theory, existing or still to be advanced, which will explain *all* the known features in terms of a completely different mechanism. However, even this alternative model, however plausible it seemed, might itself be invalidated by new observations in due course.

3.2 Required features of theories

Knowledge about the Solar System has expanded greatly with the advent of space research, with spacecraft having been sent to many solar-system objects to examine them either directly or remotely and where instruments in space, such as the Hubble Space Telescope, have made detailed observations never previously possible. By and large this new knowledge is concerned with the detail of the system; its major features have been known from Earth-based observation for a very long time. Here we are going to divide the features of the system into different categories as far as theories are concerned and for any theory to be taken at all seriously it is essential that it should explain the major and gross features of the system. There will be other kinds of feature which are less gross and, at the extreme, some fine details of the system which most theories do not even try to address. If there are competing theories of equal merit in terms of explaining the grosser features of the system then it is these finer details which may be important for distinguishing the more from the less plausible theories.

Although it may seem reasonable to most people that the viability of theories should be tested against some objective set of criteria, not everyone is so convinced. Stephen G Brush, an American science historian, has concluded that the lists of 'facts to be explained' presented by theorists do not provide a serious basis for choosing the best theory (Brush 1996). He points out that the present-day features of the Solar System may not be original and may be the result of dissipative processes and perturbations over the lifetime of the system. He also asserts that different theorists produce different lists of 'important features' with the suggestion that each author produces the list with which his own theories can best deal. That may well be so. Despite Brush's view it does seem to be useful to put forward a set of features that can be used to judge theories of the origin *and evolution* of the Solar System. If not, then by what other criteria can judgements be made? The list given here is a union of all the features suggested by various workers, as listed by Brush, plus some others which have arisen from very recent observations.

3.2.1 First-order features

A basic simple description of the Solar System, at the lowest resolution, is that it consists of the Sun plus a family of planets in almost-circular direct orbits. The Sun, with 99.87% of the mass of the system, is spinning very slowly on its axis so that it contains less than 0.5% of the angular momentum of the system, the rest being in the orbital motion of the planets. The abundance of light elements, lithium, beryllium and boron, on Earth indicates that the Earth, and probably the other planets, were formed by 'cold' material meaning that it has not been derived from inside a star, otherwise the light elements would have been destroyed by nuclear reactions. Based on this rather crude picture of the Solar System we now give a number of first-order features to be explained by any plausible theory. It should be emphasized that because these are designated as first-order features it neither means that they are the most difficult for a theory to deal with nor that satisfying them is a sufficient indication that the theory is plausible. What it does mean is that any theory that *fails* to explain these features is at least unsatisfactory and any theory that is inconsistent with these features is almost certainly wrong. In that sense these features can be thought of as a first filter to categorize theories. The first-order features are:

(1) the distribution of angular momentum between the Sun and the planets;
(2) a mechanism for forming planets in some way on short enough time-scales;
(3) planets to be formed from 'cold' material which has not come from within a hot star;
(4) direct and almost coplanar orbits for the planets.

We shall see that for many theories (1) does turn out to be a rather difficult feature to explain and is usually ascribed to the evolutionary phase of solar-system development rather than an initial creation feature. The formation of planets, feature (2), also presents difficulties. If planets are to be formed from the material in circumstellar discs then, as was indicated in section 2.8, this would need to take place on a time-scale of order 10^7 years or less, for this is the observed lifetime of the discs. On the other hand (3) and (4) present few problems for most theories and (4) might even come about as an evolutionary feature from an almost random initial configuration of planets.

3.2.2 Second-order features

The division into terrestrial and major planets is a very obvious feature of the Solar System and the difference in structure between the two types of planet is usually put down to the difference of locations in which they formed. It seems intuitively obvious that a planet formed in a higher temperature region close to the Sun would be much less able to retain volatile material, especially hydrogen, than those formed further away. This may well be true although from table 2.4 it appears that giant planets can exist even closer to a star than Mercury is to the Sun. However, the Solar System could have evolved in a very different way from the other systems (if they *are* systems) that have been observed. It has been suggested for the other systems that the large planets may have formed further out, where acquisition of a massive volatile envelope over a substantial iron-silicate core was possible, and then drifted inwards (section 2.7). It is likely that a giant planet could *exist* close to the Sun but not actually form there, especially if a planetary core of sufficient mass was not available. In the case of Jupiter, with an assumed silicate-iron core of ten Earth masses, it would just about be possible for it to acquire its hydrogen–helium envelope in Mercury's orbit.

Examining the Solar System in a little more detail we note that the major planets all have substantial satellite families, but that only the Earth and Mars of the terrestrial planets have satellites—the Moon which is very large in relation to the Earth and the two tiny Martian satellites. Pluto, the very small outermost planet, also has a comparatively large satellite companion. There are two possible general ways in which a planet may acquire satellites, either as part of the process by which the planet forms or by capture after the planet had formed. The division of satellites into regular and irregular categories (section 1.4.1) suggests to many theorists that the regular satellites, with circular direct orbits in the equatorial plane of the planet, may be associated with planetary formation but that irregular satellites are captured bodies. Of particular interest are the large irregular satellites, the Moon, already mentioned, and Triton, which has a circular but retrograde orbit around Neptune. However, since satellites are so widespread they must be regarded as an essential feature of the system and produced in some systematic way.

We have already noted the partitioning of the magnitude of the angular momentum in the Solar System but looked at in more detail a new feature is seen. The angular momentum vector of the Sun is inclined at 6° (usually misquoted as 7°) to the normal to the mean plane of the system. This is an intriguing feature, throwing out challenges to every kind of theory. Theories are of two main kinds— *cogenetic*, where the planets are produced from the same material that produced the Sun; and *dualistic* where the Sun forms first and the planets are formed later from a different source of material. For cogenetic theories the problem is to explain why the spin axis of the Sun is so *far* from the normal to the mean plane of the system. For dualistic theories the problem is to explain why the spin axis is so *close* to the normal to the mean plane. Related to the last point, the probability of two directions being collinear to within 6° just by chance is about 0.0027.

Another probability-related characteristic of any theory is to do with its prediction of the proportion of stars that would acquire planets. Until 1992 there was no evidence that any star of any kind had a planetary-mass companion and theories could not be subjected to this kind of test. From the number of stars that have now been shown to have an attendant planet it has been estimated that 3–5% of stars may possess planetary systems. This estimate could be badly wrong in either direction and, as previously stated, it is not at all certain that the detection of a single large planet is an indication that a whole family is necessarily present. Nevertheless it seems likely that the Solar System is not unique and that other planetary systems exist, albeit that they may not resemble the Solar System in a detailed way. Thus any theory that depends on a very unlikely event must be suspect and greater credence should be given to those theories that predict that planetary systems are common, at least at the 1% or so level. This brings us to the second list of features to be explained:

(5) the division into terrestrial and giant planets;
(6) the existence of regular satellites;
(7) the existence of irregular satellites;
(8) the 6° tilt of the solar spin axis;
(9) the existence of other planetary systems.

3.2.3 Third-order features

We have already commented on the planarity of the Solar System as a feature to be explained but we also indicated that it might be an evolutionary feature. It is not difficult to find dissipation mechanisms which could flatten a system. However, the system is not *completely* planar with the greatest departures from orbital planarity being shown by the extreme members Pluto and Mercury (orbits at 17° and 7° to the ecliptic, respectively). Any theory which predicted a highly coplanar system would need to invoke an evolutionary mechanism for making it less coplanar, and such mechanisms must depend on interactions between individual members of the system or some external influence. Again, the spin axes of many planets are inclined at large angles to their orbits, with Venus and Uranus having retrograde spins—a feature that needs to be addressed, especially by any theory predicting coplanar formation of non-interacting planets.

Some theorists make much of the progression of orbital radii expressed in the form of Bode's law (section 1.2.2) while others ascribe little importance to it. There are also regularities of a Bode-like form linking satellite orbital radii although these are much less convincing. What are more convincing, for both planets and satellites, are the commensurabilities linking their orbital periods (section 1.2.3). There are evolutionary mechanisms available for producing commensurabilities so not much should be expected of a theory at the creation stage of the Solar System. Nevertheless there are some theories that make a point of explaining Bode's law and, since this is so, it must be included in a comprehensive list of features of interest.

The physical division between terrestrial and major planets, and the Bode's law gap between them, is filled by the asteroids that mostly occupy the region between Mars and Jupiter. Asteroids are interpreted either as material at an intermediate stage in the process of forming planets which did not go to completion or as products of some catastrophic event involving the break-up of one or more large bodies in the system. Comparisons of infrared spectra of asteroids and meteorites indicate that they are related, a generally accepted conclusion, so that investigations of meteorites are also telling us something about asteroids. Like meteorites, asteroids have different compositions, either iron, stone or a mixture and also may have been extensively heated, and sometimes re-heated, at some stage.

Laboratory investigations have shown many compositional and structural properties and features of meteorites that relate to either the origin or the evolution of the Solar System. Of particular importance are the presence of isotopic anomalies in meteorites which indicate nucleosynthetic events which could have been either of a catastrophic nature or have continuously taken place due to bombardment of solar-system material by high-energy particles. These features of meteorites will be dealt with in some detail in sections 11.2 to 11.6. It would clearly be of interest if a theory could address the origins of these properties and features in a convincing way.

There is some debate about the distinction, if any, between comets and asteroids. In a statistical sense they occupy different regions of the Solar System, with comets being further out than asteroids although with some overlap. In particular the vast number of comets in the Oort belt, at distances of tens of thousands of AU and stretching halfway to the nearest stars, mark the practical boundary of the Solar System. The origin of the Oort cloud, and the mechanisms for its survival against outside perturbations, are possible features for theories to address. Asteroids and comets are usually thought of as having different compositions, so that comets contain large inventories of volatile material while asteroids are largely inert. Measurements on the comet *Hale–Bopp* have shown the coexistence of high-temperature silicate material with ices that could not exist at high temperatures. This clearly has some significance for the origin of comets. A number of researchers believe that some asteroids are simply spent comets, which have exhausted their volatile components. The observational evidence is not too clear in this area. Nevertheless there is no doubt that a legitimate concern of a theory should be the presence and formation of asteroids and comets. An important feature of a theory would be whether or not it distinguishes the two classes of objects and can explain their apparently different physical properties and distributions.

Finally, we come to a number of smaller objects, not previously considered, some of which have been discovered only recently. The object, Chiron, believed to be cometary in nature, is some 90 km in diameter and orbits the Sun mainly between Saturn and Uranus. Other small objects have also been detected in this region. Of special interest are the Kuiper-belt objects that move mostly outside the orbit of Neptune. Many of these are known and some of them share with Pluto a 3:2 periodic commensurability with Neptune. There is a school of thought that Pluto is just the largest member of this group of objects which, because of its size, was the easiest and the first to be detected. However, Pluto is distinguished by having a satellite so that it may be a unique object unrelated to the others.

This brief survey of fine-detail characteristics of the Solar System, many of which are described more fully in other chapters, gives the following list of third-order features for theories to address:

(10) the Solar System has significant departures from planarity;
(11) the variable directions of the planetary spin axes;
(12) Bode's law or commensurabilities linking planetary and satellite orbits;
(13) asteroids—their existence, compositions and structures;
(14) comets—their compositions and structures;
(15) the formation and survival of the Oort cloud;
(16) the physical and chemical characteristics of meteorites;
(17) isotopic anomalies in meteorites;
(18) Kuiper-belt and other small objects;
(19) Pluto and its satellite, Charon.

It may be expecting a great deal for any theory to deal with all these matters. Theories under development that are still in the process of explaining some of the very basic characteristics of the system—e.g. the angular momentum distribution and the formation of planets—would hardly be concerned with most of the list of third-order features given here. However, a well developed theory should be putting forward ideas about at least some of these detailed features and a theory which can explain large numbers of them with the fewest possible assumptions would deserve to be taken seriously. Conversely a theory which struggles to solve the basic problems would have to be regarded rather circumspectly—although the possibility that such a theory might eventually solve its problems cannot completely be discounted.

PART 2

SETTING THE THEORETICAL SCENE

Chapter 4

Theories up to 1960

4.1 The historical background

The first problem facing those who wished to understand the Solar System was to determine the true nature of the system, how the bodies moved and what were the forces that governed that motion. Once that had become established then sensible theories could be advanced about the origin of the system. To provide this background it is only necessary to follow solar-system studies up to the time of Newton for, after Newton, the essential structure of the system and the laws that governed its behaviour were well known and understood.

4.1.1 Contributions of the ancient world

The earliest civilizations, which arose in China and the Middle East, became involved in studies of the heavens, partly for their own intrinsic interest but also because they were of practical use in such activities as navigation at sea and farming. Four thousand years ago the Babylonians planted their seeds when the Sun was in the direction of the constellation Aires although, because of the precession of the Earth's axis, spring now begins when the Sun is in Aquarius. Early observers found that, while most points of light in the sky remained in fixed patterns, there were others that wandered around, called *planetes* (meaning *wanderers*) by the Greeks who realized that the Sun, the Moon, the Earth and the planets were part of a separate system. There were early suggestions that the Earth and the Sun were flat but Pythagorus (572–492 BC) suggested that all heavenly bodies were spheres, mainly for the reason that the sphere was regarded by the Greeks as the 'perfect' shape.

Aristarchus of Samos (310–230 BC) began what could be called scientific measurements of the Solar System. He understood that when a half-Moon is seen then the Sun–Moon–Earth angle is 90° and he attempted to measure the Moon–Earth–Sun angle at this time to find the ratio of distances from the Earth of the Sun and the Moon. The angle was far too close to 90° to be measured accurately

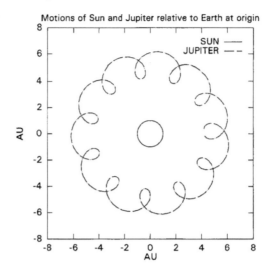

Figure 4.1. The apparent motion, with respect to the Earth at the origin, of the Sun and Jupiter.

and his assessment of the ratio of distances was 19 ± 1 compared with the true value of 390. Aristarchus also proposed that the Earth moved round the Sun in a circular path—the first heliocentric theory.

The Alexandrian Greek, Eratosthenes (276–195 BC), found a way of measuring the size of the Earth. At mid-day at the beginning of summer the sun shone straight down a deep well at Syene (modern Aswan) so that the Sun was straight overhead at that time. By measuring the length of the shadow of a column in Alexandria, which is due north of Syene, on that day he deduced that the difference of latitudes of the two locations was 1/50th of a complete revolution. Hence, from the distance between Alexandria and Syene (determined by pacing), the circumference of the Earth was found. His estimate was only 13% greater than the currently accepted value.

Since there is no sensation of movement associated with being on the Earth it seemed natural to accept that the Earth was at rest and that every other astronomical body moves relative to it. This was the basis of the geocentric model of the Solar System put forward by Ptolemy (c.150 AD), another Alexandrian Greek, which was to be dominant for the next 1400 years. The problem for a geocentric theory is that while the Sun and the Moon clearly describe circular paths around the Earth the motion of the planets is less regular, moving through a series of loops, although dominantly in an eastward direction (figure 4.1). To explain this observation Ptolemy had two components of the motion of the planets. The first component was motion in a uniform circular orbit, with a period of one year, of a point called the *deferent*. The second was an *epicycle*, uniform motion in

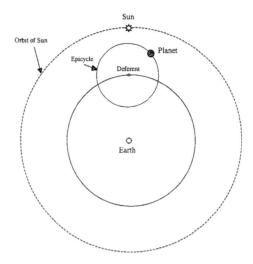

Figure 4.2. The deferent and epicycle for an inferior planet according to Ptolemy.

a circular path centred on the deferent (figure 4.2). To explain what was seen, the deferents of Mercury and Venus, the inferior planets, had always to be on the Earth–Sun line. For the superior planets, those outside the Earth, the line joining the deferent to the planet had to remain parallel to the Earth–Sun direction. Although this was a complicated and seemingly arbitrary system it did explain the motions of the planets quite well. With no theories available to explain how bodies should move relative to each other it served its purpose well enough at that time.

4.1.2 From Copernicus to Newton

Nicolaus Copernicus (1473–1543) was the next important figure to influence thinking about the Solar System. He was a Catholic cleric, educated both in the sciences and arts, and he spent some time in Rome as a professor of mathematics. He was also very interested in astronomy, especially the Solar System, and he assembled tables on the motion of the planets. From this material he deduced that the motions of the planets were much more simply described in a Sun-centred rather than in an Earth-centred system. He was attached to the idea of circular orbits although his data indicated that planetary orbital angular velocities would have to vary slightly and he therefore proposed that the orbits were circular but not concentric. This model did not completely fit the data he had available so he did add epicycles to the motion of the planets—although these were much smaller than those given by Ptolemy.

Copernicus made his work known and the Pope encouraged him to develop his ideas which were eventually published in a treatise dedicated to Pope Paul

III, *De Revolutionibus Orbium Coelestium* (The Revolutions of the Heavenly Spheres). *Revolutionibus* was actually published when Copernicus was on his death-bed and he died knowing that his great work was accepted by the Church, the major depository of knowledge at that time.

Later the attitude of the Church changed. Towards the end of the 16th century an Italian philosopher, Giordano Bruno, suggested that the stars were all like the Sun and all had planetary systems inhabited by other races of men. This conflicted with the Church's belief in the central role of mankind in the universe and, because Bruno refused to retract his ideas, he was eventually burnt at the stake for heresy in 1600. Thereafter *Revolutionibus* was treated as a potentially seditious work and in 1616 was placed on the *Index Liborium Prohibitorum*, the list of books that Catholics were forbidden to read. It should be said that the Catholic Church was not alone in condemning the heliocentric theory and, if anything, the Lutheran Church was even more vehement in its condemnation.

Not all astronomers accepted the heliocentric model and Tycho Brahe (1546–1601), a Danish nobleman and the most effective astronomical observer of the pre-telescopic period, supported a hybrid model where the Sun and the Moon orbited the Earth but all other planets orbited the Sun. This model had the merit that it accurately described the motion of all bodies relative to each other. In 1576, with support of the Danish king, Frederick II, Brahe built a substantial observatory on the island of Hven, in the Baltic Sea. He began a programme of measurements with line-of-sight instruments based on very large circles for measuring angles (figure 4.3) and his measurements of the positions of stars and the motions of the planets were much more accurate than any that had been made hitherto. When the king died Brahe's support dried up so he was forced to leave. In 1596 he went to Prague as the Imperial Mathematician at the court of Rudolph II of Bohemia where he spent his last years compiling tables of planetary motion with the help of a young assistant, Johannes Kepler (1571–1630).

After Brahe died, Kepler set out with the help of Brahe's accurate observations to try to refine the Copernican model of the planetary orbits. Most of his effort was spent on Mars, the eccentric orbit of which made it the most difficult to understand and he eventually deduced that the orbit was not a circle but an ellipse. After many years of theoretical work he was able to put forward his three laws of planetary motion:

(1) Planets move on elliptical orbits with the Sun at one focus.
(2) The radius vector sweeps out equal areas in equal times.
(3) The square of the period is proportional to the cube of the mean distance from the Sun.

The Copernican description of the orbital shapes as circles was not unreasonable. An ellipse in Cartesian coordinates is

$$\frac{x^2}{a^2} + \frac{y^2}{b^2} = 1 \tag{4.1}$$

Figure 4.3. A mural depicting Tycho Brahe's quadrant that had a radius of 2 m. Tycho is shown both at the right edge as an observer and also as the large figure at the centre of the mural.

in which a and b are the semi-major and semi-minor axes respectively. The foci are displaced from the geometric centre of the ellipse by $\pm ae$ in the x direction where e is the eccentricity of the ellipse. Finally the semi-minor axis is related to the semi-major axis by

$$b = a(1 - e^2)^{1/2}. \tag{4.2}$$

The eccentricities of most planets are quite small. Of the ones known to Kepler the largest eccentricity was that of Mercury but that planet is not easy to observe. Mars, on the other hand, is very easy to observe and has $e = 0.093$ which gives $b = 0.991a$. For the other planets known at the time, except Mercury, the ratio of b to a is even closer to unity. It took Tycho Brahe's accurate measurements to distinguish that the orbits were ellipses rather than circles, although previous measurements were good enough to indicate that the Sun was not at the geometric centre of the orbits. Kepler wrote a number of books, the most relevant to his solar-system work being the *Epitome of the Copernican Astronomy*, published in three parts between 1618 and 1621. It soon became another entry in the *Index Librorum Prohibitorum*.

Galileo Galilei (1562–1642), a professor of mathematics in Pisa, was a contemporary of Kepler and was in frequent contact with him. He was interested in mechanics, in particular the motions of planets, and through Kepler's publications he became convinced of the essential truth of the heliocentric theory. However, he was also a very devout man and at first he kept his views to himself. In 1608 he made a telescope, which had been invented in Holland shortly before, and started a programme of astronomical observations. He observed the Moon and made es-

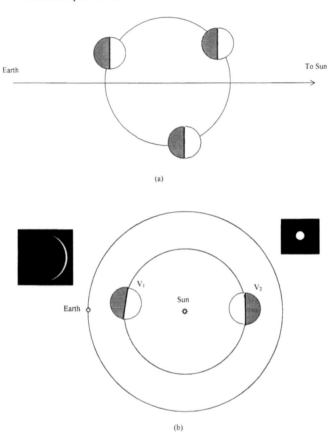

Figure 4.4. The motion of Venus relative to the Earth for: (*a*) Ptolemy's theory for which Venus is always between the Earth and Sun so that it always shows a crescent phase. (*b*) The Copernicus theory for which Venus shows all possible phases.

timates of the heights of lunar mountains. He discovered the large satellites of Jupiter, now known as the Galilean satellites, and in 1610 he saw the rings of Saturn, although he did not recognize them for what they were. However, his most important observation vis-à-vis the heliocentric versus geocentric models was that of the phases of Venus. According to Ptolemy Venus is always close to a point between the Earth and the Sun and so only cresent phases should be seen (figure 4.4). For the heliocentric theory all possible phases can be seen and, what is more, the angular size of the disc will be larger in the crescent phase than in full phase, as Galileo actually observed.

Galileo was in a quandary—he wished to respect the Church but his reason was telling him that, in respect to this astronomical question, the Church was wrong. He tried to present his views in a way that would not offend the Church by writing the *Dialogue on the Two Chief World Systems* in which two individuals, Simplicio and Salviati, discuss the relative merits of the geocentric and heliocentric models. The bias of the work was clear to all, including the Inquisition, and the book was suppressed and Galileo was put on trial. He was neither imprisoned nor ill treated in any way, although he was essentially put under house-arrest, but he was made publicly to recant and to forswear the Copernican 'heresy'.

In the year of Galileo's death Isaac Newton (1642–1727), arguably the greatest scientist of all time, was born. He formulated the laws of dynamics known as 'Newton's laws of motion' and his scientific work included the study of light, the development of calculus and aspects of hydrostatics and hydrodynamics. His great contribution to astronomy was his proof that a force between two bodies, proportional to the product of their masses and inversely proportional to the square of the distance between them, led to Kepler's laws of planetary motion. Newton's theory of gravity, with its astronomical implications, was an important part of his famous publication, the *Principia*, which took 15 years to write and appeared in 1687.

For all his scientific virtuosity Newton spent his later years studying subjects which command little respect these days—astrology and alchemy. He was irascible and quarrelsome but he also had good friends, including Edmond Halley (1656–1742) who not only persuaded Newton to publish the *Principia* but also paid the cost of the publication. Halley, who was a good all-round astronomer, is best remembered for the comet that bears his name It was he who established comets as full members of the solar-system family that could only be seen when they came into the inner part of the system (section 1.7).

Newton's work was a watershed in science, in particular in understanding the essential structure of the Solar System. With his work, and the work of others before him, the basis was established for creating scientific theories of the origin of the Solar System. We now examine a range of the major post-Newtonian theories up to 1960. These include both monistic theories, in which the Sun and planets are formed from the same pool of material, and dualistic theories where the Sun and planets originate from different sources of material and at different times. These earlier theories bring into sharp focus the problems that need to be solved. They will be considered approximately in historical order, which will help to show how ideas have developed as the subject has progressed.

4.2   Buffon's comet theory

One of the very earliest ideas on the origin of the Solar System was suggested in 1745 by Georges comte de Buffon (1707–1788) a French naturalist. This theory, both dualistic and catastrophic, postulated a grazing collision with the Sun's surface by a comet that ejected solar material from which the planets then formed at various distances from the Sun. Buffon had no idea of the nature of comets and clearly assumed that they are much more massive than they are in reality; from the Shoemaker–Levy collision with Jupiter it is now known that the Sun would be little affected by such an event.

The theory was criticized in 1796 by Pierre Laplace in his work *Exposition du Système du Monde*. Laplace argued that the ejected material would be on closed orbits around the Sun so that, since they started at the Sun's surface, they would eventually return to the surface and be reabsorbed. Laplace conceded that the mutual perturbations of the ejected material might negate the conclusion that material would return to the surface but he also raised doubts about the rather eccentric orbits that would result from Buffon's model. Laplace was convinced that a plausible model had to give circular orbits, an outcome from his own nebula theory that will now be described.

We cannot leave this account of Buffon's collision hypothesis without noting that other collision scenarios for the origin of the Solar System have since been advanced although all have involved collisions between two stars (Arrhenius 1901, Jeffreys 1929).

4.3 The Laplace nebula theory

4.3.1 Some preliminary ideas

René Descartes (1596–1650), a French philosopher, physicist and mathematician, was a contemporary of Galileo and Kepler and was familiar with their work. He believed in the heliocentric theory and he put forward an idea of how the Solar System might have formed. Of course, the necessary mathematical background, provided by Newton, was not available so the model was rather vague and qualitative. Descartes assumed that space was filled by a 'universal fluid' of an unspecified nature and that this formed vortices around stars. Planets were formed in eddies within a vortex and around the planets new vortices formed with smaller eddies giving rise to satellites. There was no scientific basis for the model, although it was based on observations of fluid motion, but it did address the heliocentric nature of the Solar System, its planarity and the direct orbits of the planets and the known satellites. Descartes wrote up his ideas in a book *Le Monde* but, with knowledge of the Church's treatment of Galileo, he was afraid to publish it and it appeared posthumously in 1664. A century later, in 1755, the German philosopher, Immanuel Kant (1724–1804), showed by the use of Newtonian mechanics that a cloud of gas contracting under gravity would flatten into a disc-like form; this idea has echoes of Descartes' vortices except that the universal fluid now becomes a gas.

During the 18th century the art of making telescopes had much improved and a notable maker and user of a very good instrument was the German-born, but British, astronomer William Herschel (1738–1822). In common with other observers of his time he had observed fuzzy nebulae which he assumed to be unresolved collections of stars, as indeed most of them probably were. However, an observation he made in 1791 seemed to him clearly to show a single star sur-

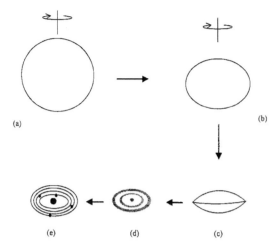

Figure 4.5. An illustration of Laplace's nebula theory. (*a*) A slowly rotating and collapsing gas-and-dust sphere. (*b*) An oblate spheroid form as the spin rate increases. (*c*) The critical lenticular form. (*d*) Rings left behind in the equatorial plane. (*e*) One planet condenses in each ring.

rounded by a luminous halo and he was drawn to the idea that stars are produced from nebulae and the halo he saw was just a residue of an original nebula.

4.3.2   The nebula model of Solar System formation

The work of Descartes, Kant and Herschel sets the scene for the first well formulated theory of the origin of the Solar System put forward by the French physicist and astronomer, Pierre Laplace (1749–1827). Laplace's model is illustrated in figure 4.5. The starting point is a slowly spinning spherical cloud of gas and dust which is collapsing under gravity (figure 4.5(*a*)). As it collapses so, to conserve angular momentum, it spins more quickly and flattens along the spin axis (figure 4.5(*b*)). Eventually material in the equatorial region is in free orbit around the central mass at which stage the cloud takes on a lenticular form (figure 4.5(*c*)). Thereafter further contraction leads to material being left behind in a disc-like form as the central mass continues to contract. Laplace postulated that the release of material into the disc would be spasmodic so instead of a uniform disc there would be a series of annular rings (figure 4.5(*d*)) and in these rings material would clump together through the action of gravity. There would be several clumps in each ring but their orbital velocities would not be precisely the same, so that faster ones would catch up slower ones and amalgamate. Eventually there would be one condensation in each ring (figure 4.5(*e*)) so giving the planetary system. The inner material of the collapsed cloud, the majority of its mass, would form the Sun. Smaller versions of the same mechanism, operating on the collapsing planetary clumps, would give rise to satellite systems. This theory was published by Laplace in 1796 in *Exposition du Système du Monde* and it had a mixed reception with both strong supporters and detractors.

4.3.3   Objections and difficulties

One strong scientific criticism of Laplace's nebular model was due to the work of Clerk Maxwell (1831–1879). While still a student at Cambridge, Maxwell had written an Adams Prize essay in which he showed that the rings of Saturn could only be stable if they consisted of small solid particles and that they could not possibly be gaseous. The same analysis could be applied to Laplace's gaseous rings. The rings could not condense into planets under self-gravitation because they would be disrupted by inertial forces due to the differential rotation between the inner and outer parts of each ring. Indeed the rings would have to have been hundreds of times more massive than the planets they produced if they were to resist this disruptive process. While Maxwell offered this criticism of Laplace's theory he did not imply that the nebula hypothesis was necessarily invalid.

Another, and fairly obvious, difficulty of the theory concerns the distribution of angular momentum because there seemed to be no obvious mechanism by which the small proportion of the material forming the planets could take up nearly all the angular momentum. Most of the angular momentum should reside in the central body and it is not difficult to show that if the intrinsic angular momentum (i.e. angular momentum per unit mass) of outer material of the original sphere is sufficient to produce, say, Uranus (Neptune was not discovered at that time) then the central condensation would not be able to collapse at all to form the Sun. This difficulty can be expressed in various forms. One way is to assume that the whole nebula was spinning at the angular speed of Uranus in its orbit when the Uranus ring formed. The angular momentum of the whole nebula would then be

$$H = \alpha M_\odot R_U^2 \omega_U \qquad (4.3)$$

in which R_U and ω_U are the radius and angular speed of Uranus in its orbit and α, the moment-of-inertia factor, is a constant of order unity (two-fifths if the nebula was uniform and spherical). Inserting numerical values, this gives $H = 4 \times 10^{46} \alpha$ or $4 \times 10^4 \alpha$ times the estimated value for the Sun. No reasonable value of α can reconcile this discrepancy. The other approach is simply to accept the estimate for the total angular momentum in the present Solar System and imagine that it is possessed by a nebula reaching out as far as Uranus. In that case the material would have had such a low angular speed that it could not have detached itself as Laplace's theory requires to form a ring. Difficulties of this kind were noted by many, but particularly by the French physicist Jacques Babinet (1794–1872) who, nevertheless, still believed in the general validity of the nebula theory.

4.4   The Roche model

4.4.1   Roche's modification of Laplace's theory

Another French scientist, Edouard Roche (1820–1883), who was working on the configurations of stars, suggested in 1854 that Laplace's original cloud could have had a high central condensation so that most of the mass was close to the spin axis and would thus have little associated angular momentum. This would require a very tiny value of α in (4.3) and would almost correspond to a 'Roche model' where the star has a finite size but such a high central condensation that virtually all the mass is at the centre. Later, in 1873, Roche produced a mathematical analysis of Laplace's theory in which the system being analysed was described as 'the Sun plus an atmosphere', clearly implying high central condensation. Here we give a resumé of Roche's analysis, not precisely in the form he gave it but bringing out the essential arguments.

The Sun-plus-atmosphere model originally extends well beyond the range of the planets and it is collapsing as it cools. The atmosphere is taken as co-rotating with the Sun but, for a given angular speed, ω, this can only happen out to a distance R_L given by

$$R_L^3 = \frac{GM_\odot}{\omega^2}.$$
(4.4)

Any part of the atmosphere that is further out than R_L must go into free orbit around the central mass.

As the system collapses so the value of ω increases to conserve angular momentum and the value of R_L reduces. If the form of collapse is such that R_L diminishes more rapidly than the effective radius of the atmosphere, then all the atmosphere beyond a distance R_L will form a ring of radius a where

$$a^3 = \frac{GM_\odot}{\omega^2}$$
(4.5)

and a will be the semi-major axis of the planet forming from the ring material.

Roche examined in detail the way in which material from the atmosphere would contribute to the ring. He used analysis he had developed in his work on the profiles of spinning stars. The cross-section of the atmosphere through the rotation axis is shown in figure 4.6(*a*). The forces acting on material at the point P on the surface are the gravitational field due to the central mass, the centripetal acceleration along SP and the pressure of the atmosphere. If these three forces are in balance then the pressure exactly balances the other two forces. Since the pressure force acts normal to the surface of the atmosphere, that normal must be along the direction given by

$$-\frac{GM_\odot}{r^3}r - r\sin\theta\omega^2\hat{i}$$
(4.6)

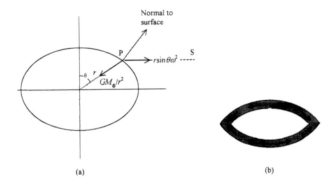

Figure 4.6. (*a*) The relationship of the forces due to gravity and spin, and the normal to the surface as described by Roche. (*b*) Two critical lenticular cross-sections at two different times. The material between them flows towards the equator and is shed from the atmosphere of the star.

in which $\hat{i}$ is the unit vector along the x direction. Roche showed that the equation defining the cross-section of the atmosphere is

$$\frac{2}{r} + \eta r^2 \sin^2 \theta = c \tag{4.7}$$

in which $\eta = \omega^2/GM_\odot$ and c is a constant chosen to give the correct volume for the atmosphere. If the equatorial radius of the atmosphere is r_e ($\theta = \pi/2$) and the polar radius is r_p ($\theta = 0$) then, from (4.7),

$$\frac{r_e - r_p}{r_p} = \frac{1}{2}\eta r_e^3. \tag{4.8}$$

The left-hand side of (4.8) is a measure of the flattening of the profile and this is seen to decrease as the material leaves the atmosphere to form rings, since r_e will clearly be reducing. Between two different values of ω, and two different values of c (because some of the atmosphere is lost), the material indicated by shading in figure 4.6(*b*) will be released into a ring in the equatorial plane with shaded material flowing from the poles towards the equator.

4.4.2 Objections to Roche's theory

Roche sidestepped the angular momentum problem by postulating a very high central condensation, even perhaps a fully condensed Sun, but many other problems remained. Firstly, with such a diffuse atmosphere it is unlikely that there will be sufficient viscous coupling to ensure co-rotation of the atmosphere with the Sun. Secondly, the British astronomer James Jeans (1877–1946) showed that with the nebula distribution required by Roche the outer material would have been

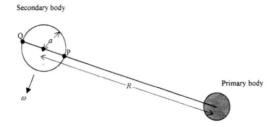

Figure 4.7. A satellite in circular orbit around a primary body.

so tenuous that it could not have resisted the tidal forces due to the central mass and so could not have condensed. This argument is based on work done by Roche himself who derived the so-called *Roche limit*. Consider a spherical satellite of mass m and radius a moving in a circular orbit at distance R ($\gg a$) from a body of mass M ($\gg m$) as shown in figure 4.7. The angular speed is given by

$$\omega^2 = \frac{GM}{R^3} \tag{4.9}$$

that gives no net force along the radius vector at O, for material at the centre of the satellite. If the satellite is tidally locked to the primary body then the point P, closest to the primary, is also moving at angular speed ω. Thus the net acceleration (force per unit mass) at P along OP is given by

$$A_{\mathrm{OP}} = \frac{GM}{(R-a)^2} - (R-a)\omega^2 - \frac{Gm}{a^2}.$$

Substituting for ω^2 from (4.9) and making the usual approximations this becomes

$$A_{\mathrm{OP}} = \frac{3GMa}{R^3} - \frac{Gm}{a^2}. \tag{4.10}$$

There will be an equal and opposite acceleration at the point Q of the satellite, furthest from the primary. If A_{OP} is positive then the satellite will be disrupted, unless it is materially strong enough to resist the tidal stretching forces. From (4.10) the condition that the satellite will not be disrupted can be expressed as

$$\rho_{\mathrm{S}} \geq 3\rho_{\mathrm{mean}} \tag{4.11}$$

where ρ_{S} is the density of the satellite and ρ_{mean} is the mean density of the system within a sphere of radius R. Another way of expressing the condition is to give the critical distance, the Roche limit R_{L}, within which the satellite will be disrupted in the form

$$R_{\mathrm{L}} = \left(\frac{3\rho_{\mathrm{P}}}{\rho_{\mathrm{S}}}\right)^{1/3} r_{\mathrm{P}} \tag{4.12}$$

in which ρ_P is the density of the primary body and r_P its radius. Other analyses for finding the Roche limit, for example with comparable values of a and R, give similar expressions but with different numerical constants.

Jeans (1919) pointed out that for material in a ring shed from the collapsing nebula to condense, it would need to have a density comparable to, but greater than, the mean density of the system. This would give an atmospheric mass of magnitude similar to that of the central mass, which restores the angular momentum problem.

By the beginning of the 20th century the Laplace theory, even as modified by Roche, no longer commanded much support; it was a pleasingly simple model that spontaneously and simultaneously produced the Sun and its attendant planets but it simply did not work. However monistic theories were by no means completely discredited and other ideas about evolving nebulae were to appear in due course.

4.5   The Chamberlin and Moulton planetesimal theory

4.5.1   The planetesimal idea

For solar-system cosmogony the last decade of the 19th century and the first decade or so of the 20th century were dominated by two Americans, Thomas Chamberlin (1843–1928) and Forest Moulton (1872–1952). They were in constant communication, exchanging and testing ideas on each other, but the great majority of their publications were done separately.

In the 1890s Chamberlin was considering the difficulties associated with the nebula theory, in particular relating to the formation of planets from gaseous material. He introduced what he saw as a solution of that problem by suggesting that what actually accumulated to form planets, or at least planetary cores, were small bodies that condensed out of the nebula material. Later these postulated bodies became known as *planetesimals* and they have become an essential ingredient of later theories. Chamberlin was initially concerned with the way the planetesimals would collect together as it seemed to him that a natural outcome would be a planet with retrograde spin, which is the exception rather than the rule for the actual planets. The basis of his concern is shown in figure 4.8(*a*); if the planetesimals within the region shown coalesced to form a planet then, since the inner particles travel faster than the outer particles, the body would have a retrograde spin. Later he was able to resolve the difficulty by taking into account that the planetesimals would be on elliptical orbits. As an example, the planetesimals coming together in the shaded region of figure 4.8(*b*) contribute retrograde spin in the dark grey shaded region and direct spin in the black shaded region. Since the direct-spin region is the larger of the two then an overall direct spin should result. The conclusion was suggestive rather than definitive but satisfied Chamberlin that planet formation by planetesimals was a viable possibility.

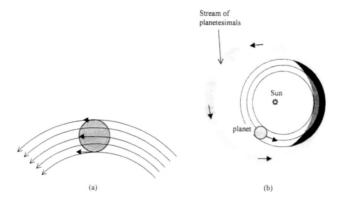

Figure 4.8. (*a*) Chamberlin's argument that, because of their differential Keplerian speeds, planetesimals joining a proto-planet would give retrograde spin. (*b*) Chamberlin's later argument that the eccentric orbits of planetesimals can give direct spin. The stream of planetesimals shown, all moving faster than the growing planet, give retrograde spin in the dark grey region but prograde spin in the larger black region.

4.5.2 The Chamberlin–Moulton dualistic theory

Although this planetesimal hypothesis was produced in the process of examining the nebula hypothesis, from 1900 Chamberlin and Moulton began to develop an alternative scenario for planet formation (e.g. Chamberlin 1901, Moulton 1905). Observational astronomers were taking photographs of spiral nebulae, whose true nature was not known at that time, and Chamberlin and Moulton were interpreting these as ejection of material from a star. They considered the idea that such material could form planets that would then go into orbit around the parent star but they eventually rejected this idea on the grounds that the resulting orbits would be far too eccentric. Chamberlin next turned his attention towards solar eruptions and he considered the possibility that the observed spiral nebulae were the result of a disruptive interaction between a star in the process of erupting and another star, which prevented the erupted material from returning to the parent star. This idea gradually evolved into a model for the formation of the Solar System.

The model requires a very active Sun with massive prominences and a passing star exerting its tidal influence. The action of the star had to be just sufficient to retain prominence material outside the Sun but not enough to remove material from the Sun itself as only 1/700th of a solar mass was required for planet formation. This required the Sun–star distance to be just greater than the Roche limit for the Sun, and a massive star, a few times the mass of the Sun, was postulated. In the tidal field of the star, prominence material would be pulled out on either side, both towards the star and away from it, forming two spiral arms out to about the distance of Neptune (figure 4.9). The outer spiral-arm material would have originally come from the surface of the Sun while the inner material would have

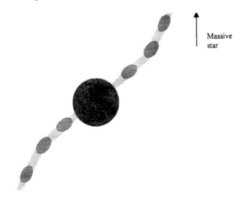

Figure 4.9. The Chamberlin and Moulton mechanism. The bunching of density in each stream is due to the loss of solar material in a spasmodic way.

come from deeper in the Sun and so have a larger inventory of higher-density material—which would explain the general difference in the composition of the terrestrial and major planets.

The prominences were taken to have left the Sun in the form of irregular pulsations and these would appear as high-density regions in the spiral arms. Within these regions rapid cooling would lead to the formation of liquid or solid planetesimals which would then accumulate to form the planets. Smaller collections of planetesimals in orbit around the planets would give satellite systems. It was claimed that retrograde satellites would tend to acquire more eccentric orbits and so plunge into the planet leading to the conclusion that retrograde orbits could only exist in the outer reaches of the satellite system. The existence of Pheobe, the outermost and retrograde satellite of Saturn, and the then newly discovered retrograde satellites of Jupiter seemed to support this idea. A final detail of the model was that since the prominences had been ejected by the Sun they would be in the Sun's equatorial plane and the small 7° (*sic*) tilt of the solar spin axis would be due to the influence of the passing star.

4.5.3   Objections to the Chamberlin–Moulton theory

The Chamberlin–Moulton theory was always described in a rather qualitative way, for example the way that spiral arms would develop due to the action of the passing star was described by diagrams without any detailed dynamics being involved. In that sense, being just descriptive, it was not easy to criticize although, as stated in section 3.1.1, this is a feature of an unsatisfactory theory. The theory had fairly wide acceptance in the United States but less acceptance elsewhere. A damaging attack on the theory by a German astronomer, Friedrich Nölke, in 1908 was practically unknown outside Germany. His criticisms were that:

(i)   stellar interactions would be too rare to explain the observed numbers of spiral nebulae;

(ii) if the passing star was so close to the Sun that it was inside the Roche limit and disrupted the Sun then the spiral arms would be highly asymmetrical—a feature not observed in spiral nebulae;

(iii) the periods of the planets are such that, if the inner part of a spiral arm corresponded to Mercury and the outer part to Neptune, the spirals would quickly distort;

(iv) the accretion of planetesimals would not always give direct spins;

(v)   retrograde satellites would not have more eccentric orbits and so tend to be absorbed by planets;

(vi) the passing star would be so distant that it could not be expected to disturb the system sufficiently to explain the tilt of the solar spin axis.

Naturally, once the true nature of spiral nebulae became clear, one of the fundamental observational foundations of the Chamberlin–Moulton theory was removed. Support for it was fairly solid up to about 1915 but thereafter steadily declined.

4.6   The Jeans tidal theory

4.6.1   A description of the tidal theory

The next theoretical development stemmed directly from the Chamberlin–Moulton model and was due to the British astronomer James Jeans (1877–1946). At the beginning of his scientific career, in the first few years of the 20th century, he tackled a number of astronomical problems to do with the stability of nebulae but then went on to do some very basic and important work in kinetic theory and radiation physics. Around 1916 he returned to astronomy and quite quickly made the contribution to solar-system cosmogony for which he is best known.

Jeans took up the idea that the Solar System was a product of the interaction between another star and the Sun but his model was substantially different from that of Chamberlin and Moulton. The most important difference was that solar prominences were not involved. Instead the main body of the Sun was tidally influenced by a massive star passing within the Roche limit to give solar disruption. The tide drawn up on the Sun was so great that material escaped from it in the form of a filament (figure 4.10(a)). This filament was gravitationally unstable and broke up along its length into a series of blobs (figure 4.10(b)) each of which condensed to become a proto-planet. These proto-planets were influenced gravitationally by the retreating star and so given enough angular momentum to go into orbit around the Sun (figure 4.10(c)). Finally the tidal effect of the Sun on the proto-planets when they made their first perihelion passage gave a smaller-scale version of the planet-forming process to give natural satellites.

The model had its superficial attractions. Like the Chamberlin and Moulton model it gave a planar system of planets in direct orbits and the problem of the

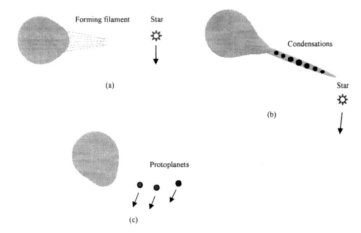

Figure 4.10. An illustration of Jeans' theory. (*a*) The escape of material from the tidally-distorted Sun. (*b*) Proto-planetary condensation in the ejected filament. (*c*) Proto-planets attracted by the retreating massive star.

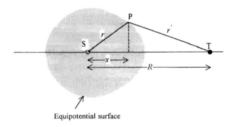

Figure 4.11. An equipotential surface of a tidally distorted star.

slow spin of the Sun was side-stepped because the Sun was already formed prior to the process of planetary formation. The slow solar-spin problem was still there but did not have to be addressed by the tidal theory. On the other hand, without prominences, there was no ready explanation of the tilt of the solar spin axis since the Sun's spin and the plane of the Sun–star orbit would have been completely randomly oriented with respect to each other.

What made this theory quite different from all previous theories is that Jeans was an accomplished mathematical theorist and the important processes in his model were subjected to analysis and shown to be valid. We shall now consider these analyses, all of which still have some relevance in the consideration of astronomical problems.

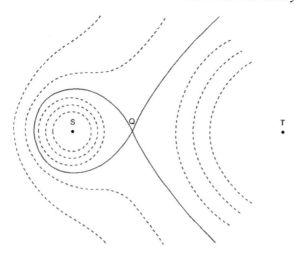

Figure 4.12. Equipotential surfaces around star S due to its own mass and that of star T.

4.6.2 The tidal disruption of a star

In figure 4.11 the star S, which we identify as the Sun, is approached by another star, T, of mass M_*. The material of S is highly centrally condensed and it is taken as a Roche model so that it has a fixed volume but has all its mass concentrated at its centre. If star T approaches slowly so that there is no inertial lag in the reaction of S to its field then the material of S will be distorted so that its boundary is an equipotential surface with respect to S. That is to say that the difference of potential between all points on the boundary and the point S is the same at any time. Jeans gave this difference of potential for the point P, in figure 4.11, as

$$\Omega_P = \frac{GM_\odot}{r} + \frac{GM_*}{r'} - \frac{GM_* x}{R^2} \tag{4.13}$$

where r, r', x and R are shown in the figure. This expression, the negative of that usually given, can be verified by finding $\partial\Omega_P/\partial x$ and $\partial\Omega_P/\partial y$ and seeing that they are the components of the acceleration of P with respect to S. The contours corresponding to a cross-section of these surfaces is shown in figure 4.12 for the case $M_*/M_\odot = 3$.

The behaviour of S as T approaches can now be determined. When T is distant and has little effect on S, the volume of S can be contained within an equipotential surface that is little different from spherical. As T approaches so the system of surfaces shrinks in proportion to the distance ST and the volume of S occupies increasingly distorted surfaces. Eventually the distance ST is reached where the volume of S occupies the largest closed equipotential surface, marked with the full line in figure 4.12, and for any further decrease in the distance ST material from S escapes in a stream from the point Q.

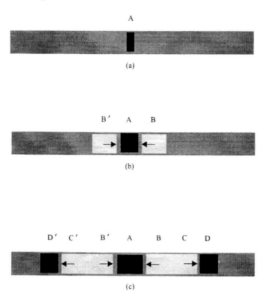

Figure 4.13. (*a*) A filament with a density excess at A. (*b*) Material at B and B′ is attracted towards A. (*c*) Material at C and C′ moving away from the depleted regions B and B′, so creating higher-density regions at D and D′.

4.6.3   The break-up of a filament and the formation of proto-planets

The stream of material flowing out of the Sun would have been in the form of a gaseous filament, with some solid component, and each part of it would have been at some particular density and temperature. In what follows it will be assumed that the stream had a uniform density and temperature but this will not substantially affect the conclusions that are drawn.

What Jeans showed was that such a stream would be unstable and break up into a series of blobs. The physical basis of this is illustrated in figure 4.13. In figure 4.13(*a*) there is a small density excess in region A of a long uniform stream of material. Because of an imbalance of forces, material in the vicinity of A experiences an attraction towards A which creates two lower density regions, B and B′, on either side of A (figure 4.13(*b*)). Material further out than B or B′, at C and C′ say, now experience outward accelerations and produce high-density regions at D and D′ (figure 4.13(*c*)). In this way the stream eventually breaks up into a series of blobs.

It is possible to analyse this model in terms of the properties of the gas to find the distance, l, between the condensations but the general form of the expression can be found from dimensional analysis. The rate at which a disturbance moves along the filament will obviously be related to the speed of sound in the gas given by

$$c = \sqrt{\frac{\gamma k \theta}{\mu}} \qquad (4.14)$$

in which γ is the ratio of specific heats of the gas, k is the Boltzmann constant and θ and μ are, respectively, the temperature and mean molecular weight of the gas. The other factors influencing l are the gravitational constant G and the density of the gas ρ. The relationship found by Jeans, with the numerical constant not given by dimensional analysis, is

$$l = \left(\frac{\pi}{\gamma G \rho}\right)^{1/2} c = \left(\frac{\pi k \theta}{G \rho \mu}\right)^{1/2}. \qquad (4.15)$$

In the Jeans analysis of this problem the stream actually takes up a wave-like form so that l is the wavelength of the disturbance. What the analysis does not include is the line density of the filament, σ, i.e. the mass per unit length, for it is this which decides whether or not the density maxima in the filament will continue to collapse into proto-planets.

Although the blobs are not necessarily spherical they would probably be roughly so and the fate of the blobs will be decided by whether they are greater or less in mass than the Jeans critical mass given by (2.22). If σl is greater than the right-hand side of (2.22) with the θ, ρ and μ of the filament material inserted then a proto-planet will form.

4.6.4 Objections to Jeans' theory

The analysis that accompanied the tidal theory of Solar System formation was very persuasive and for a number of years the theory enjoyed a great deal of support. During the early days of the development of the theory one of the chief supporters was the British geophysicist Harold Jeffreys (1891–1989). In fact Jeffreys, who was very mathematical in his approach to geophysical problems, was so active in the field that the tidal theory is sometimes called the Jeans–Jeffreys theory although the two individuals never worked together. Jeffreys criticized the Chamberlin and Moulton model on the grounds that colliding planetesimals would vaporize each other rather than give planets, and also made numerous contributions on the implications of Jeans' theory for geology.

Despite his early support of the tidal theory, the first important criticisms of it actually came from Jeffreys (1929). The first objection was of a probabilistic nature, that massive stars were rare and that the probability that one would pass by the Sun within the required distance was extremely small. The objection was valid but not very damaging; there was no way of knowing that the Solar System was not unique in the universe and it could be argued that, because an extremely improbable event actually happened, so mankind was around to consider its implications. The second argument was very mathematical and may not have been well understood by many who read it. Jeffreys argued that Jupiter and the Sun have similar mean densities and that this would have been the density of material drawn from the Sun to form Jupiter. The material of the Sun would have been characterized by its *circulation*, $\nabla \times v$, where v is the local velocity of the material, and this is related to the spin period of the Sun, which is of the order of 27 days. Jupiter should have the same circulation, but Jupiter's spin period is under 10 hr which is inconsistent with a solar origin. This argument too is somewhat suspect; if the material from the Sun forming Jupiter had, say, 10% or so of the mean density of the Sun and had then collapsed to form Jupiter the problem would not exist. Whether or not it was damaging to the tidal theory, what this argument *did* do was to persuade Jeffreys to abandon the tidal model. Instead he began to consider the effect of a grazing collision of a star with the Sun—indeed a return to Buffon but with a much more massive colliding body. This would have induced extra spin into the ejected material and hence solve the 'circulation problem' if it was a valid one. Although Jeffreys had argued against the tidal theory on the grounds of probability, and a direct collision would be even less probable, there were many astronomers who were much more convinced by the Jeffreys collision model than the Jeans tidal version.

The next objection was much more cogent and it was put forward in 1935 by the American astronomer Henry Norris Russell (1877–1957). For an elliptical orbit of semi-major axis a and eccentricity, e, the semi-latus rectum, $p = a(1 - e^2)$, cannot be greater than twice the perihelion distance, $a(1 - e)$. For material pulled out of the Sun the perihelion distance cannot be greater than the Sun's radius and hence this imposed a tight constraint on the intrinsic angular momentum, H, which can be imparted to the ejected material since

$$H = (GM_\odot p)^{1/2}. \tag{4.16}$$

Even allowing for some extra angular momentum imparted to the ejected material by the gravitational attraction of the passing star, it is not possible to explain the orbits of the outer planets convincingly. In fact to obtain ejected material in orbit around the Sun at *any* distance there must be some appreciable disturbance of the initial ejecta orbits, otherwise it will be reabsorbed by the Sun.

Russell's argument presents an angular momentum problem but in a completely different form to that occurring in the nebula theory. By assuming a pre-existing Sun the tidal theory has no need to explain its slow spin but what it now cannot do is provide enough angular momentum to put the planets where they are needed.

Another telling objection was made in 1939 by the American Lyman Spitzer (1914–1997) who made use of Jeans' own critical-mass formula (2.22). If the Sun had been in much its present condition when the Jupiter material was pulled from it, then it would necessarily take material from a depth such that the density would equal the Sun's average density and the temperature would be about 10^6 K. Inserting these values into (2.22) gives a minimum mass of 2×10^{29} kg, or one hundred times the mass of Jupiter. It thus seems clear that Jupiter could not condense directly from such material.

To these arguments could be added that concerning the light elements lithium, beryllium and boron which are consumed by nuclear reactions at solar temperatures and would not be sufficiently present in very hot solar material to explain the terrestrial abundance. However, by 1919 Jeans had changed his model and was suggesting that when the stellar passage took place the radius of the Sun was equal to that of the orbit of Neptune. This negated almost all the objections that were subsequently raised and applied to the original form of the theory. There would be no difficulty in getting material out to a sufficient distance—it was already there. The material was cool, thus satisfying the light-element requirement while at the same time, with an estimated temperature of 20 K, the Jeans critical mass would be 10^{27} kg or one-half the mass of Jupiter. The Russell and Spitzer objections now only applied fully to Jeffreys' collision model which considered the Sun to be in its present state. However, while solving some problems, the extended Sun introduced others, in particular how the newly formed planets moving towards the perihelion would have interacted with the collapsing Sun. In any case the very extended Sun would be rather like a nebula with all its attendant difficulties.

Jeans himself was aware of the problems of the tidal theory and wrote: 'The theory is beset with difficulties and in some respects appears to be definitely unsatisfactory'. At the end one was left with a hypothesis that could not be sustained but also a body of theoretical analysis that was sound and unchallenged and might still have a part to play in the future.

4.7 The Schmidt–Lyttleton accretion theory

4.7.1 The Schmidt hypothesis

The demise of the two dualistic theories, those of Chamberlin and Moulton and of Jeans, which had dominated cosmogony for the first three to four decades of the 20th century did not see an immediate end to them. The next interesting idea, a quite different form of dualistic theory, was proposed in 1944 by a Soviet planetary scientist, Otto Schmidt (1891–1956). Telescopic observation shows the existence of cool dense clouds in the galaxy, revealed by the way in which they block out the light from stars behind them. Schmidt argued that from time to time a star would pass through such a cloud and might, after its passage, acquire an envelope of gas and dust. From this cloud the planets could form. However, Schmidt was convinced by an argument that two-body capture could not take place, which is certainly true for point masses. If two point masses approach each other from infinity, gravitationally interact and do not collide then they must end up an infinite distance apart. Schmidt assumed that this would apply to a star and a cloud and therefore postulated a third body in the vicinity of the stellar passage through the cloud in order to take away some of the energy of the two-body system. The need to have a third body certainly made the idea seem very improbable.

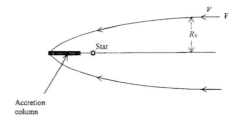

Figure 4.14. Bondi and Hoyle accretion. Material interaction on the axis destroys the component of velocity perpendicular to the axis, leaving material with less than escape speed.

4.7.2 Lyttleton's modification of the accretion theory

In 1961 the British astronomer Ray Lyttleton (1911–1995) took up Schmidt's idea but showed that the third body was not necessary. He proposed that a mechanism of line accretion described by Bondi and Hoyle (1944) would enable cloud material to be captured by the star. The passage of the star through the cloud is illustrated in figure 4.14 where cloud material moves relative to the star at speed V. The cloud material is gravitationally focused and interacts along the axis, cancelling out the component of motion perpendicular to the axis. If the horizontal component of velocity is less than the escape speed from the star then it will eventually be captured. If the point F corresponds to the largest distance from the axis at which material can be captured then a fairly straightforward analysis (Appendix IV) shows that the distance from the axis of the associated approaching material is

$$R_V = \frac{2GM_*}{V^2}. \tag{4.17}$$

To allow for the temperature of the cloud Bondi (1952) suggested a modification of (4.17) of the form

$$R_B = \frac{2GM_*}{V^2 + c^2} \tag{4.18}$$

where c is the speed of sound in the cloud material.

The star would capture a cylinder of cloud material of radius R_B and of length l equal to that of its path through the cloud. This would have a mass

$$m_C = \pi R_B^2 l \rho \tag{4.19}$$

where ρ is the density of the cloud material. Lyttleton (1961) assumed that $2R_B \ll l$ so that the captured material was derived from a long thin rod-like region of the cloud. Taking the intrinsic angular velocity of the cloud material as ω he was thus able to estimate the angular momentum of the captured material as

$$H = \tfrac{1}{3} m_C l^2 \omega. \tag{4.20}$$

Lyttleton then considered various numerical values for the parameters of the model such that, with the Sun as the star, the mass and angular momentum of the captured material would agree with that of the planets. He started by assuming that the cloud was in thermal equilibrium with background galactic radiation at 3.18 K and that it consisted of molecular hydrogen. This gave $c^2 = 3.86 \times 10^4$ m^2 s^{-2}. He took the density of the cloud to be 10^{-20} kg m^{-3} and ω as the intrinsic angular velocity of the galaxy, 10^{-15} s^{-1}. By taking $m_C = 3 \times 10^{27}$ kg and $H = 4 \times 10^{43}$ kg m^2 s^{-1}, approximate values for the planetary system, Lyttleton deduced that $l \approx 10^{16}$ m, $R_B \approx 5 \times 10^{15}$ m and $V \approx 0.2$ km s^{-1}.

4.7.3 The problems of the accretion theory

Lyttleton's numerical conclusions have several unsatisfactory features. The condition that $2R_B \ll l$ does not hold although that does not necessarily invalidate the theory. More difficult to accept are the very low temperature of the cloud and the very low speed of passage of the Sun through the cloud. A temperature in the range 10–100 K is much more acceptable for a cool interstellar cloud. In addition, the proper speed of the Sun in the galaxy is quite typical for stars of its type, ~ 20 km s^{-1}, and it is unlikely to approach a cloud at such a low speed as Lyttleton suggests. In fact since the cloud and Sun accelerate towards each other due to their mutual gravitational attraction the minimum speed of contact is about 0.3 km s^{-1} and is likely to be much greater.

Lyttleton's capture mechanism was critically assessed by Aust and Woolfson (1973). They pointed out that there would be considerable tidal distortion of the cloud and that a filament of matter might be drawn out of the cloud and captured. Aust and Woolfson suggested different parameters that would make Lyttleton's model more acceptable. They assumed that not all the captured material would go into planet formation—some material might be lost from the system or captured by the Sun—which would allow a somewhat larger value for m_C. Some parameters suggested by Aust and Woolfson were: density of cloud 10^{-19} kg m^{-3}, temperature of cloud 20 K and the Sun reached the boundary with free-fall speed (corresponding to the unlikely possibility of zero relative speed at infinity). This then gives $m_C = 5.6 \times 10^{27}$ kg and $H = 1.9 \times 10^{44}$ kg m^2 s^{-1} which are reasonable if some allowance is made for loss of captured material.

It cannot be pretended that this theory is a very convincing one for a number of reasons but, in particular, since it depends on such an unlikely approach scenario for the Sun and the cloud. Given a pre-existing Sun it appears to solve the angular momentum problem although it requires that planets must form somehow from the initially diffuse mixture of gas and dust. This is a requirement of several theories and will be considered in more detail in section 6.4.3. Perhaps the greatest criticism of the Lyttleton theory as far as it has been developed is that it is rather vague in that it just provides a suitable set of ingredients for making the Solar System. However, it is not very specific in describing mechanisms that will form a system consisting of planets and satellites, a very basic requirement for

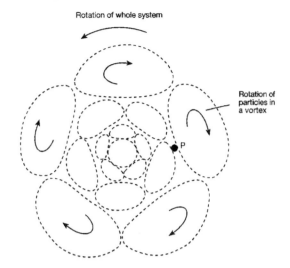

Figure 4.15. The von Weizsäcker configuration of vortices. The combination of rotation of the whole system and rotation within vortices enables individual portions of matter to be in Keplerian motion about the central mass.

any plausible theory.

This theory was the last in a line of dualistic theories, at least for a time, and it was contemporaneous with a new monistic theory that will now be described and was the harbinger of a longer-lasting and extant theory that appeared a quarter of a century later.

4.8   The von Weizsäcker vortex theory

4.8.1   The basic model

In 1944 Carl von Weizsäcker (1912–) revisited the idea of a proto-planetary disc but introduced a new idea that within the disc a pattern of turbulence-induced eddies was set up as shown in figure 4.15. A suitable combination of clockwise rotation of each vortex with anti-clockwise rotation of the whole system can lead to individual elements of the disc moving around the central mass in Keplerian orbits. Thus there would be very little dissipation of energy due to the overall motion of the system but material would be colliding at high relative velocity at the boundary between vortices, as shown at the point P. According to the von Weizsäcker model, in such regions small roller-bearing eddies would form and in these regions, where matter was heavily interacting, material would coalesce to give condensations. The condensations would form in rings and once all condensations in a ring had come together there would be a family of planets. If there were five vortices to a ring then von Weizsäcker showed that the orbital radii would give something similar to Bode's law.

4.8.2   Objections to the von Weizsäcker model

This model has been very heavily criticized, especially by Jeffreys (1952) in a Bakerian lecture to the Royal Society. He and other critics have argued that turbulence is a phenomenon associated with disorder and would not spontaneously produce the highly ordered structure required by the theory. A natural outcome for a disc with turbulence is to produce a quietly rotating system with all parts in circular orbit around the central mass. The viscosity of the system would then lead to material moving inwards and outwards but preserving an axially symmetric system. In this natural evolutionary pattern viscosity would give a loss of energy and therefore a residual low-energy system; on the other hand the von Weizsäcker system of vortices is a high-energy configuration which would not be stable and therefore could not actually form at all.

The basic problem of all disc theories is not tackled by the von Weizsäcker model. It has nothing to say about the angular momentum problem and it gives no mechanism for producing a slowly spinning Sun. It is also a poor theory in the sense that it only deals with the formation of planets, and that unconvincingly, and has no suggestions to make about satellite formation or other very basic characteristics of the Solar System.

4.9   The major problems revealed

The theories up to 1960 were representative of the two main types—first, monistic, which includes Laplace's theory, together with its Decartes and Kant antecedents and the von Weizsäcker model; and second, dualistic, which includes the ideas of Buffon, Chamberlin and Moulton, Jeans, Jeffreys and Schmidt and Lyttleton. All the theories have been able to explain some aspects of the Solar System, no matter how qualitatively, but all of them have had weaknesses that were deemed to be fatal to their acceptance. Since 1960 there have been a number of advances in theory which could possibly resolve some of these weaknesses. What we shall do here is to summarize the successes and failures of the older theories in terms of what was understood in 1960. Where a theory has been successful in explaining some aspect of the Solar System we shall identify the favourable factor and so produce a list of what might be desirable features of a successful theory.

4.9.1   The problem of angular momentum distribution

The monistic theories have all struggled unsuccessfully to resolve the problem of how a single nebula could evolve spontaneously to give virtually all the angular momentum to a tiny fraction of the material. The first attempt to solve this problem, that of Roche by postulating a very highly condensed nebula, was first, very unrealistic in the degree of central condensation it required but, second, required such a diffuse nebula in its outer parts that a proto-planetary condensation could not possibly form there. Another approach has been to postulate that the disc was not very diffuse but, just the opposite, so dense that it might have accounted for an appreciable fraction of the total mass of the nebula—anything from 10–50% of the mass of the Sun. The greater part of the disc would then be lost leaving the planets behind. The partitioning problem is thus numerically changed but is still a very challenging one. If the original mass of the disc was, say, 10% of the mass of the Sun, and if the intrinsic angular momentum of the lost material was the same as that of the disc as a whole, then the partitioning required is that 10% of the material (the original disc) should have contained 99.993% of the total angular momentum. This model then poses another problem, that of providing the energy to dispose of most of the disc. If the Sun, as part of its early evolution, passed through a T-Tauri stage then it could have lost almost 10% of its mass over 10^6 years in the form of a very energetic solar wind. It is believed by some workers that this could have been the instrument for sweeping away most of the mass of the disc. There are considerable doubts about this explanation, especially as the disc would be a fairly flat structure and would only have interacted with a small fraction of the T-Tauri emission.

The dualistic theories of the two-star interaction type have been no more successful than the monistic theories in handling angular momentum. While they avoid the problem of a slowly spinning Sun by assuming its pre-existence, they cannot find mechanisms to remove material to a sufficient distance from the Sun or, in other words, to give it enough angular momentum. Any attempt to use solar material to produce planets, with the Sun in its present form, seems doomed to failure. However, the Schmidt–Lyttleton dualistic accretion theory does resolve the angular momentum problem by capturing material in a spread-out form which possesses the right amount of angular momentum to explain the planetary motions at the time of its capture.

4.9.2   Planet formation

The Schmidt–Lyttleton accretion theory plus the monistic theories have as starting points for planet formation a disc of material, so an important part of such theories must be the mechanism by which such material accumulates to form planets. If a condensation is going to form spontaneously due to gravitational instability of the disc then the density and temperature of the material must be able to satisfy Jeans' criterion (2.22) for the condensation of a planetary mass, and also be outside the Roche limit given by (4.12). The total mass within the disc may then be very large. There will be a tendency either for many planetary condensations to form, giving an embarrassing abundance of planets, or for there to be a massive gaseous residue which must be disposed of in some way, which requires energy from somewhere, presumably the Sun.

The material ejected from the Sun by Jeffreys' collision theory would presumably be in a fairly restricted region of space so that we might consider it similar in that respect to the Chamberlin and Moulton theory and Jeans' theory. Chamberlin and Moulton introduced the idea of planetesimals as small solid condensations within the material coming from the Sun. There are good theoretical grounds for believing that as solar material cooled some of the more refractory components would condense and form solid particles, probably in the form of grains, and accumulations of these could then be the planetesimals envisaged by Chamberlin and Moulton. Up to 1960 no coherent and detailed theory had been advanced for the formation of planets given planetesimals as a starting point but we shall see that, since then, considerable effort has been made on this topic.

The process of planet formation in Jeans' theory is the most straightforward, depending as it does on gravitational forces and the operation of the Jeans criterion (2.22) which, in its turn, depends on the well-founded Virial Theorem. However, as Spitzer (1939) so convincingly demonstrated, it is not possible for solar material in its present condition to satisfy (2.22) for Jupiter, let alone for the less massive planets. By expanding the Sun out to the orbit of Neptune, that is by considering a proto-Sun in an early stage of its development so that material was cool, Jeans was able to deal with this difficulty. This then brought up an extra problem in explaining how the planets so formed would interact with such an extended Sun.

4.9.3 Implications from the early theories

From the various difficulties thrown up by the early theories it is possible to postulate the circumstances which would negate them. However, in performing this exercise it must again be emphasized that it is being done on the assumption that only the theoretical treatments of angular-momentum transfer and planet formation available in 1960 are taken into account. There have been many theoretical developments since 1960 that have been specifically directed towards the angular-momentum and planetary-formation problems and these will be dealt with and assessed in chapters 5 and 6.

The first implication is that the angular-momentum problem is so intractable for the early nebula theories that it is desirable to assume a slowly spinning pre-existing Sun. Finding an explanation of why the Sun spins so slowly seems intrinsically easier to do if the Sun is to be produced alone rather than trying to explain it in the context of both planet and Sun formation for which the simultaneous antagonistic partitioning of mass and angular momentum are required.

The one theory which seems to overcome the angular momentum problem is the Schmidt–Lyttleton theory in which planetary material comes from a body other than the Sun and this seems to be a good scenario for explaining the angular momentum of the planets. This suggests the first two desirable features:

(1) *There is a slowly spinning pre-existing Sun.*
(2) *Material is captured from a body other than the Sun.*

The formation of planets from very diffuse and low-density material requires non-gravitational processes, for example, to give the aggregation of planetesi-mals. On the other hand if the density is sufficiently high, and the temperature sufficiently low, so that Jeans' criterion (2.22) is satisfied and formation is outside the Roche limit then gravity can take over to give planets without any theoretical difficulties. Jeans' first attempt to explain planets by pulling them out of the Sun was soundly based theoretically but the material was not in a form that could di-rectly form planets. On the other hand, with a greatly expanded Sun planets could be produced but then would have to plough through the Sun's material. These considerations of planetary formation suggest that

(3) *the Sun should be in a condensed form;*
(4) *the material forming the planets should be dense and cool.*

If the planetary material is to come from a disc that has a high density then there is either the problem of mass disposal or that of producing planets too pro-lifically. To limit the amount of material from which the planets are to form while, at the same time, having a high density suggests that the Chamberlin and Moulton or Jeans models are preferable so that

(5) *the material from which the planets form should be of as low mass as possible and be concentrated in some way, e.g. in two steams or a single filament.*

In the following chapter we shall examine newer ideas of Solar System for-mation and see the extent to which either the desirable features identified here have been incorporated or how other ways have been proposed by which the dif-ficulties of the older theories could be resolved.

PART 3

CURRENT THEORIES

Chapter 5

A brief survey of modern theories

5.1   The method of surveying theories

In looking at theories of solar-system formation up to 1960 what has become clear is that it is not an easy task to find a plausible theory—none up to that time could be so described—and also that there are some basic problems that are very difficult to solve. Since 1960 four new approaches, or developments of old approaches, have been put forward that may be regarded as extant theories—in the sense that they each have active adherents and are not universally regarded as implausible. These are: the Proto-planet Theory (McCrea 1960, 1988), the Capture Theory (Woolfson 1964), the Solar Nebula Theory (Cameron 1973) and the Modern Laplacian Theory (Prentice 1974). We shall also refer in subsequent chapters to one older theory, Schmidt's Accretion Theory (section 4.7) first proposed in 1944 and later developed by Lyttleton.

In this chapter we shall be giving a broad-brush description of the four most recent theories—the Accretion Theory has already been described—without any mathematical detail but sufficient to give an idea of their basic structures and how, if at all, they relate to older theories. In addition we shall indicate their critical features and any potential problems which they have and need to resolve, but without at this stage commenting on the extent to which the problems are, or are not, actually resolved. In later chapters the characteristics of the Solar System will be considered, one after the other, and the extent to which the five theories under review have been successful or otherwise in explaining these characteristics will be considered in some detail.

The presentation of the five theories may seem rather uneven. This reflects the different states of development of the theories rather than any deliberate bias and may also be related to the complexity of the model under discussion; complex mechanisms require more explanation than simple ones.

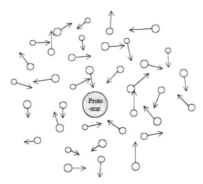

Figure 5.1. A region of the forming stellar cluster and individual proto-planets, some in aggregations.

5.2   The Proto-planet Theory

As we have already seen one of the basic problems to be addressed in cosmogony is why the Sun and other late-type stars spin so slowly. An explanation for this has been offered in section 2.6.2.6 in relation to a theory of star formation but Mc-Crea suggested that star formation and planetary formation should be considered together. Following this idea a theory should explain the formation of a system, planets plus a star, and show how the star has most of the mass while the planets have most of the angular momentum.

   McCrea's starting point is an interstellar cloud of gas and dust that is eventually to form a galactic cluster. About 1% of the mass of the cloud is in the form of grains and the remainder is the normal cosmic mix of hydrogen and helium. The assumption is made that the cloud is in a state of supersonic turbulence. Due to the collisions of turbulent elements nearly all the mass of the cloud consists of compressed regions of gas that moved around in random fashion within a lower-density background (figure 5.1). McCrea's general approach is to feed into his model a minimum number of assumed parameters and then to use the derived quantities which describe the resultant Solar System as a test for the plausibility of the model. In the original 1960 form of the theory the compressed regions were called floccules and had masses about three times that of the Earth. In a later revised version of the theory (McCrea 1988) the masses were increased to just over 100 Earth masses, approximately the mass of Saturn, and were re-designated as proto-planets. Here we shall use the term 'blobs' to describe these objects and so distinguish them from 'proto-planets' the entities which will of themselves ultimately collapse to form planets.

   From the assumed total mass and radius of the cloud it is possible to find the root-mean-square speed of the blobs by an application of the Virial Theorem (Appendix II). Then, from another assumed parameter, the density of the blobs, their radii could be found and hence the rate at which they collided. The reasonable assumption was made that collisions between blobs would be inelastic so that they would coalesce and gradually build up into larger aggregates. When, by chance, one aggregate in a particular region became appreciably larger than its neighbours then, because of gravitational focusing, its collision radius with blobs would increase so giving an accelerating rate of absorption of other bodies. In this way one body in each region would become dominant and absorb most of the original blobs within it, hence giving a proto-star starting its evolution towards the main sequence.

From the number and mean random speed of the original blobs it is possible to estimate the expected total angular momentum associated with the original cloud but very little of this angular momentum is associated with the spin of the proto-stars which form. The addition of blobs to the growing proto-star is from random directions so that the final angular momentum is a random-walk addition of the individual contributions. The missing angular momentum is accounted for by having several of the blobs, or small aggregations of them, in orbit around each proto-star and McCrea showed that the number required was small and similar to the number of observed planets.

The blob mass is just larger than that of Saturn so it appears that Jupiter, at least, was formed from a number of the original blobs. McCrea assumes that all the initial proto-planets were more massive than the residual planets we now see and that some form of mass loss occurred. In the process of collapsing, the proto-planets would have become rotationally unstable and broken up into two unequal parts with a mass ratio of order 8:1 (Lyttleton 1960). Part of the original spin angular momentum of the proto-planet would then appear as relative motion of the two fragments around the centre of mass (figure 5.2) with the less massive portion having a speed relative to the centre of mass eight times that of the main portion. Escape speeds from the Solar System are less in its outer regions and McCrea proposed that the small portions were lost in the regions of the major planets, leaving the major portions rotationally stable although with quite rapid spins, such as are observed. Another result of the fission of the original proto-planets is that in the 'neck' between the two portions small droplets would have condensed and been retained by the retained major portion as a system of regular satellites (figure 5.2).

The disruption process took a different form in the inner part of the system. First, it is assumed that the rotational disruption took place after there had been some segregation of material and that the rotational instability occurred in a dusty core. Thus the bodies produced would have consisted of solids and the smaller body would not have sufficient speed to escape, since it was so close to the Sun. McCrea suggested that the pairs of bodies Earth–Mars and Venus–Mercury were each the product of such a dusty-core disruption.

When the theory is examined in more detail it will be found to give very satisfactory numerical agreements of predicted and observed quantities derived from the assumed input parameters. The important features of, and potential problems

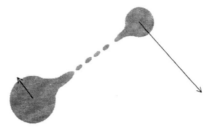

Figure 5.2. The fission of a rapidly-spinning proto-planet with the formation of proto-satellite droplets.

faced by, this theory are:

(i)   It is a monistic theory that deals with both the partitioning of mass and angular momentum.
(ii) It must be shown that the input parameters are reasonable.
(iii) The formation of blobs of suitable mass by turbulent collisions must be demonstrated.
(iv) The blobs must be stable, at least for the time required for them to combine into proto-stars or small aggregations.
(v)   The predicted solar spin must be shown to be compatible with that at present.
(vi) It should be demonstrated that the 'missing' angular momentum after the proto-stars form is taken up by proto-planets in orbit and is not taken up in some other way.
(vii) An explanation is required for the near alignment of the solar spin and planetary orbital angular momentum vectors.
(viii) The basic mechanism does not give a planar system of planets in circular orbits so how this comes about must be explained.
(ix) A reason is required for why the total proto-planet disrupts in the outer system but only the core in the inner system.

5.3   The Capture Theory

The demise of Jeans' tidal theory (section 4.6), which had so much acceptance and for such an extended period of time, convinced most cosmogonists that there was no plausible theory to be derived from that direction. In 1964 Woolfson introduced a new variant of the tidal theory, which called on many of the mechanisms investigated by Jeans, which were generally accepted as being theoretically sound, but which was different from the Jeans model in a number of important respects.

In section 2.6.2 a theory of star formation in a galactic cluster was described which gave results consistent with observations of young clusters. In particular the first stars produced had somewhat more than one solar mass ($\sim 1.4 M_\odot$) and

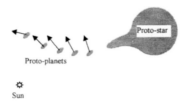

Figure 5.3. The initial motion of proto-planetary condensations in the Capture Theory.

subsequently stars of lesser mass were formed. In the rather dense environment of a young cluster, stellar interactions will be quite common—indeed it is the interactions that give individual stars sufficient energy to escape from the cluster and which, hence, gradually disperse it and provide field stars. The form of interaction considered by Woolfson, consistent both with the star-forming model and observation of young clusters, is one between a condensed solar mass star, identified as the Sun in relation to the formation of the Solar System, and a proto-star of lesser mass which was newly formed and so in a diffuse state.

In the interaction considered the proto-star moved on a hyperbolic orbit relative to the Sun and passed within the Roche limit for disruption. The consequent behaviour of the proto-star was similar to that shown in figure 4.10; a filament was drawn out of the proto-star at the extremity of the tidal bulge and the gravitationally unstable filament broke up into a series of condensations. The line density of the filament was sufficiently high for each blob to have had a mass exceeding the Jeans critical mass and so the blobs became contracting proto-planets. The essential differences between this model and the Jeans are:

(a) The material which comes from the proto-star was captured by the condensed star—that is the aspect giving the theory its name.
(b) The material forming the planets was cool, thus removing many of the objections to Jeans' original tidal theory.
(c) At the time of the interaction the proto-star had a radius of about 20 AU and the aphelion distance of its orbit was of the order 40 AU. It is this latter distance which governed the scale of the Solar System, and the intrinsic orbital angular momenta of the planets, which Jeans' theory could not reproduce.

The proto-planets were produced on highly eccentric orbits, with eccentricities in the range 0.7–0.9 and with aphelia ranging out to more than 100 AU. The initial motions from the time the proto-planetary blobs separate from the filament were towards aphelia, as illustrated in figure 5.3; hence the proto-planets had from tens to hundreds of years to condense before they had to survive the tidal forces of a perihelion passage. While this enabled the proto-planets to condense to give major planets they *were* influenced by solar tidal forces to the extent that they became distorted and the outer material, especially that in the tidal bulge, acquired spin angular momentum (figure 5.4(a)). The form of collapse of a distorted proto-star,

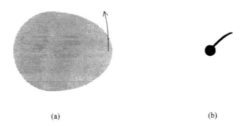

(a)   (b)

Figure 5.4. (*a*) A tidally distorted proto-planet with spin induced in the tidal bulge. (*b*) The form of the proto-planet after considerable collapse showing the formation of a filament.

especially with an induced spin as described, is that the tidal bulge material gets progressively left behind as the main body collapses, as shown in figure 5.4(*b*). The material in the bulge thus took on the form of a filament and the Jeans mechanism of gravitational instability occurs, with individual blobs condensing to form a family of regular satellites. This mechanism for satellite formation was that proposed by Jeans in his original tidal model, although he did not give a detailed description,.

The transition of material from the proto-star tidal filament to proto-planets also gave some material in a diffuse form around the Sun, so providing a resisting medium. This rounded off the planetary orbits that, together with various interactions between the planets as the rounding-off process took place, gave the orbits we see today.

Features and potential problems of the Capture Theory are:

(i)   It is a dualistic theory and requires a separate mechanism to produce a slowly spinning Sun.
(ii) It must be shown that condensations in the proto-star filament were captured by the condensed star and not retained by the proto-star.
(iii) The proto-star filament must have had a sufficient line density to form stable proto-planetary condensations.
(iv) The proto-planets must have condensed on a time-scale sufficiently short that they could have survived the first perihelion passage.
(v)   The form of collapse of a tidally distorted proto-planet to give a filament and satellites must be demonstrated.
(vi) It must be shown that, with a reasonable mass of resisting medium, planetary round-off takes place within the lifetime of the medium.
(vii) An explanation is required for the near alignment of the solar spin and planetary orbital angular momentum vectors.
(viii) It must be shown that the probabilities of capture-theory interactions in evolving stellar clusters are sufficiently high to explain the inferred frequency of extra-solar planets.

5.4   The Solar Nebula Theory

It became evident in the 1960s that many features of meteorites could be understood in terms of their condensation from a hot vapour. A number of theoretical studies of condensation sequences from material of Solar System composition cooling at various temperatures and pressures were published (e.g. Larimer 1967, Grossman 1972) and the idea was reinforced that material in the early Solar System had been in a hot gaseous form. In addition Safronov (1972) published an influential paper on the formation of planets starting with very diffuse material. Although Safronov's results indicated long time-scales for planetary formation he did present a well-structured model that it was felt could be modified to resolve the time-scale problem. This background led to a revival of the original dualistic Laplace idea of the spontaneous formation of the Sun and planets from a slowly spinning sphere of gas and dust (section 4.3)—but with the difference that new theoretical advances would enable the problems of the original theory to be overcome. This began an avalanche of investigation into the Solar Nebula Theory that became the dominant paradigm of cosmogony in the last decades of the 20th century, accounting for much more than 90% of all the work in this field.

The model is still under active development, and has not yet completely settled into an agreed sequence of events leading to the present Solar System, but there are some dominant ideas and these will now be described. A very early and major contribution was made by Cameron (1973). One of his early conclusions (Cameron 1978) was that

> At no time, anywhere in the solar nebula, anywhere outwards from the orbit of Mercury, is the temperature in the unperturbed solar nebula ever high enough to evaporate completely the solid materials contained in interstellar grains.

Although one of the basic motivating influences for re-examining this type of theory was removed, if Cameron's comment was valid, nevertheless the impetus of the work was completely unaffected. A new surge of confidence in the essential correctness of the Solar Nebula Theory was generated by the detection of circumstellar discs (section 2.8) which would be a necessary observable feature of the theory.

A major problem of Laplace's theory was that of angular-momentum distribution so that, as the nebula develops, angular momentum is transferred from the inner condensing core material to the disc that is forming in the equatorial plane. A number of possible mechanisms have been suggested and explored for this angular momentum transfer, involving one or more of

(a) turbulent viscosity within the disc,
(b) gravitational effects due to the formation of spiral arms in the disc,
(c) interactions between ionized matter leaving the central region and a magnetic field generated within it and

(d)   angular momentum transport by waves generated within the disc.

In the original Laplace theory the separation of the disc from the central condensation was in the form of a series of concentric rings, each forming the basis of a single planet. In the Solar Nebula Theory the starting point for planet formation was a disc of mostly gaseous composition with 1 or 2% of solid material and at a temperature that fell with increasing distance from the centre. In some earlier versions of the theory the disc was quite massive, of about one solar mass, and the density and temperature within it was such that Jupiter-mass regions satisfied the Jeans criterion. The disc would have thus been gravitationally unstable and giant planets would have formed spontaneously. Such a model has no difficulties in producing planets but produces so many planets that the theorist is then left with a major disposal problem. For that reason a less massive disc is now favoured, with mass between 0.01 and $0.1 M_\odot$, which is consistent with observations and requires the planets to be formed by an accretion process.

The accretion of terrestrial planets and the solid cores of the giant planets is assumed to have taken place in two stages. The first involved the formation of planetesimals, a term first used by Chamberlin and Moulton (section 4.5.1), and here taken to be bodies anywhere in the size range from hundreds of metres to tens of kilometres. An essential first step in planetesimal formation was the settling of the dust to form a thin layer in the mean plane of the disc. This would have been gravitationally unstable (Goldreich and Ward 1973) and the solid condensations within it would have provided the planetesimals. There is a minority view (Weidenschilling *et al* 1989) that planetesimal formation required solid material to stick together when it came into contact if a dust disc was to form in a sufficiently short time. Regardless of the way in which planetesimals were formed, the way that they would have come together to form planets was described by Safronov in some detail. Within each annular region of the nebula one body would have become increasingly dominant, as its capture cross-section increased with mass, until eventually it absorbed all the planetesimals in the region. Once the cores of the giant planets were produced they would then attract the nebula gas so forming the gas giants, a process that would have happened on a very short time-scale of 10^5 years or so.

No specific theory is available for natural satellite formation for the Solar Nebula Theory except that it is a small-scale version of the mechanism for producing planets. Thus a collapsing proto-planet would have developed a disc in its equatorial plane and proto-satellite condensations would have formed within it.

The critical features and potential problems of the Solar Nebula Theory are as follows:

(i)   It is a monistic theory that simultaneously deals with the partitioning of mass and angular momentum.
(ii) Some mechanism, or combination of mechanisms, must be shown to transfer sufficient angular momentum from the condensing Sun to the disc.

(iii) It must be shown that planets will be produced on a time-scale compatible with the observed lifetimes of circumstellar discs ($<10^7$ years).

(iv) The surplus disc material, after the planets formed, must be disposed of.

(v) Since the model appears to predict a highly planar system, the tilt of the solar spin axis requires explanation.

5.5 The Modern Laplacian Theory

The Solar Nebula Theory is a derivative of the original Laplace nebula model although the proposed pattern of development of the nebula was somewhat different—for example, the material in the disc was not in the form of a set of annular rings. An idea put forward by Prentice (1974) followed the Laplacian picture more closely and so is called the Modern Laplacian Theory.

In an attempt to rescue the Laplace model, Roche (1854) suggested that the nebula had a very high central condensation so that if it was rotating uniformly the majority of the mass would have had very little of the angular momentum (section 4.4). This would certainly have helped the angular-momentum problem but it also made it impossible to solve the problem of producing planets since the density in the outer part of the nebula would then have been too low to condense within the tidal field of the central mass. Prentice's contribution was to propose a mechanism that, at the same time, gave a high degree of central condensation, corresponding to a moment-of-inertia factor of 0.01, and also concentrated the material in the outer regions where planets were to form.

The first stage of the process was the formation of a condensed proto-Sun from a cold dense cloud in such a way that when its radius was about $10^4 R_\odot$ it contained only 1% of the original intrinsic angular momentum of the cloud material from which it formed. Prentice adopted a suggestion of Reddish and Wickramasinghe (1969) that in the interiors of cold dense clouds, temperatures are so low (≤ 5 K) that grains of solid molecular hydrogen may form, containing some helium as an impurity. These grains became part of a collapsing cloud but for most of the collapse, until the linear dimension of the grain cloud was reduced by a factor of ten, the grains were coupled to the cloud by dynamical friction. In this way the final cloud retained only 1% of the initial angular momentum of its material. The gravitational energy of the collapse vaporized the solid H_2 grains so that by the time the cloud radius reached $10^4 R_\odot$ the proto-Sun was a collapsing gaseous sphere. A degree of central condensation had been attained at this stage due to the more rapid fall of denser CNO grains towards the centre of the cloud. Again, modelling shows that a collapsing gaseous cloud is extremely non-homologous in its collapse and forms a dense core (Larson 1969).

When the proto-Sun reached a radius equal to that of Neptune's orbit there was a balance of centrifugal and gravitational forces so that the equatorial material was in free orbit around the central mass. This is similar to Laplace's model of the situation and would have led to low-density equatorial material being left behind

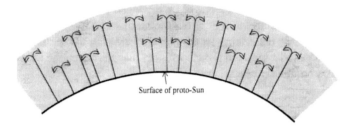

Figure 5.5. Needle-like elements due to supersonic turbulence. Material in the shaded region slowly falls back to the surface of the proto-Sun.

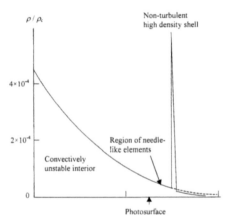

Figure 5.6. The density inversion beyond the boundary of the evolving proto-star due to the mechanism shown in figure 5.5. Eventually the high-density region splits off to form a ring.

in the equatorial plane as the proto-Sun collapse continued. It is here that Prentice introduced another new idea, that of turbulent stress within the collapsing cloud. If the proto-Sun was strongly convective, and turbulent supersonic motions were generated within it, then material in regions of lower density could have been propelled by buoyancy effects beyond the normal surface at supersonic speeds. This created a swarm of needle-like elements that were ejected at high speed from the surface but returned to the surface in the form of a much lower speed stable flow (figure 5.5). The effect of this is that the density in the surface regions of the proto-Sun was very much increased, as illustrated in figure 5.6. Another effect of the supersonic turbulence is that it would have had its greatest effect in the outer parts of the proto-Sun, acting as an extra source of pressure that exclusively extended the outer regions and so further increased the central condensation.

The enhancement of the density in the surface regions directly influenced

Figure 5.7. A sequence of cross-sections of the evolving turbulent contracting proto-Sun and its concentric system of orbiting gaseous rings.

the density of the material left behind by the collapsing proto-Sun. Prentice also showed that an instability occurred in the equatorial shedding process so that the material in the equatorial plane was in the form of a series of annular rings, much as Laplace had postulated (figure 5.7). In the Modern Laplacian Theory the rings all had very similar mass, about $10^3 M_\oplus$, with temperatures at the time of ring detachment falling off slightly more slowly than the inverse of the ring radius. Prentice postulated that several rings were formed within the orbit of Mercury but that their material was completely vaporized. For the terrestrial planets the rings would have contained solid silicate and metal grains, of total mass $4M_\oplus$, while in the major planet rings there would also have been ice grains giving a total mass of solids of $11\text{--}13M_\oplus$ within each ring.

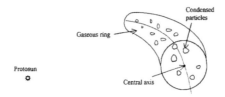

Figure 5.8. Condensed particles within a ring in the Modern Laplacian Theory.

Prentice gave arguments to show that the solid material in each Laplacian ring would have fallen towards the ring axis, as shown in figure 5.8. A hierarchical sequence of Jeans' instabilities caused bunching of the material until all the solid material along the ring axis had collected together. The proportion of the total solid material within the whole ring that collected together depended on the balance between the rate of settling towards the axis and the rate at which axial solid material accreted. Finally, in the major planet region, the cores were sufficiently massive, and the local gas sufficiently cool, for gaseous mantles to be accreted.

Prentice assumed that the contraction of the atmospheres of the newly formed major proto-planets followed the same pattern as that of the proto-Sun with supersonic turbulent stress being generated and the shedding of rings to give regular satellite families.

The Modern Laplacian Theory is certainly the most complex of the modern theories in terms of the physical phenomena it invokes and the detailed mathematical analysis of the processes involved. The critical features and potential problems of the model are as follows:

(i) It is a monistic theory which simultaneously deals with the partitioning of mass and angular momentum.

(ii) It critically depends on a loss of 99% of the original angular momentum of the proto-Sun by formation from H_2 grains—which requires an extremely low cloud temperature.

(iii) The several rings postulated as having formed within Mercury would have had a total angular momentum several hundreds times that of the present Sun. It must be shown that the energy was available to have disposed of virtually all of this mass.

(iv) Would the solid material in each ring have fallen to the ring axis as described and would the gaseous rings have had a sufficiently long lifetime for this to happen?

(v) The model would have given a very planar system. How is the tilt of the solar spin axis explained?

(vi) Would collapsing gaseous proto-planets have provided sufficient energy for supersonic turbulent stress and would problem (iv) have been even more pressing in the smaller system?

5.6 Analysing the modern theories

As previously noted the various modern theories have been developed to very different degrees but they all deal with, or imply, the first-order features described in section 3.2.1, viz. the distribution of angular momentum, planet formation, a cold origin for planets and direct planetary orbits. To these we may add the formation of the Sun, or stars for planetary systems in general, and satellite formation. Any theories which convincingly explain these features without obviously insuperable problems must be regarded as *prima facie* plausible and any ranking of such theories would then depend on either the application of the Occam's razor principle or on how well more subtle features of the Solar System are explained. The following chapter deals with these augmented first-order features.

Chapter 6

The Sun, planets and satellites

6.1 Surveying extant theories

As was indicated in section 3.1.1 there is no such thing as a correct theory but, nevertheless, theories can be judged on the basis of their plausibility. In this and succeeding chapters the characteristics of the Solar System, some of them major through to others much more subtle, will be considered in relation to the extant theories which were described in outline in chapter 5. These are the Proto-planet Theory, the Capture Theory, the Solar Nebula Theory and the Modern Laplacian Theory together with occasional references to the Schmidt–Lyttleton Accretion Theory described in section 4.7. Not all the theories have been developed to the same extent so not all characteristics can be discussed in relation to all theories. Although the discussion will be mainly centred on the Solar System the existence of other planetary systems will not be ignored and, for appropriate topics, the relevance of the ideas to other planetary systems will be discussed.

In this chapter we shall be considering the most basic characteristics of the Solar System—the Sun, the planets and their regular satellites—the explanations for the origin of which must be regarded as requirements for any plausible theory. This can be regarded as a first sieve to identify those theories which are either implausible or, at least, less plausible. To repeat what was stated in section 5.1, it may seem that in dealing with various topics the depth and extent of treatment between one theory and another may seem rather uneven. This will reflect the level of development of different aspects of the different theories as presented in published work, although sometimes a detailed treatment developed in relation to one theory can be carried over to another.

6.2 Formation of the Sun: dualistic theories

In section 4.1.2 the division of theories into the categories of monistic and dualistic was introduced, dependent on whether the Sun and planets were produced

as products of the same process or of different processes. The Modern Laplacian Theory and the Solar Nebula Theory are clearly in the monistic category and
the Accretion Theory is clearly dualistic. The Proto-planet and Capture Theories
are also dualistic although, for them, there is an intimate connection between the
star and planet-forming processes. For the Proto-planet Theory the same material
in the same form produces either the Sun or a planet, depending on how it happened to move in the system. In the Capture Theory the same environment which
enabled stars to form also enabled interactions between stars to take place.

For the Accretion Theory and the Capture Theory a separate model for star
formation is required and one such model is described in chapter 2. There may
be other plausible mechanisms but to have at least one is reassuring. On the other
hand the Proto-planet Theory provides a precise mechanism for star formation
and explains the slow spin rate of Sun-like stars by the random-walk addition of
angular momentum contributions as proto-planetary blobs join the growing proto-
star. Here the analysis we shall follow is a slightly simplified form of that given
by McCrea but one which gives similar results except for unimportant numerical
factors.

The starting point is an accumulation of a few blobs to form the nucleus of
the growing proto-star. If its radius is r and a blob of mass m joins it when moving
at speed V then the magnitude of its contribution to the angular momentum will,
on average, be approximately

$$h = \tfrac{1}{2}mVr. \tag{6.1}$$

Although r and V will depend on the extent to which the proto-star condensation
has grown we may regard them as some kind of average quantities during the
whole process. If N blobs form the final body then the total angular momentum
will be the result of a random-walk addition of individual contributions

$$H^* = \sqrt{N}h = \frac{1}{2}\sqrt{N}mVr = \frac{MVr}{2\sqrt{N}} \tag{6.2}$$

where M is the final mass of the proto-star. The mean radius of the proto-star is
assumed to be close to its final radius, which McCrea takes as 7×10^8 m, close to
the present radius of the Sun, and V will be of the order of the escape speed from
the proto-star so that

$$V^2 = \frac{2GM}{r}. \tag{6.3}$$

From this, with some factor of uncertainty of order unity,

$$H^* = \left(\frac{GM^3r}{2N}\right)^{1/2}. \tag{6.4a}$$

By a more detailed approach McCrea found

$$H^* = \frac{2}{3}\left(\frac{GM^3r}{3N}\right)^{1/2}, \tag{6.4b}$$

which is just over one-half of (6.4a). One of McCrea's assumed parameters was
that the blob masses were 6.7×10^{26} kg, approximately the mass of Saturn, so that
$N = 3000$ was necessary for M to be the mass of the Sun, 2×10^{30} kg. Inserting
the assumed and derived values into (6.4b) gives $H^* = 4.3 \times 10^{42}$ kg m^2 s^{-1}.
The time-scale for the growth of the Sun by this process can be estimated from
the probable sizes of the blobs and of the growing Sun and is in the range 10^5–
10^6 years, which is acceptably short. The estimate of angular momentum of the
Sun from the Proto-planet Theory, H^* is some 30 times its present value. This
may be compared with the estimate from the Woolfson (1979) star-formation theory, as given in section 2.6.2.6, that is anything from just over one to 20 times the
present value. Angular momentum enhanced by this factor presents no problems
for the Sun forming as a compact body. Nevertheless a reduction in angular momentum by a factor up to 30 is the main problem that needs to be resolved for
these theories of star formation and we shall now look at a plausible mechanism
for reducing initial spin by the required amount.

6.2.1   The magnetic braking of solar spin

For dualistic theories the Sun is produced as an isolated body although there may
be mechanical coupling to a surrounding medium. If such coupling occurs then it
will involve viscous drag that will slow down the spin. However, the first stage in
the collapse of a proto-star is quite rapid, approximately free-fall, until the opacity
of the proto-star increases to the point where energy is trapped within the body
and internal pressure opposes and slows down the collapse. For this reason the
duration of the viscous drag stage is probably too short to have an appreciable
effect on the star. As a worst-case assumption, the Sun is taken as a condensed
body at the Kelvin–Helmholtz contraction stage (section 2.3.1) with up to 20 or so
times its present angular momentum. The most plausible mechanism for slowing
down the spin thereafter involves the coupling of ionized material moving out of
the Sun with the solar magnetic field. Charged particles leaving the Sun, in the
form of a solar wind, will spiral along field lines in the vicinity of the Sun where
the field is strong. Since the magnetic field rotates with the Sun, so the escaping
material will co-rotate with the Sun while moving outwards and hence remove
angular momentum. The important factors governing this process are the nature
and strength of the magnetic field and the rate at which ionized material is lost
from the Sun.

    First we consider the mechanism by which the charged particles are initially
coupled to field lines and later become decoupled from them. The condition that
governs whether or not the charged particles couple to field lines depends on the
relative strengths of the magnetic pressure (energy density) given by

$$P_B = \frac{B^2}{2\mu_0} \tag{6.5}$$

in which B is the field and μ_0 the permeability of free space, and the total gas

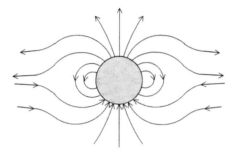

Figure 6.1. Magnetic field lines round a star with a strong stellar wind of ionized particles.

pressure

$$P_g = nk\theta + nmv^2 \qquad (6.6)$$

in which θ is the temperature and n the number density of particles of mean mass m and bulk flow velocity v. The first term in (6.6) is the normal gas pressure and the second term is the dynamic pressure due to the bulk flow. Only if the magnetic pressure exceeds the total gas pressure can the motion of the charged particles be controlled by the field and equality of the two pressures gives the approximate conditions under which decoupling of the particles from the field may be considered to take place.

The form of the solar magnetic field is quite complex because of the rapid flow of the solar wind. At larger distances from the Sun, where the field is weaker, the wind is more-or-less unconstrained by the field but, on the other hand, the magnetic field becomes frozen into the plasma and field lines take on the directions of the local flow. The net effect on the field lines is shown schematically in figure 6.1. The result is that close to the equatorial plane, and at large distances, R, from the Sun, the fall-off in field varies as between R^{-1} and R^{-2} rather than as the R^{-3} that is expected for a dipole field. By carrying over results from measurements of the Jovian magnetic field Freeman (1978) suggests that the solar field is of dipole form to within 25% or so out to $30R_\odot$ and gradually moves towards R^{-1} dependence thereafter. For theoretical purposes he suggested a form of field given by

$$B = D_\odot \left\{ \frac{1}{R^3} + \frac{1}{(30R_\odot)^2 R} \right\} \qquad (6.7)$$

in which $D_\odot$ is the magnetic dipole moment of the Sun.

In finding the distance at which the magnetic and gas pressures are in balance we may make two simplifying assumptions. The first is that for protons as the charged particles, with a typical solar-wind speed of 500 km s^{-1}, $mv^2 > k\theta$ unless θ is of order 10^7 K so that the first term on the right-hand side of (6.6) can be ignored. The other assumption, which the results eventually justify, is that out to distances where decoupling takes place the field is effectively of dipole form.

Equating the magnetic and gas pressures with these simplifications

$$\frac{D_\odot^2}{2\mu_0 r^6} = nmv^2. \tag{6.8}$$

The quantity nm is the local density of the ionized material and, in terms of the rate of mass loss from the Sun, dM/dt, assuming that all lost material is ionized,

$$nm = \frac{dM/dt}{4\pi r^2 v}. \tag{6.9}$$

Inserting (6.9) into (6.8) we find that co-rotation of ionized material will persist out to a distance

$$r_c = \left(\frac{2\pi D_\odot^2}{\mu_0 v \frac{dM}{dt}}\right)^{1/4}. \tag{6.10}$$

With present values for the Sun, $dM/dt = 2 \times 10^9$ kg s^{-1}, $D_\odot = 8 \times 10^{22}$ T m^3 and $v = 5 \times 10^5$ m s^{-1}, (6.10) gives $r_c = 3.4 R_\odot$. This means that the lost mass takes from the Sun $(3.4)^2$ times the angular momentum it had when it was part of the Sun. At the present solar-wind rate the total mass loss of the Sun over its lifetime would have been about 1.4×10^{-4} of the initial mass and the loss of angular momentum about 0.16% of the original angular momentum.

It is generally assumed that the early Sun was far more active than it is now which would have given both a greater rate of mass loss and also a higher early magnetic field. The early field was considered by Freeman (1978) from various points of view. We shall see in section 9.4 that the remnant magnetism of lunar rocks indicate that there may have been a local magnetic field early in the Moon's history much higher than the present one and Freeman estimates this as about 2×10^{-6} T. To obtain such a local field from the Sun would have required a magnetic dipole moment of $D_0 \approx 7 \times 10^{27}$ T m^3 or 10^5 times its present value. This dipole moment would have given a surface field for the Sun of 20 T, which is two orders of magnitude greater than that which is observed for magnetic stars. If, when this dipole moment existed, the Sun had a radius four or five times larger than at present then the surface field is brought down to a value compatible with observations. The other aspect of an early Sun is that the rate of loss of mass was almost certainly far higher than at present. The assumption is sometimes made that the Sun went through a T-Tauri stage when the loss of mass was of order $10^{-7} M_\odot$ year^{-1} (some 10^7 times the present rate) sustained for a period of 10^6 years. It is found that with such a total loss of mass and with the extremely high original field deduced from lunar-rock measurements the Sun would have virtually no residual spin at all, which is unrealistic. There are the following arguments against such a scenario:

(i)   There is no evidence of such high magnetic fields for T-Tauri stars.
(ii) The material from T-Tauri stars gives a strong spectrum of neutral hydrogen suggesting that the material is only lightly ionized and hence will not be strongly coupled to the field.

Table 6.1. The fraction of the initial angular momentum remaining after 10^6 years with different combinations of magnetic dipole moment and rates of mass loss.

	D (T m^3)		
$\mathrm{d}M/\mathrm{d}t$ (kg s^{-1})	8×10^{23}	8×10^{24}	8×10^{25}
2×10^{13}	0.9907	0.9107	0.3925
2×10^{14}	0.9708	0.7437	0.0517
2×10^{15}	0.9093	0.3867	0.0001

(iii) Observational evidence suggests that material is moving both away from and towards T-Tauri stars. This may mean that the material is not lost from the star but is ejected from it in the form of large *prominences*, such as are seen for the Sun but on a much larger scale.

(iv) Most late-type stars, of solar mass or less, spin slowly and also require some braking mechanism. Stars of much less than $1M_\odot$ certainly do not have a T-Tauri stage so proposing such a stage as an essential ingredient for solar braking is tantamount to suggesting that there are two mechanisms. This is not impossible but seems unlikely.

Returning to the original requirement of reducing the angular momentum of the Sun to a few per cent of its original value, the magnetic braking effect is able to do this without any outlandish assumptions. If the moment of inertia of the Sun is always of the form $\alpha M R_\odot^2$, where M is the changing mass and the radius is assumed not to change, then the rate of change of angular velocity per unit mass loss is given by

$$\frac{\mathrm{d}\Omega}{\mathrm{d}M} = \frac{r_c^2 \Omega}{\alpha M R_\odot^2}.$$

Integrating this gives

$$\Omega = \Omega_0 \left(\frac{M}{M_0}\right)^{r_c^2 / \alpha M R_\odot^2}. \tag{6.11}$$

The effect of different combinations of rate of loss of mass, between 10^4 and 10^6 the present rate, and magnetic dipole moment, between 10 and 1000 times the present value, is shown in table 6.1 with $\alpha = 0.055$ (Woolfson 1979). It is clear that combinations of rate of loss and dipole moment towards the upper ends of the ranges considered are capable of giving the required reduction in angular momentum. It is concluded that, assuming an active early Sun with plausible values of rate of mass loss and magnetic field, there are no problems with the early Sun models predicted by the Proto-planet Theory and Woolfson's star-formation theory as far as the present spin rate is concerned.

6.2.2   The solar spin axis

A very significant feature of the Sun in relation to the system as a whole is the 6° tilt of the solar spin axis relative to the angular momentum vector for the total system. This angle of tilt is, on the one hand, too large to be ignored and requires explanation but, on the other hand, is too small to be confidently accepted just as a matter of chance. The probability that two vectors in randomly chosen directions are parallel to within 6° is about 0.0027. One way of describing the angular momentum associated with the Sun's spin is to equate it to that of one-quarter of Jupiter's mass orbiting at its equator. This provides the key to explaining the tilt of the spin axis for the dualistic theories. Taking the Capture-Theory model, the planets formed within a filament drawn from a passing proto-star, as described in section 5.3, but not all the filament material was converted into planets. Much of it ended up surrounding the Sun and acted as a resisting medium within which the early planets moved. This medium was dissipated over time due to the action of solar radiation; radiation pressure acted on gaseous material pushing it outwards, as did the ram pressure associated with the solar wind which may have been more violent than at present and would also have expelled very small solid particles. For example, with a T-Tauri strength of emission the pressure force on a 10 μm particle would exceed the gravitational force on it and so propel it outwards. However, larger solid particles would react differently and would spiral in towards the Sun under the action of the Poynting–Robertson effect (Appendix V). A 100 μm silicate particle in the vicinity of Jupiter would, with the present solar luminosity, take about five million years to join the Sun. The total lifetime of the surrounding medium would probably have been a few million years, a conclusion supported by observation of circumstellar discs (section 2.8). During this time, with an active Sun, the medium would have dispersed, with some small part of it being absorbed by the Sun.

The Capture-Theory model gives no correlation between the original spin axis of the Sun and the Sun–proto-star orbital plane, which approximately defines the plane of the planetary orbits. However, material drawn into the Sun by the Poynting–Robertson effect would have pulled the spin axis towards the normal to the mean plane of the system. It is impossible to specify a particular scenario for the development of the solar spin axis but it is possible to specify plausible scenarios. If the active Sun's spin had been greatly reduced in the period before it interacted with the proto-star, and the solar activity had subsequently declined, then the absorption of a mass equal to one-quarter that of Jupiter, together with the small residue of the original spin, would readily give the 6° tilt. If the Sun was still more active than now then somewhat more than one-quarter of Jupiter's mass could have been absorbed since the resultant spin could have been reduced further after the absorption of material. To provide one-quarter of Jupiter's mass of solid material would require a total mass of material between 12 and 25 Jupiter masses, assuming that 1 or 2% of the proto-star material was in the form of solid grains. This is a perfectly feasible mass for the medium surrounding the Sun. A very similar argument also applies to the accretion-theory model. It is likely that by the time the Sun passed through an interstellar cloud it would be quite mature and have a very slow spin, again with the assumption of an early active stage. If most of the solid content of the accreted cloud was absorbed by the Sun then, again, the tilt would be explained.

The final dualistic theory, the Proto-planet Theory, has its early configuration much less clearly defined. There is no reason why the proto-planets that are captured by the central star should be co-planar or even have orbits in the same sense. Some process is required to flatten the system and to eliminate any retrograde objects. One possibility is that if several of the original proto-planets orbited in a contrary sense to the majority then these would preferentially suffer collisions and the material from the collisions would create an envelope around the Sun. Interactions within the envelope would give an orderly spinning system with a well-defined spin axis and the planets orbiting in this medium would be perturbed to settle down into the mean plane. At the same time the solid component would be absorbed into the Sun so pulling its spin axis towards the normal to the mean plane. There are no clear time-scales attached to the Proto-planet Theory, or readily deducible ones, so the plausibility of this scenario cannot be tested. However, it is a *possible* scenario so the theory cannot be refuted just on the grounds of the tilt of the Sun's spin axis.

6.3   Formation of the Sun: monistic theories

In the monistic theories there are two main problems in explaining the Sun in its present state. The first is actually to form the Sun by somehow transporting most of the angular momentum from the central regions of the collapsing nebula and the second, once the Sun had formed, is to slow down its spin to the present observed level. The second problem is similar to that faced by the dualistic theories, albeit in a rather more extreme form, but we shall now consider these problems separately.

6.3.1   Removing angular momentum from a collapsing nebula

If a collapsing cool dense cloud in which stars may form reaches the comparatively high density of 10^{-17} kg m^{-3} while still coupled to the galactic rotation with intrinsic angular speed 10^{-15} s^{-1}, then a spherical volume with mass equal to that of the Sun would possess angular momentum 1.0×10^{46} kg m^2 s^{-1}. This is about 700 times that of the Solar System as a whole or more than 100 000 times that of the Sun. Thus the first problem is to remove most of the angular momentum and here we look at the way this is achieved by the Modern Laplacian Theory according to Prentice (1978).

6.3.1.1   The Modern Laplacian Theory

As explained in section 5.5, Prentice assumes that solid hydrogen grains exist in cool dark clouds and these grains, collapsing within the cloud, remain coupled to it until something like 99% of their original angular momentum had been removed. A problem with this scenario is that there is no evidence for temperatures within interstellar clouds low enough to produce solid hydrogen grains and usual estimates of their temperatures are in the range 10–100 K. The continuous heating effect of cosmic radiation would seem to ensure higher temperatures with thermal equilibrium becoming established by a balance between the heating and cooling processes, which depend on material being in a gaseous form (section 2.2.2).

The energy released by the faster fall of the denser CNO grains towards the centre would give rise to a dense luminous core of radius $3R_\odot$ together with a very tenuous envelope. The very small moment-of-inertia factor of the configuration at this stage implies a very small density in the outer part of the nebula with all the associated difficulties for producing planets which Roche failed to resolve. Prentice claims that the process of supersonic turbulence by which he envisages the build-up of a high-density ring is supported by observations of T-Tauri stars. Material is ejected at a speed which is high but less than the escape speed from the star. In his model Prentice assumed that material is supersonically expelled at around 200 km s^{-1} but returns to the star much more slowly— at about 10 km s^{-1}. This gradually builds up a dense shell outside the main body of the slowly-shrinking Sun. The loss of material from the main part of the nebula results in a constant re-adjustment of the nebula material to restore quasi-equilibrium, and the large intrinsic angular momentum associated with the forming ring is derived from the material in the main body of the star. When a ring becomes sufficiently massive it detaches from the main body of the Sun and the process starts again. Prentice describes the formation of planets down to Mercury in this way but the model also requires the presence of several rings, and hence potential planets, within Mercury. The reason these planets did not form is put down to the ambient temperature that was too high to allow material to condense. Nevertheless this material existed, whether it formed planets or not, so the assumption must be made that the material of these rings, or at least their angular momentum, was removed in some way.

There are many points of uncertainty in this model. It seems very contrived and requires a high degree of turbulence in a slowly contracting proto-Sun. For example, Prentice estimates that it would take 10^5 years for the proto-Sun to shrink from a radius of $10^4 R_\odot$, at which stage the temperature was 20 K, to $10^3 R_\odot$ (~4.7 AU) where the temperature was 150 K. The release of gravitational energy is only slightly more than enough to explain the heating of the material so, when radiated energy is taken into account, it must be wondered where the energy comes from to provide and sustain the supersonic turbulence. Certainly energy from nuclear reactions is not available at that stage. Once an internal energy source is available then the concept of supersonic turbulence is feasible and

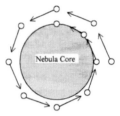

Figure 6.2. Material spiralling in towards a growing nebula core.

becomes a possible explanation of the T-Tauri observations. Another point, re-lated to timing, is that the predicted time-scale for the collapse of the nebula in the Modern Laplacian Theory, 4×10^7 years, disagrees badly with the observed lifetime of circumstellar discs.

A final difficulty of the model is that it offers no explanation of the tilt of the solar spin axis. It should lead to a highly planar system with planets being formed in almost circular orbits so that there would be no possibility of interactions dis-turbing the planarity.

6.3.1.2 The Solar Nebula Theory

Possible mechanisms by which the material in a flattened solar nebula could lose angular momentum and so join the central star-forming region have been reviewed by Larson (1989). However, his main emphasis seemed not to be that of produc-ing a slowly rotating star but rather on removing angular momentum on a short-enough time-scale because observation shows that circumstellar discs survive for up to 10^7 years at most.

Mechanisms that lead to material losing angular momentum by slowly spi-ralling in to join the central body are not in themselves going to solve the angular momentum problem. This can be illustrated by considering the Sun being built up by material spiralling in within the equatorial plane and then flowing, with little delay, to build up a spherically symmetric body (figure 6.2). For this approximate calculation the assumption is made that the Sun grows in concentric layers with uniform density out to its present radius, $R_\odot$, and the angular momentum it would then have is calculated. This can be multiplied by 0.055/0.4 ($\sim$0.14), the ratio of the moment of inertia factor of the actual Sun to the model uniform Sun, to obtain an extreme underestimate of the angular momentum of a Sun built up in that way. Actually if the Sun somehow accumulated from material that spiralled inwards it would have a very much larger initial radius. A simple analysis shows that the predicted angular momentum is

$$H_\mathrm{P} = 0.14 \times 0.6 M_\odot \sqrt{GM_\odot R_\odot} = 5 \times 10^{43} \text{ kg m}^2 \text{ s}^{-1} \tag{6.12}$$

or about 400 times the actual value. A better approximation is given in sec-tion 7.3.2. Given the extreme assumptions in the calculation, all in the direction of reducing the final angular momentum, the true factor must be at least 10^4 and probably much greater than that. Since all the mechanisms described by Larson give a spiralling-in motion to the material it is concluded that, without some other way of removing angular momentum, the Sun may not be able to form from a nebula at all. A possible mechanism may be the magnetic braking described in section 6.2.1 although it would require magnetic fields and rates of loss of mass for which there is no direct evidence. It would also require the mechanism to operate while the Sun was forming.

The four basic mechanisms suggested by Larson will now be described. Each of them addresses the important question of time-scale for incorporating material into the central condensation.

Turbulent viscosity

If the collapsing nebula, in the form of a disc, had some kind of turbulence within it then a theory by Lynden-Bell and Pringle (1974) suggests that angular-momentum transport could occur. Since turbulence quickly dissipates, without an input of energy the nebula disc would quickly settle down into quiet rotation with the only relative motions of material due to Keplerian shear. Cameron (1978) has suggested that material falling on to the disc from outside, or gravitational insta-bility within a disc, could generate the required turbulence. Lin and Papaloizou (1980, 1985) suggested that once the central body began to generate heat, convec-tion would drive the turbulence. However, analysis of all these suggestions show that either the effects they gave would be too weak or that they would only occur for a very short time in the early stages of formation of the nebula. The basis of the Lynden-Bell and Pringle mechanism is that, for a rotating system in which energy is being lost but angular momentum must remain constant, inner material will move further inward while outer material will move further outward. This amounts to a transfer of angular momentum from inner material to outer. That this is so can be shown very simply. Consider two bodies of equal mass in circu-lar orbits, with radii r_1 and r_2 around a central body of much greater mass. The energy and angular momentum of the system are

$$E = -C\left(\frac{1}{r_1} + \frac{1}{r_2}\right) \tag{6.13a}$$

and

$$H = K(\sqrt{r_1} + \sqrt{r_2}), \tag{6.13b}$$

where C and K are two positive constants. For small changes in r_1 and r_2 the changes in E and H are:

$$\delta E = C\left(\frac{1}{r_1^2}\delta r_1 + \frac{1}{r_2^2}\delta r_2\right) \tag{6.14a}$$

and

$$\delta H = \frac{1}{2} K \left(\frac{1}{\sqrt{r_1}} \delta r_1 + \frac{1}{\sqrt{r_2}} \delta r_2 \right). \tag{6.14b}$$

If angular momentum remains constant then, from (6.14b),

$$\delta r_1 = -\sqrt{\frac{r_1}{r_2}} \delta r_2 \tag{6.15}$$

and substituting this in (6.14a) gives

$$\delta E = \frac{C}{\sqrt{r_2}} \left(\frac{1}{r_2^{3/2}} - \frac{1}{r_1^{3/2}} \right) \delta r_2. \tag{6.16}$$

Given that δE is negative then it is clear that if $r_2 < r_1$ then δr_2 must be negative, that is to say that the inner body moves inwards and hence, from (6.15), the outer body moves outwards.

Larson concluded that for this mechanism to give the required time-scale the maintained turbulent velocities must be at least one-tenth of the thermal velocities and that convective turbulence might marginally satisfy this condition.

Gravitational torques

If an evolving nebula, which will be a rotating and flattened system, develops trailing spiral arms then gravitational torques occur which transfer angular momentum outwards (Larson 1984). This may be readily seen in figure 6.3 where the gravitational interaction between material at the points P and Q is seen to add angular momentum to P and to subtract it from Q. It is also clear that the cumulative action of the whole spiral will have a similar effect at both points. The torque will be smaller if there are many spiral arms since, in the extreme case with an infinite number of arms, the effect will be that of a continuous disc which will not add or subtract angular momentum at either point.

Larson has shown that this mechanism would be very effective for a massive disc of about $1M_\odot$ around a central condensation of $1M_\odot$ giving an accretion rate of about $10^{-5} M_\odot$ year^{-1}. However, for discs which have only one-tenth of the central mass the effect is negligible. Since less massive discs tend to be favoured by solar nebula theorists the gravitational torque mechanism is not likely to be relevant to the accretion process.

Magnetic torques

If the region within which the nebula is evolving possesses a magnetic field and if the material of the nebula is highly ionized then it is possible that the nebula can be coupled to the field as it collapses and so lose angular momentum. This effect is not likely to be important in practice as fields of the necessary strength and the degree of ionization required will not be present to control the motion

Figure 6.3. The gravitational effect of a massive trailing spiral arm is to add orbital angular momentum at P and subtract it at Q.

of the nebula, especially when its density exceeds about 10^{-9} kg m^{-3}, which is quite early in its development.

Wave transport

If a wave generating in a moving medium has an energy E per unit mass of the medium in which it moves then it is transporting momentum at a rate E/v per unit mass, in which v is the phase velocity relative to the moving medium. This can also be written $E/(v_p - v_m)$ in which v_p is the local speed of the wave and v_m the speed of the medium. We now consider all motions as approximately circular about a spin axis; this would apply if the wave motion or the material of the medium was gradually spiralling inwards or outwards. If the material is in the form of a disc then the angular momentum per unit area of the disc can be represented by

$$H_a = \frac{E\mu r}{v_p - v_m} = \frac{E\mu}{\Omega_p - \Omega_m}, \qquad (6.17)$$

where μ is the surface density of the disc. It will be noticed that if $\Omega_p > \Omega_m$ then the wave contributes positive angular momentum locally, otherwise the contribution is negative.

In a disc with Keplerian speeds the differential motion causes a wave motion to be wound into a trailing spiral pattern. Waves that move inwards have $\Omega_p > \Omega_m$ so that if there is dissipation and they deposit their properties in the local medium they will contribute negative angular momentum. Conversely, waves that move outwards deposit positive angular momentum. The net effect is that angular momentum is transported outwards; angular momentum is removed from inner material and added to outer material. Larson demonstrated that wave disturbances from Jupiter could clear the inner Solar System on a time-scale of about 10^6 years but, of course, this presupposes that Jupiter has itself been produced very quickly if this result is to be consistent with observation.

Although all the mechanisms are either marginal in their ability to meet Larson's self-imposed 10^6 year observational limit for the existence of the nebula, or fail to meet it, the observations of circumstellar discs do seem to allow a few times 10^6 years. Thus there seems to be no theoretical difficulty in bringing material in towards the central condensation on a suitable time-scale. However, just bringing material inwards is not enough to ensure that a star will be formed. The large amount of angular momentum associated with the material will prevent it from forming a compact body. Until it *is* a compact body it would be unable to generate a large magnetic field or a flow of ionized material to remove angular momentum. What seems to be required is that angular momentum should be removed from inwardly acting material causing it to plunge inwards to join the central body rather than to spiral in gradually. To produce this effect requires the condition of both having material moving *inwards* to form a central body while at the same time having ionized material coupled to a magnetic field moving *outward* to remove angular momentum. This is difficult to envisage.

Despite the many mechanisms that have been proposed for transporting angular momentum Miyama (1989) has stated:

> The formation of a slowly spinning Sun starting with the solar nebula is a central problem and one that has still not been satisfactorily solved.

The other major problem is to produce planets from the nebula within the time of its existence and we now move on to the topic of planet formation to see how well the various theories under review deal with this problem.

6.4 Formation of planets

The five theories we are considering present four different scenarios for planet formation, the two presenting similar mechanisms being the Accretion Theory and the Solar Nebula Theory. We shall now discuss these scenarios in detail and comment on their strengths and weaknesses.

6.4.1 Planets from the Proto-planet Theory

In the final form of the Proto-planet Theory (McCrea 1988) the planets are formed by the spontaneous fragmentation of an interstellar cloud into planetary-size units. An input parameter into the theory was the temperature of the cloud, taken as 40 K. Assuming that the mass of the proto-planet blobs was the Jeans critical mass then an estimate could be made of the mean density of the material within which the blobs formed. From the Jeans critical mass formula (2.22)

$$\rho = \left(\frac{5k\theta}{G\mu}\right)^3 \frac{3}{4\pi m^2}. \tag{6.18}$$

Inserting the values of m and θ suggested by McCrea gives a mean density of cloud material of 10^{-6} kg m^{-3}. The formation of planets, or indeed any other astronomical bodies, by spontaneous gravitational collapse given the correct combination of temperature and density presents no theoretical problems; depending as it does on the Virial Theorem the validity of the Jeans criterion in the formation of condensations cannot be doubted. There are possible objections to the McCrea assumptions. The first is that the model described in section 2.6.2, in which the initial conditions are based on observational evidence, concludes that the first condensations produced in a turbulent collapsing cloud are greater than a solar mass, rather than of planetary mass. A second objection is that the Proto-planet Theory would give a highly non-planar system—at least initially.

The Proto-planet Theory predicts the number of resulting planets from the fact that they have almost all the angular momentum of the system contained in their orbits. If the spherical region forming the Sun has the mass of the Sun then from its mean density we find that its radius, R, is 7.8×10^{11} m, which is also the radius of the orbit of Jupiter. McCrea assumes that planets could go into orbit at radii up to a few times this distance, with a mean orbital distance kR, which allows for the existence of the outer planets. If there are p planets at a mean distance kR in circular orbits then the total orbital angular momentum will be approximately

$$H_P = pm\sqrt{GMkR} = \frac{p}{N}\sqrt{GM^3kR}. \tag{6.19}$$

The mean square speeds of the original N individual blobs may be estimated from the Virial Theorem by

$$MV^2 = \frac{3GM^2}{5R}$$

in which the left-hand side is twice the kinetic energy and the right-hand side is the magnitude of the gravitational potential energy. The expected total angular momentum of the randomly moving blobs is then

$$H = \frac{1}{2}mVRN^{1/2} = \frac{1}{2}\sqrt{\frac{3GM^3R}{5N}}. \tag{6.20}$$

Equating this to H_P in (6.19), since most of the angular momentum of the system is in the orbiting planets, one finds

$$p = \sqrt{\frac{3N}{20k}}. \tag{6.21}$$

Taking $k = 3$ and $N = 3000$ gives $p \approx 12$, a reasonable if somewhat high value.

Since there is no reason why the planets should have been coplanar and in prograde orbits then this leads to the expectation of more predicted than actual planets. If a minority had retrograde orbits then they could have been eliminated by collisions since they would be more likely to interact with the majority prograde bodies. The question is whether the residual planets could then have been brought nearly into a common plane by the action of a circumsolar resisting medium. Results given in section 7.4.5 suggest, but do not show convincingly, that this could be so.

6.4.2   Planets from the Capture Theory

The Capture-Theory process produces proto-planets within a filament and there are four stages that have to be verified:

(i)   The formation of a filament was suggested by Jeans' analytical work but needs to be confirmed by numerical work.

(ii) The filament must have a sufficiently high line density for proto-planet condensations that satisfy the Jeans criterion (2.22) to form.

(iii) It must be shown that the proto-planetary condensations are captured by the Sun.

(iv) The proto-planets must collapse to high density on a short enough time-scale for them to resist disruption when they make their first perihelion passage.

6.4.2.1   A first model of the basic Capture-Theory process

The first model of the Capture Theory was described by Woolfson in 1964. At that time computers were not very powerful and Woolfson used a two-dimensional point-mass model of a proto-star with forces between the points to simulate the effects of gravity and gas pressure. The outcome of this early computational work is shown in figure 6.4 as a sequence of configurations of the model proto-star culminating at a stage just after perihelion. The distortion of the proto-star, as derived theoretically by Jeans, is seen to happen; the lag in the direction of the tidal bulge, due to inertia in the system, was also an effect predicted by Jeans. In the final configuration the formation of a filament is clear and the subsequent motions of the points in the filament, under the gravitational influence of the Sun and proto-star, were found until the energy with respect to the Sun became constant. A few of the points escaped from the Sun, marked H for *hyperbolic orbit* in figure 6.4, but many of them were retained. The captured points mostly had orbits of high eccentricity and their perihelion distances varied from 1.4 to 38.4 AU, a very satisfactory range in relation to the Solar System.

This early model served the purpose of confirming the validity of the Capture-Theory concept and was the precursor of more detailed modelling to come.

6.4.2.2   The limacoid model

The first three-dimensional computational model of the Capture Theory was described by Dormand and Woolfson (1971). Their approach avoided the need directly to determine inter-point forces which meant that they were able to incorporate more points in their model proto-star. They noted that the profile of a

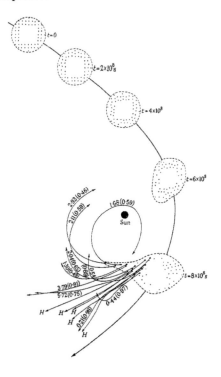

Figure 6.4. The disruption of a model proto-star. Captured points are marked with their orbital perihelion distances (10^{12} m) and eccentricities (in brackets). Escaping points are marked H (hyperbolic orbits). Initial proto-star parameters: mass $= 0.15 M_\odot$, radius $= 2.2 \times 10^{12}$ m, perihelion $= 6.67 \times 10^{12}$ m (from Woolfson 1964).

tidally-distorted body would be similar to a limaçon, a shape described in polar coordinates as

$$r = \frac{a(k \cos \theta - 1)}{k - 1}, \quad \text{for} \quad k > 1 \quad \text{and} \quad \cos \theta \leq 1/k. \quad (6.22)$$

The quantity a is the length of the axis and k the *eccentricity* of the limaçon. Limaçons for three values of k are shown in figure 6.5; a circle corresponds to $k = \infty$ and as k approaches unity the shape becomes more and more elongated. Rotation of a limaçon about its axis produces a surface termed a *limacoid*.

In the 1971 modelling the proto-star was initially simulated by two concentric shells of points distributed at the vertices and on surface normals to two regular polyhedra. A separate limacoid was 'fitted' to the points corresponding to each shell which involved finding values of k and a to give a least-squares match to the points. A match of limacoids to points is shown in figure 6.6 for the proto-star just before its perihelion passage. The gravitational field of a limacoid at a point can be calculated as a function of distance from the origin and angle of the

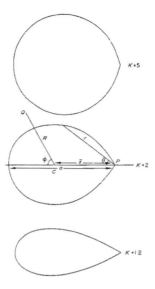

Figure 6.5. The limaçon $r = a(k\cos\theta - 1)/(k - 1)$, $\cos\theta \le 1/k$ for $k = 5, 2$ and 1.2.

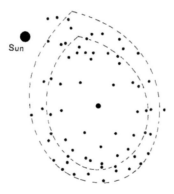

Figure 6.6. A configuration of proto-star points fitted to two limacoids. Outer shell $k = 2.27$, $a = 37.2$ AU; inner shell $k = 2.44$, $a = 28.3$ AU.

radius vector to the axis and this was used to calculate the field at each mass point due to the two limacoids. The proto-star was taken to have a central condensation so three-quarters of its total mass was concentrated at the centre.

A central repulsive force was also added to the model to simulate gas pressure and to prevent the proto-star from collapsing faster than the Kelvin–Helmholtz rate. Any points outside the outer limacoid were treated separately and unnatural 'pooling' of these points was prevented by a process of velocity sharing between neighbouring points. This added to the model the characteristics of a

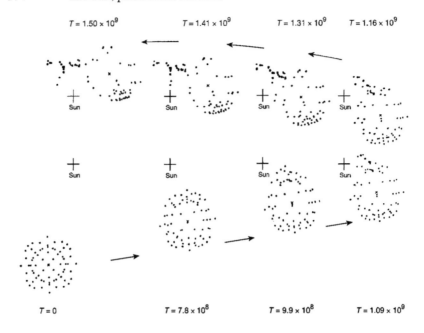

Figure 6.7. A Sun–proto-star encounter from the limacoid model.

fluid with viscosity.

Although the model was computationally very efficient it did impose a symmetry which would not actually be maintained in realistic situation. For this reason the simulation could only be continued until the distribution of points departed significantly from a limacoidal form.

The result of one limacoid simulation is shown in figure 6.7 and the orbits of some captured points are illustrated in figure 6.8. As for the 1964 model, the scale of these orbits was similar to those in the Solar System, although the eccentricities are much greater. The mass distribution is shown in figure 6.9 for four separate limacoid simulations together with the smoothed-out distribution for the actual Solar System. The computed distributions bracket the true distribution and have the desirable property of having a central peak. The peak is related to the rate of loss of the material by the proto-star which is greatest when it is close to perihelion. The orbits of the captured mass points are shown for one of the limacoid models in table 6.2. The total mass of the points represents the mass contained in the filament, only part of which ends up within the planets.

The limacoid results indicated that the line density of the filament, and the density of its material, would be sufficiently high for proto-planets to form. What can also be inferred from figure 6.8 is that the first motion of the points is away from the Sun following capture, which clearly has implications for the ability of proto-planets to condense and survive.

Figure 6.8. The orbits of some captured mass points from the limacoid model.

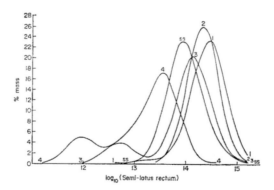

Figure 6.9. The mass distribution for captured material from four simulated Sun–proto-star encounters. The dashed line shows the smoothed-out distribution for the Solar System.

6.4.2.3 The SPH model

The smoothed particle hydrodynamics approach (Appendix III) is ideally suited to investigate the Capture-Theory model since it realistically models gravitation,

Table 6.2. Orbits of captured mass points from the limacoid model.

Point	Semi-latus rectum (AU)	Eccentricity	Mass (Jupiter units)
1	16.0	0.938	1
2	15.5	0.979	1.5
3	11.3	0.755	0.5
4	10.5	0.756	3
5	6.3	0.784	1
6	7.7	0.920	1
7	7.7	0.730	1
8	6.5	0.780	2.5
9	0.3	0.995	2

Proto-star parameters: Mass $M_\odot/4$, radius 16.7 AU, eccentricity of orbit 1.5.

pressure forces and viscosity. An encounter between the Sun and a proto-star of mass $M_\odot/5$ and initial radius 20 AU is shown in figure 6.10 (Dormand and Woolfson 1988). The Sun–proto-star orbit had eccentricity 1.4 with perihelion 10 AU. At the beginning of the simulation the proto-star was collapsing at 25% of the free-fall velocity.

Almost 20% of the mass of the proto-star is captured by the Sun and the distribution in the orbital semi-latera recta is shown in figure 6.11. These extend only slightly beyond the orbit of Mars so are unsatisfactory for describing the Solar System. However, it is clear from other modelling that more extensive systems can be produced with different parameters.

6.4.2.4 Initial orbits

Planets from the Capture Theory are produced with elliptical orbits close to the plane of the Sun–star orbit. Departure from strict planarity is due, first, to the spin of the proto-star providing the planetary material and, second, to close interactions between planets during the period they were rounding off. The order of magnitude of departure from planarity due to the spinning star can be estimated without dependence on very precise parameters for the Capture-Theory model. Taking the proto-star at the time of disruption with a mass of $M_\odot/4$ and a radius of 15 AU then the maximum equatorial speed is of order 4.5 km s^{-1}, beyond which it would disrupt. If the proto-star orbit had a perihelion distance of 20 AU and an eccentricity of 2.0 then the orbital speed at perihelion would have been of order 11.5 km s^{-1}. This suggests a maximum possible inclination of the path of the escaping material from the plane of the star orbit, when the equatorial motion is perpendicular to the star orbit, of about $\tan^{-1}(4.5/11.5)$ or 21°. An inclination close to that value, which depends on extreme conditions in the proto-star, is very

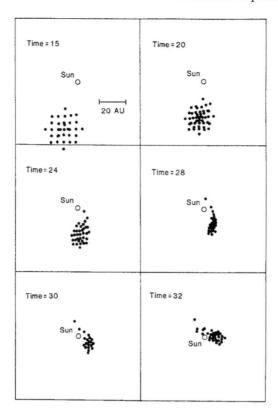

Figure 6.10. Configurations in a tidally-disrupted SPH proto-star. Times are in years from the start of the simulation.

unlikely but variations of, say, up to 3° are plausible.

The conditions in the proto-star have not been uniquely defined in the Capture Theory but the theory given in section 2.6.2 suggests an initial density of 3×10^{-13} kg m^{-3} and temperature 20 K when it first formed, corresponding to a radius of about 500 AU. Subsequently it would collapse, following a path similar to that shown in figure 2.15 but modified for a lesser-mass star. It would increasingly depart from free-fall collapse as it became more opaque. Based on grain opacity, assuming grains of 100 μm in the stellar material, the optical thickness of the proto-star would reach unity when its radius was about 75 AU following which its collapse would first accelerate more slowly and then finally decelerate. By the time its radius reached 10 AU it would have a centrally condensed structure but with a mean density of 4×10^{-8} kg m^{-3}. Capture-Theory modelling indicates that of the order of one-fifth of the proto-star was removed by the tidal action of the Sun. This means that the mean density in the filament would also be of order 4×10^{-8} kg m^{-3} with an initial temperature of order 200–300 K,

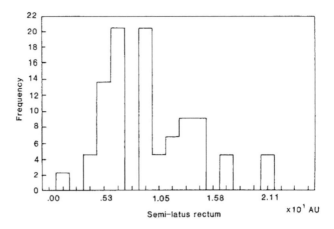

Figure 6.11. Distribution of semi-latera recta of captured elements from the SPH simulation in figure 6.10.

assuming that about one-half of the energy of collapse of the proto-star was still contained as heat energy within it. However, the proto-planet filament would have an optical thickness appreciably less than unity and so would quickly cool. Since the initial paths of the condensations within the filament are on highly eccentric orbits with motions *away* from the Sun their temperatures can fall to low values.

6.4.2.5 *The collapse of a proto-planet in isolation*

Schofield and Woolfson (1982a) set up a very detailed model to study proto-planet evolution under the conditions suggested by the Capture Theory. The first step was to model an isolated planet which was spherically symmetric and thus, essentially, one-dimensional. In addition no account was taken of rotation, turbulence, magnetic fields or any external gravitational field. This simplification enabled the solution of the equations of hydrodynamics and radiative heat transfer to be modelled conveniently and realistically. For the computation the model was in the form of a number of concentric spherical shells of equal mass, each characterized by pressure, density, temperature and chemical composition. Since the model started with very low density, heat flow could be taken as dominated by radiation, dependent on the opacity of the material, with negligible contributions from conduction and convection. In the later stages of planetary development considerable convection would occur so the model was only valid for the very early stages of the planet's evolution. This was sufficient for the purpose of demonstrating that proto-planet material was capable of collapsing into a dense planetary body. The problem of the way the planet would evolve into its final form is not one of prime importance to the cosmogonist.

The initially homogeneous composition of the planet was taken as 74% hy-

Table 6.3. Schofield and Woolfson's proto-planet model.

Initial mass	2×10^{27} kg (1 Jupiter mass)
Initial radius	2.28×10^{11} m ($\sim$1.5 AU)
Initial density	4×10^{-8} kg m^{-3}
Composition	
Hydrogen	74%
Helium	24%
Other elements	2%
Mean molecular weight	3.88×10^{-27} kg
Number of shells	16
Initial temperature giving	
(a) Jeans' critical mass	32.8 K
(b) 1.5$\times$ Jeans' critical mass	24.6 K

drogen, 24% helium with 2% of heavier elements and the material behaves like a perfect gas with the appropriate mean molecular weight. The characteristics of the model, which was considered to be a proto-Jupiter, are given in table 6.3

The ratio of the actual model mass to the Jeans critical mass was varied by changing the temperature as seen in cases (a) and (b) in table 6.2. With the lower temperature, gravity forces would dominate over pressure and so the collapse would be more rapid. In other experiments with the model the planet was embedded in a medium exerting a small constant pressure over the surface of the body.

The model used a Lagrangian frame in which the motion of boundaries was determined; between adjacent boundaries there was a fixed quantity of material. The equation of motion of a spherical boundary of radius r is described by

$$\frac{\partial^2 r}{\partial t^2} = -\frac{GM(r)}{r^2} - 4\pi r^2 \frac{\partial P}{\partial M(r)} \tag{6.23}$$

where $M(r)$ is the total contained mass within a surface of radius r and P is the pressure.

Under the conditions considered by the model, heat flow is dominated by radiation and a standard Rosseland mean opacity $\kappa_R(\rho, T)$ was used which was an amalgamation of results given by Cameron and Pine (1973), Hayashi and Nakano (1965) and Cox and Stewart (1970). In optically thick regions of the proto-planet the diffusion approximation for radiative transfer was used, with the form

$$L(r) = -\frac{16}{3} \frac{\sigma r^2}{\kappa_R(\rho, T)\rho(r)} \frac{d(T^4)}{dr} \tag{6.24}$$

in which σ is Stefan's constant and T is temperature. In optically thin regions the luminosity is given by

$$L(r) = 4\pi\sigma(I + E + E_\mathrm{S})$$

in which the contribution from material interior to r comes from

$$I = \int_0^r x^2 \kappa(x)\rho(x)T^4(x) \exp\left\{-\int_x^r \kappa(x')\rho(x')\,\mathrm{d}x'\right\}\mathrm{d}x \qquad (6.25a)$$

that from material exterior to r comes from

$$E = \int_r^R x^2 \kappa(x)\rho(x)T^4(x) \exp\left\{-\int_x^r \kappa(x')\rho(x')\,\mathrm{d}x'\right\}\mathrm{d}x \qquad (6.25b)$$

and that from exterior sources comes from

$$E_\mathrm{S} = R^2 T_\mathrm{E}^4 \exp\left\{-\int_r^R \kappa(x')\rho(x')\,\mathrm{d}x'\right\} \qquad (6.25c)$$

where R is the radius of the planet and T_E the external temperature. The heat flow equation using the luminosity was taken in the form

$$\frac{\partial u}{\partial t} + P\frac{\partial v}{\partial t} + \frac{\partial L}{\partial M} = 0 \qquad (6.26)$$

in which u is the intrinsic internal energy and $v = 1/\rho$ is the specific volume.

The material was assumed to behave as a perfect gas with molecular hydrogen as 25% p-H_2 (para-hydrogen) and 75% as o-H_2 (ortho-hydrogen) and above 1500 K the dissociation of hydrogen molecules is taken into account. The heat capacity of molecular hydrogen and the thermodynamic properties of dissociated hydrogen are given by Vargaftik (1975) as is the heat capacity of helium. The contribution to the heat capacity of evaporating H_2O grains were taken into account but not that of grains of refractory material since they evaporate at about 1500 K at which stage the dissociation of molecular hydrogen dominates the heat capacity.

The results of running the higher-temperature model (a), where the mass equalled the Jeans mass, are shown in figure 6.12. The outer regions expand while the central regions collapse. After nine years the disturbance from the boundary reaches the centre causing a bounce and a subsequent re-expansion for about 70 years. There is then a slow contraction with some fluctuations lasting about 500 years at the end of which the central density and temperature are about 3.4×10^{-6} kg m^{-3} and 180 K respectively. At this stage water ice evaporates so reducing the opacity and absorbing latent heat which leads to another sustained collapse during which H_2 dissociates at a temperature of about 1500 K. This leads to even more rapid collapse that is virtually free-fall in the central regions. After 720 years the central density and temperature reach 24 kg m^{-3} and 6000 K, respectively, at which point the model is terminated. The outer shell reaches escape velocity and so is completely lost.

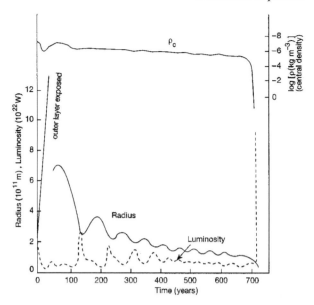

Figure 6.12. The radius, luminosity and central density of an evolving proto-Jupiter (Schofield and Woolfson 1982a).

With starting temperature (b), corresponding to having 1.5 times the Jeans mass, a similar density and temperature is reached in just over 20 years. Intermediate temperatures between (a) and (b) in table 6.3 give a range of times to reach an internal temperature of 6000 K, as illustrated in figure 6.13. As expected the addition of an external pressure, as might be provided by an external medium, makes the collapse more rapid for any given initial temperature and also prevents the loss of the outer shell.

The model lacks the effect of external influences such as the gravitational effects of the Sun and the proto-star and also that of radiation due to the Sun. Schofield and Woolfson tested the extreme assumption that the whole of the surface of the proto-planet was constantly bathed in solar radiation (to preserve the spherical symmetry) with the present solar luminosity. This increased the dispersion of the material but the cooler model still collapsed on a time-scale that was short compared with its orbital period.

The Capture-Theory model produces proto-planets on highly eccentric orbits moving towards aphelion. One of the necessary conditions for proto-planets not just to form but also to survive is that they should have become sufficiently compact to survive complete tidal disruption at the first perihelion passage. The estimated initial orbital period for the proto-Jupiter is of order 100 years so it is clear that the higher temperature in table 6.3 would not lead to an enduring planet. On the other hand at the lowest temperature the planet could form and survive.

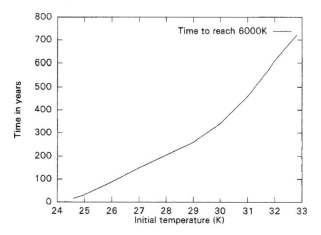

Figure 6.13. The time for the central temperature of an evolving proto-Jupiter to reach 6000 K as a function of the initial temperature of the material.

The final factor, an important one which has not been included, is that of the gravitational effects due to the Sun and the proto-star and these we now consider.

6.4.2.6 *Planetary condensation in a gravitational field*

In order to introduce the effects of the gravitational fields of the Sun and the proto-star Schofield and Woolfson (1982b) first produced a point-mass model with forces between the points simulating gravity and viscosity. These forces were so arranged that the behaviour of an isolated body matched well with that of the physically realistic model from section 6.4.2.5. The model was then exposed to external forces to see how it would behave. One of the computational difficulties that this introduces is that the time steps for solving the differential equations are controlled by the closest approach of point masses so that if just two of them approach closely the calculation was drastically slowed down or, alternatively, the accuracy was severely impaired. To solve this problem Schofield and Woolfson devised a redistribution scheme whereby from time to time the mass points were redistributed on a regular grid. This is illustrated in figure 6.14; the masses and velocities of the regularized points could be so arranged as to conserve the energy, momentum and angular momentum of the system.

An example of a simulation produced by this technique is shown in figure 6.15 where it is seen that a considerable degree of condensation is achieved before the proto-planet reaches aphelion. There is a strong indication that planets can form in the presence of the solar tidal force and the angular momentum contained in the condensed planet was similar to that in the Jovian system—which is relevant to satellite formation. However, there are some faults in the simulation, in particular that the regularization was always carried out within an ellipsoidal

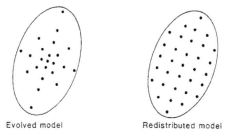

Evolved model   Redistributed model

Figure 6.14. The redistribution scheme used in the computational model of a proto-Jupiter evolving in the tidal field of the Sun (Schofield and Woolfson 1982b).

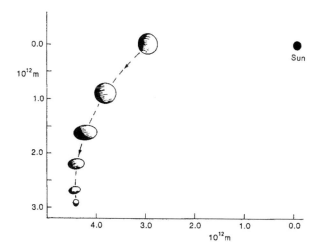

Figure 6.15. Proto-planetary evolution starting from the semi-minor axis ($a = 20$ AU, $e = 0.77$) (Schofield and Woolfson 1982b).

envelope so that no other shape for the proto-planet was possible. Again, the non-gravitational forces were physically based only in that they caused the proto-planet to behave as expected in isolation but it would clearly be better if some more physically-based model could be used in the presence of the gravitational forces. Such modelling is possible through the use of SPH (Appendix III).

6.4.2.7 Simulation of planetary collapse by SPH

The SPH method has been applied by Allinson (1986) to study the behaviour of a globular body of three Jupiter masses with initial radius 2 AU corresponding to a density of 5×10^{-8} kg m^{-3}, slightly greater than that used by Schofield and Woolfson. This body, that could be regarded as a section of a filament from the proto-star, had an orbit with semi-latus rectum 8.5 AU and eccentricity 0.77. The

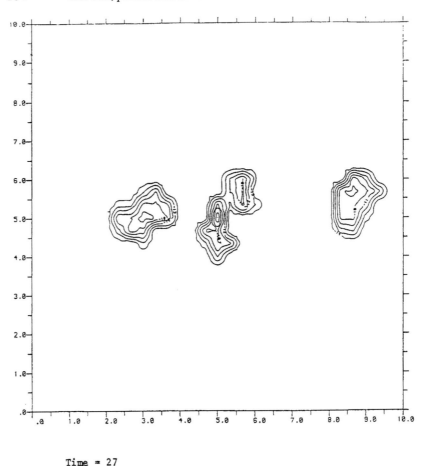

Time = 27

Figure 6.16. An SPH simulation of the tidal break-up into three main fragments of a blob of matter corresponding to part of a proto-planet filament (Allinson 1986).

proto-planet started within the Roche limit and was disrupted into three proto-planets of lesser mass; the density contours after 27 years (figure 6.16) indicated that the three condensing blobs would be able to survive the following perihelion passage.

6.4.3   Planets from the Solar Nebula Theory

The formation of terrestrial planets, or the cores of major planets, from a diffuse nebula is a topic that has been subjected to a great deal of analysis. Many different models have been advanced, almost too many to deal with exhaustively, but they can be put into three main categories which depend on the mass of the nebula within which the planets form. These are:

(i)   within a nebula so massive that proto-planets can form directly by gravitational instability;
(ii)   within a low-mass nebula through slow accretion processes; and
(iii) by more rapid accretion, described as runaway growth, requiring an intermediate-mass nebula.

We shall consider these categories in order and indicate the range of ideas within each category and the strengths and weaknesses of each of the models.

6.4.3.1   A massive nebula

One very obvious way of producing planets is to have a sufficient quantity of material at a high enough density and a low enough temperature for planets to form directly via the Jeans instability criterion (2.22). This, after all, is the way in which planets are assumed to form in the Proto-planet and Capture Theories and there are no particular difficulties with this process. Several workers have suggested such a mechanism (e.g. De Campli and Cameron (1979)) and here we can indicate the kinds of parameters that enable the process to work.

We consider a disc nebula of mass $1M_{\odot}$, radius 40 AU and thickness 5 AU. The mean density of the material in it would be 2.4×10^{-8} kg m^{-3} and for a temperature of 50 K the Jeans critical mass, assuming atomic hydrogen, would be just over three times the mass of Jupiter. Small variations of the parameters, plus the possibility of having molecular rather than atomic hydrogen, could give Jupiter-mass condensations or less. The problem with this model is not that of producing planets but rather of producing too many planets with an associated disposal problem. Nearly 99.9% of the original nebula has to be disposed of in some way, which can only be either inward or outward. Inward disposal presents the problem that as well as providing mass it also provides angular momentum that, in its turn, must be eliminated. Outward disposal brings up the question of finding a source of energy to remove such a large mass to infinity.

Because of the serious problems it raises, including the lack of observational evidence for such massive nebulae, models of this kind are out of favour and accretion processes have been the main topic of investigation since about 1980.

6.4.3.2   Accretion in a low-mass nebula: planetesimal formation

In a quietly rotating disc the dust component would eventually settle down into the mean plane. If the dust was in the form of particles a few micrometres in extent then the time-scale for it to settle would be a million years or more, taking up all or at least a considerable fraction of the observed lifetimes of nebulae. On the other hand if it is assumed that dust particles would stick together when they came into contact then they would build up into larger particles and fall more quickly towards the central plane. This is a perfectly reasonable assumption as laboratory experiments under ultra-high vacuum condition show that small particles cold-weld together on contact. If the final particles are of the order of centimetres in size then the settling time is only a few tens of years.

The formation of a dust disc is a very useful first step in planetary formation as it brings closer together the essential material for the formation of terrestrial planets and the cores of the major planets. The process of forming these solid bodies is broken down into two stages. The first of these is the formation of planetesimals, typically from hundreds of metres to 100 km in diameter, depending on where they form and which theoretical approach is followed. The second stage is the accumulation of planetesimals to form planets or cores. Finally, where giant planets are concerned, the planetary cores capture gaseous material.

There is a strong argument for the initial planetesimal stage in the aggregation of nebula material into planets. Most solid bodies in the Solar System show signs of bombardment by projectiles with sizes corresponding to those of planetesimals and, indeed, the larger planetesimals may still be visible as asteroids or even, perhaps, comets. As previously indicated, there is no universally accepted theory of planetesimal formation and there are two main ideas—the first through gravitational instability within the dust disc and the second by the sticking together of solid particles when they collide. Solid material in the dust disc will tend to clump together through mutual gravitational attraction but this tendency will be opposed by disruptive solar tidal forces. This is a Roche limit problem and by an analysis similar to that which gave (4.12), but without tidal locking, we find that the clumping process will be possible if the mean density of solid material in the vicinity, ρ_S, satisfies

$$\rho_S > \frac{3M_\odot}{2\pi R^3} = \rho_{cr} \tag{6.27}$$

where R is the distance from the Sun and ρ_{cr} is the critical density for clumping. Safronov (1972) showed that a uniform disc would be unstable to density perturbation and that a two-dimensional wave-like variation of density would develop. For a disc with a density greater than the critical density, regions with a diameter equal to some critical wavelength of disturbance would separate out into condensing clumps and Safronov showed that, at the critical density, the wavelength would be about eight times the thickness of the disc.

Assuming that a nebula of mass $0.1M_\odot$ with a solid component of 2% is spread into a uniform disc of radius 40 AU the mean densty per unit area of solids, σ, is 35 kg m^{-2}. If the critical density is assumed then this indicates the thickness of the dust disc, h, and hence the volume of material clumping together is approximately $60h^3$. Deductions concerning the masses and dimensions of the resulting condensations, assuming the solid material has a mean material density, ρ_{sol}, of 2000 kg m^{-3}, are given in table 6.4. It can be seen that this predicts planetesimals of dimensions from a few kilometres up to, perhaps, 100 km in the inner Solar System but this conclusion was challenged by Goldreich and Ward (1973). They invoked thermodynamic principles to show that the condensations would be much smaller, hundreds of metres in extent rather than the more than

Table 6.4. The masses, m_p, and radii, r_p, of planetesimals, according to Safronov (1972) in the vicinity of the Earth and Jupiter.

	Earth	Jupiter
ρ_{cr} (kg m^{-3})	2.83×10^{-4}	2.01×10^{-6}
$h = \sigma/\rho_{cr}$ (m)	1.24×10^{5}	1.74×10^{7}
$M_p = 60h^3\rho_{cr}$ (kg)	3.20×10^{13}	6.35×10^{17}
$R_p = (3m_p/4\pi\rho_{sol})^{1/3}$ (km)	1.6	42

kilometre-size bodies predicted by Safronov. Since the escape velocity from a solid body of radius 500 m is about 0.5 m s^{-1} and the velocity dispersion of the initial Goldreich–Ward planetesimals is expected to be small, $\sim$0.1 m s^{-1}, then collisions with the largest planetesimals will give accretion and growth of larger bodies. Eventually planetesimals of the size predicted by Safronov would come about, albeit by a two-stage process.

An alternative view of the way that planetesimals formed has been advanced by Weidenschilling *et al* (1989). They have argued that the presence of even a small amount of turbulence in the disc would inhibit gravitational instability. The free-fall time for the collapse of a planetesimal clump in the vicinity of Jupiter would be somewhat over one year but if the material was stirred up by turbulent motions before the collapse was well under way then the condensation would simply not form. On the other hand they state that, since the mechanism of adhesion of fine-grained material is a necessary process for forming a dust disc on a reasonable time-scale, then there is no reason why should it not also operate to form planetesimals. There is some justification for this criticism. If a planetesimal clump in the vicinity of Jupiter were to be formed into a sphere at the density given in table 6.4 then the escape speed from it would be 1.4 m s^{-1} and would be less if it was not in spherical form. Turbulent speeds of this order of magnitude would thus tend to promote disruption of the clump although whether or not the nebula would be turbulent is not generally agreed. For some purposes theorists postulate a quiet nebula, for example, to enable planetesimals to form, but then other theorists prefer a turbulent nebula, for example, as an aid to angular momentum transfer (Cameron 1978).

In the Weidenschilling *et al* model, particles are assumed to stick together whenever they come into contact. The collision cross-section for the collision of particles with radius s_1 and s_2 is $\pi(s_1 + s_2)^2$ and the rate of collisions will depend on the relative velocities of particles. For very small particles, with radius less than a few tens of micrometres, relative velocities are mainly due to random thermal motions with root-mean-square speed $(3k\theta/\mu)^{1/2}$ where θ is the temperature and μ the particle mass. Assuming that all particles grow at the same rate so that their radii are always equal this leads to a growth of s as a function of time of form

$$s(t) = s(0) + \left\{ 15\rho_p \left(\frac{k\theta}{8\pi\rho_s^3} \right)^{1/2} t \right\}^{2/5} \tag{6.28}$$

in which ρ_p is the partial density of solids in the nebula and ρ_s is the density of the material of the solid grains.

As the grains increase in size so they are less influenced by thermal motion and they begin to settle towards the mean plane of the nebula. The relative speed of particles now depends on larger particles settling more quickly than smaller ones. It can be shown that the growth pattern of a grain is now of the form

$$s(t) = s(0) \exp \left(\frac{f\Omega^2 zs}{4c} \right) \tag{6.29}$$

in which f is the mass ratio of solids to gas in the nebula, Ω the local Keplerian frequency, z the distance from the mean plane of the nebula and c the root-mean-square thermal velocity of the gas molecules. As a particle settles so it also drifts towards the mean plane, which enables it to grow to a larger size. Typically, in the vicinity of the Earth the time to reach the mean plane is of the order a few hundred years; the largest objects formed are about 10 m in diameter and the mass ratio of solids to gas in the mean plane is roughly unity. The further development of the metre-size bodies has not been investigated and it is also possible that the time-scale for settling into the mean plane could be severely underestimated. If the accretion of grains was such that they were not compact bodies but had a fractal-like structure (figure 6.17) then Weidenschilling *et al* suggest that the total time for settling into the mean plane could be 10^5–10^6 years.

The general view of solar-nebula theorists is that, while there may be some uncertainties in the formation mechanism for planetesimals, it seems likely that kilometre-size bodies will form by some process or other on a relatively short time-scale. If this is so then virtually the whole lifetime of the dusty nebula is available for the next stage of forming planets from planetesimals.

6.4.3.3 *Planets from planetesimals*

The basic theory of the accumulation of planetesimals to produce planets was developed by Safronov (1972) and most subsequent work has been developments, or variants, of it. Safronov showed that if the random relative velocity between planetesimals is less than the escape speed from the largest of them then that body will grow and eventually accrete all other bodies with which it comes into contact. When the planetesimals are first produced they move on elliptical orbits at different distances from the Sun and gravitational interactions between them, acting as elastic collisions, will increase the random motions. However, as the relative velocities of planetesimals, and the eccentricities of their orbits, increase through elastic gravitational interactions so the probability of actual collisions between planetesimals, which will be inelastic, increases. The effect of inelastic

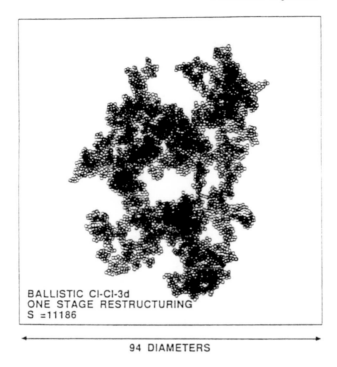

BALLISTIC CI-CI-3d
ONE STAGE RESTRUCTURING
S =11186

94 DIAMETERS

Figure 6.17. Grains forming a fractal-like structure.

collisions is to damp down the randomness in the motion. Safronov showed that
a balance between the effects which increase and decrease random motions, and
hence the relative speed of planetesimals, occurs when the mean random speed,
v, is of the same order as v_e, the escape speed from the largest planetesimal. In
general one could write

$$v^2 = \frac{Gm}{\beta r} = \frac{v_e^2}{2\beta} \tag{6.30}$$

where m and r are the mass and radius of the largest planetesimal and β is a factor
in the range 2–5 in most situations.

In a simple case where all colliding bodies adhere, the rate of growth is proportional to the collision cross-section, which must take account of the focusing
effect of the mass of the accreting body. In figure 6.18 a particle is shown approaching from infinity with a speed v in a direction such that the closest approach
to the centre of the accreting body is D. At the critical condition for accretion the
particle, moving at speed v', will intersect the body at point P. Equating angular
momentum

$$vD = v'r \tag{6.31a}$$

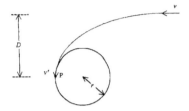

Figure 6.18. A particle deflected by the gravitational field of an accreting body in the critical path just to strike it.

and from energy considerations

$$v'^2 = v^2 + v_e^2. \tag{6.31b}$$

Combining (6.30), (6.31a) and (6.31b) gives the accretion cross-section

$$\pi D^2 = \pi r^2 \left(1 + \frac{v_e^2}{v^2}\right) = \pi r^2 (1 + 2\beta). \tag{6.32}$$

The rate of growth is proportional to the collision cross-section so the proportional rate of growth of any planetesimal, of mass m, to that of the largest planetimal, of mass m_L, is

$$\frac{\dot{m}/m}{\dot{m}_L/m_L} = \frac{r_L}{r} \frac{1 + 2\beta(r/r_L)^2}{1 + 2\beta}. \tag{6.33}$$

This ratio in (6.33) equals unity both when $r = r_L$ and $r = r_L/2\beta$ and, for values of r between those two values, the ratio is less than unity and the relative size of the two bodies diverges. Eventually when $r = r_L/2\beta$ the ratio of masses will remain constant at r_L^3/r^3 or $8\beta^3$. For the range of values of β given previously this corresponds to the mass ratio of the largest forming body to the next largest of between 64 and 1000. The rate of accumulation of mass will be

$$\dot{m}_L = \pi r_L^2 \rho_p (1 + 2\beta)v \tag{6.34}$$

where v is the speed of the planetesimals relative to the growing planetary embryo. To estimate the time-scale for the formation of a planet it is required firstly to find an expression for the density ρ_p, which can be done in terms of the surface density of the nebula, σ. For a particle in a circular orbit of radius r the speed in the orbit is $2\pi r/P$ where P is the period of the orbit. If there is a random speed up to v perpendicular to the mean plane of the system then the orbital inclinations will vary up to $\phi = vP/2\pi r$. The material at distance r will be spread out perpendicular to the mean plane through a distance $h = 2r\phi = vP/\pi$. Hence

$$\rho = \frac{\sigma}{h} = \frac{\pi\sigma}{vP}$$

and the time-scale for the formation of a planet is

$$\tau_p = \frac{m}{\dot{m}} = \frac{4r_L \rho_s P}{3\pi\sigma(1 + 2\beta)} \tag{6.35}$$

in which ρ_s is the density of the forming planet. This equation does not take into account the exhaustion of material as the planet forms, which will reduce σ with time and, since r_L also varies with time, finding the time-scale should involve some integration in the analysis. However, if a final figure is used for r_L, which increases the time estimate, and an initial figure for σ, which reduces the time estimate, then a reasonable overall estimate of the total time will be obtained for a given surface density and position in the nebula.

Several different models for the distribution of surface density and the total nebula mass have been suggested. Here, for illustration, we consider a total nebula mass of $0.1M_\odot$, which reduces but does not completely remove the problem of disposal of material, with a 2% solid fraction. If the surface density varies as R^{-1}, where R is the distance from the Sun, then this gives a surface density of solids at 1 AU of 943 kg m^{-2}. Taking $(1 + 2\beta) = 8$, $\rho_s = 3 \times 10^3$ kg m^{-3} and $r_L = 6.4 \times 10^6$ m this gives a time for forming the Earth of 1.1×10^6 years. Repeating the calculation for the formation of a $10M_\oplus$ core for Jupiter gives a time of 1.5×10^8 years for Jupiter and forming a $3M_\oplus$ core for Neptune takes 7.8×10^9 years—which exceeds the age of the Solar System. In view of the observed lifetimes of nebula discs, ten million years at most, efforts have been made to find mechanisms which will shorten the times drastically. An obvious way to do this is to increase the surface density of the disc and suggestions have been made that there were local enhancements in the regions of planetary formation which did not require the total mass of the disc to increase, with all the attendant problems. Another line has been to find ways of slowing down the relative speed of planetesimals since, from (6.32), this will increase the capture cross section of interactions and hence speed up planetary growth. The inclusion of viscous drag into the system makes a small improvement in this direction but not enough to solve the problem. Stewart and Wetherill (1988) have suggested that the random speed of a body in the planetesimal swarm may not be independent of mass but that some energy equipartition law may operate so that the larger masses move more slowly. If this is true then when larger masses approach each other there is an enhanced probability of collision. Combining this idea with local density enhancement gives what the authors call *runaway growth* with planet formation times from 3.9×10^5 years for Jupiter up to about 3×10^7 years for Neptune.

The runaway growth idea at least shows that new ideas might reduce Safronov's time-scales but the formation time problem is still not solved. Modelling a nebula fails to give the required density in the region of Jupiter by a factor of 10 or so (Boss 1988). Again modelling indicates that the planetary embryos produced in the Jupiter region have less than one Earth mass and, instead of combining to form a Jupiter core, they scatter each other into very eccentric orbits (Wetherill 1989).

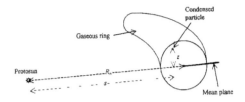

Figure 6.19. A schematic view of a gaseous ring and one of the condensed particles in the Modern Laplacian Theory.

For giant planets, once the core has been formed, it is necessary to attract nebula gas to form the total planet as it appears today. Assuming that the nebula is still present this final stage should take of order 10^5 years and presents no tight constraint on theories.

6.4.4 Planets from the Modern Laplacian Theory

In section 6.3.1.1 the formation of the Sun was described in relation to the Modern Laplacian Theory that described how material was shed from a nebula in a series of annular rings. Prentice (1978) shows that with the temperature in the rings varying from 1260 K for the Mercury ring to 26 K for the Neptune ring, with Jupiter at 122 K, rocky materials would condense out within each ring and icy materials from Jupiter outwards. The total mass of solids available for the terrestrial planets would have been about $4M_\oplus$ with $11\text{--}13M_\oplus$ for the major planets. Since the actual masses of the terrestrial planets are much less than the amount of material available it is suggested that much of it was in the form of fine dust which would remain suspended in the gas. When the condensates first form they are moving around the Sun with a Keplerian angular speed corresponding to the central core of the ring (figure 6.19) and they become decoupled from the gas in the ring. It is then proposed that particles above the mean plane would experience a component of the solar attraction bringing them down towards the mean plane. This probably implies that some degree of coupling is still present. Otherwise, if suddenly decoupled from the ring as a whole, the condensed solids would simply go into an inclined orbit that would cut through the mean plane. In addition it is stated that for condensed particles closer to the Sun than the core of the ring, the instantaneous angular speed is greater than the local Keplerian speed and given by

$$\omega_{\text{ring}}(s, z) = \omega_n \frac{R_n^2}{s^2} \tag{6.36}$$

where R_n is the radius of the nth ring, s the distance from the Sun of a point within the ring projected on to the mean plane and z the distance from the mean plane. The assumption here is that all the material within a single gaseous ring has the same intrinsic angular momentum. This means that material moving from the main nebula to form a ring retains its intrinsic angular momentum during the process of ring formation and continues to retain it after the ring has formed and separated. The implication of this is that the inner material of a ring will be moving at greater than the Keplerian angular speed so that the centrifugal force will thus be greater in magnitude than the gravitational force and produce a force towards the core. As long as the material is gaseous this force will be resisted by the radial pressure gradient in the cross-section of the ring but once condensed particles are formed the pressure ceases to be effective and they move towards the core. A similar argument is used to show that condensed particles further from the Sun than the core will move inwards towards the core. This proposed pattern of angular speed does not take into account viscous effects that will tend to slow down the faster-moving inner material and speed up the slower-moving outer material while it is in a gaseous state. However, we accept for now the argument that condensed particles are attracted towards the core of the ring.

The time for precipitation to the axis of the ring at distance R from the Sun has been estimated by Prentice as

$$t_{\text{seg}} = \frac{1.1 \times 10^4}{a\rho_s} \left(\frac{R_\odot}{R}\right)^{1/2} \text{ years} \tag{6.37}$$

with all right-hand-side quantities in SI units and where the settling particles have radius a and density ρ_s. For silicate particles of radius 1–100 μm in the vicinity of the Earth the settling times are between 3×10^5 and 3×10^7 years whereas for ice particles between 100–10^4 μm in the Jupiter region settling times are only 3×10^3–3×10^5 years. From his analysis of the evolution of the Sun, Prentice estimates that the Sun would have a peak in luminosity at 3×10^5 years that would have interrupted the process of particle segregation. This provides an explanation of why only part of the solid material went into forming planets in the inner Solar System—although it offers no explanation of why Mars has so little mass compared to the Earth.

For the next stage of planetary formation from the stream of particles moving close together on a circular path at the centre of each ring Prentice argues that the Jeans instability will cause bunching. Small particles will get together to form planetesimals and then bunching of planetesimals will produce a larger aggregation until, finally, one dominant mass absorbs all the others in the ring. At the time of maximum luminosity of the Sun, which Prentice identifies as a T-Tauri stage, the gaseous components of the ring are completely swept away in the terrestrial region, thus terminating further growth of planets there, while parts of the gas in outer rings is removed. How much is removed depends on distance from the Sun; while it would have been hotter in the inner parts of the system the material there would have been more strongly bound.

Once the solid planetary condensations have formed they are able to accrete gaseous material. The initial masses of the rings are all similar, about $1000 M_\oplus$, but in the inner part of the system the temperature is too high for the gas to be

Table 6.5. Properties of the rings at the time of their detachment.

Planet	R_n (10^{11} m)	θ (K)	ρ (kg m^{-1})	a (10^{10} m)	v_{esc}^2 (m^2 s^{-2})	v_θ^2 (m^2 s^{-2})	t_{dis} (years)
Earth	1.5	530	3.5×10^{-5}	0.76	1.13×10^6	6.65×10^6	0.13
Jupiter	7.79	122	1.1×10^{-7}	5.96	2.18×10^5	1.53×10^6	2.21
Neptune	44.96	26	2.9×10^{-10}	48.3	3.78×10^4	3.26×10^5	39.1

substantially accreted. In the outer part of the system all gas available would be accreted and the question then arises as to why it is that the masses of the major planets vary as much as they do. Prentice takes up an idea by Hoyle (1960) about the evaporation of gases from the Solar System. He estimates that nearly all the gas is lost from the vicinity of Uranus and Neptune, some 70% is lost in the region of Saturn but that only 20% is lost in the Jupiter region. The gas remaining after the evaporation provides a reasonable explanation for the final masses of the major planets.

There are a number of difficulties with planet formation for the Modern Laplacian Theory one of which, relating to angular speeds in various parts of a ring, has already been mentioned. The idea that condensed particles form at a sufficiently large size that they instantaneously decouple from the gas must be suspect and if they remain coupled or partially coupled, even if only for some time after their formation, then their subsequent behaviour would be quite different from that postulated. Another point of uncertainty is the stability of the rings after they are formed. The information in the first three numerical columns of table 6.5 is derived from a table given by Prentice relating to the properties of the gaseous rings at the time of their formation.

From the first three columns of table 6.5, which give the radius, temperature and density of the ring, it is possible to calculate the radius of the cross-section of the ring, a, given in the fourth column. The escape speed from the edge of a ring is not simple to find analytically but it will not be very different to that from a sphere of radius a at the same density. The square of the escape speed is given as v_{esc}^2 and the mean-square thermal speed is given as v_θ^2. From a study of solid bodies with and without atmospheres it appears that for an atmosphere to be stable and long-lasting the ratio v_{esc}^2/v_θ^2 must be greater than 60 or so. For the values in the table, where the ratio is less than unity, the outer material will move outwards at the speed of sound and the ring will quickly dissipate. The time of dissipation, t_{dis}, is based on the time for a sound wave to move into the centre of the ring from the outside. It is clear that, notwithstanding the uncertainty in v_{esc}^2, it is improbable that the rings could survive for the 3×10^5 years necessary for planetary formation to occur.

6.5   Formation of satellites

With problems at the level of producing planets, the Solar Nebula and Accretion Theories have not produced detailed theories of satellite formation, except to indicate that it resembles the process of planet formation but on a smaller scale. There is a tendency for cosmogonists to assume that the relationship of satellites to planets is similar to that of planets to the Sun and hence that the formation processes in these systems should just be scaled versions of each other. In 1919 Jeans put forward the view that any theory which proposed a mechanism for the origin of satellites which differed from that for producing planets would be 'condemned by its own artificiality'. Later Alfvén (1978) stated that

> We should not try to make a theory of the origin of planets around the Sun but *a general theory of the formation of secondary bodies around a central body*. This theory should be applicable both to the formation of satellites and the formation of planets.

There is, to casual observation, a great similarity between the planetary system and the various satellite systems. Indeed Galileo was confirmed in his belief in the Copernican theory when he first saw the major satellites of Jupiter. However, a closer examination of the two types of system reveals that there are also significant differences. An important characteristic is the distribution of angular momentum that has been such a stumbling block to so many cosmogonic theories. One way of looking at this is to look at the ratio of the intrinsic angular momentum (angular momentum per unit mass) of the secondary body due to its orbital motion to that of the material of the primary body at its equator due to its spin. This ratio will be

$$S = \frac{(GM_P r_S)^{1/2}}{R_P^2 \omega_P} \tag{6.38}$$

where M_P, R_P and ω_P are the mass, radius and spin angular velocity of the primary body and r_S the orbital radius of the secondary body. The values of S for various primary–secondary combinations are given in table 6.6.

Another form of comparison is to find the ratio of the total orbital angular momentum associated with the family of secondary bodies and to compare this with the spin angular momentum of the primary. This comparison brings in the additional factor of the amount of mass associated with the two types of body. In doing this what must be taken into account is that the satellites are probably the residues of larger gaseous condensations which only retained their solid components. In table 6.7 the ratios of angular momenta are given on the basis of satellite masses enhanced by a factor of 100, at the top end of estimation of the enhancement factor.

The distinction between the Sun-centred system and the planetary-centred systems is less when represented in this way but it is still very significant. If the Sun–planet ratio was of the same order of magnitude as the planet–satellite ratios there would have been no angular momentum problems for the Laplace theory or

Table 6.6. The ratio of the intrinsic angular momentum of the secondary orbit to that of the primary spin at the equator.

Primary	Secondary	Ratio S
Sun	Jupiter	7 800
Sun	Neptune	18 700
Jupiter	Io	8
Jupiter	Callisto	17
Saturn	Titan	11
Uranus	Oberon	21

Table 6.7. $R_A = \dfrac{\text{Total orbital angular momentum of secondary bodies (satellites enhanced)}}{\text{Spin angular momentum of primary body}}$.

Primary	R_A
Sun	200
Jupiter	0.6
Saturn	0.7
Uranus	0.4

the other nebula theories which followed it. The striking difference in the ratios for the two types of system, two to three orders of magnitude different, should make the idea that the formation mechanisms could have been different more acceptable.

We shall now look at the ideas put forward about satellite formation by the Proto-planet, Capture and Modern Laplacian Theories.

6.5.1 Satellites from the Proto-planet Theory

From the masses of the proto-planet blobs, 6.7×10^{26} kg, and the density of the material that formed them, 10^{-6} kg m^{-3}, it may be deduced that at the time of their formation they had radii of 5.4×10^{10} m. They would have collapsed quite quickly; the free-fall collapse time is about 2 years but more realistically, taking into account the analysis in section 6.4.2.5, a collapse time of a few tens of years is probable. If, say, three of these bodies came together to produce a Jupiter mass then the angular momentum associated with the group would be of order

$$H_J = \frac{\sqrt{3}}{2} m V r_m = \sqrt{\frac{9GM}{20R}} m r_m \qquad (6.39)$$

where r_m is the characteristic radius of the blobs at the time they combine and V is given by (6.20). Blobs are more likely to combine when they first form

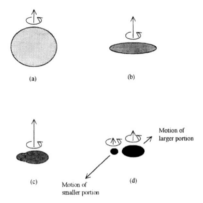

Figure 6.20. The critical break-up of a rotating and collapsing pseudo-incompressible fluid sphere. (*a*) An oblate spheroid. (*b*) A general (Jacobi) ellipsoid. (*c*) An unstable pear-shaped configuration. (*d*) Two bodies produced by fission.

and they are large but, on the other hand, there is much more time for them to combine after they have condensed. Taking $r_m = 10^{10}$ m, corresponding to partial condensation, gives $H_J = 5.9 \times 10^{40}$ kg m^2 s^{-1} which is nearly 100 times greater than the spin angular momentum of Jupiter. Even allowing for somewhat smaller values of r_m it is clear that the proto-planet condensations would probably have formed with considerably more angular momentum than that in their present spins.

The evolution of a rapidly spinning and collapsing body was first considered by Jeans (1929) and later by Lyttleton (1960). As the body collapses and spins more rapidly, at first the body takes up the form of an oblate spheroid (figure 6.20(*a*)) and then a MacLaurin spheroid (figure 6.20(*b*)) which is a general spheroid with three unequal axes. Further collapse brings it to a pear-shaped form (figure 6.20(*c*)) that is unstable and finally it breaks into two parts (figure 6.20(*d*)) in which the ratio of mass of the two parts is 8:1. If this happens to a proto-Jupiter then, as illustrated in figure 5.2, the less massive component will be moving faster relative to the centre of mass than the more massive component. In an outer region of the Solar System, it would be moving fast enough to escape from the system entirely. Much of the original angular momentum will now appear in the relative motions of the two components leaving the more massive part with only a fraction of the original angular momentum. It is envisaged that droplets forming in the neck between the components as they moved apart would be retained by the more massive components and become regular proto-satellites.

In the inner part of the system there would be two differences in the scenario. First, it is envisaged that the fission took place in the dusty core of the condensing original proto-planet, so that the bodies were of terrestrial composition and, second, the escape speed in the inner Solar System was high enough to retain the faster moving smaller component of the disruption. McCrea considered that the pairs Earth–Mars and Venus–Mercury were produced in this way although the mass ratios are 9:1 and 15:1 instead of the expected 8:1. McCrea points to the similar spin periods of the Earth and Mars as supporting evidence although Venus and Mercury will have been too greatly affected by the tidal effects of the Sun to provide similar support.

6.5.2   Satellites from the Modern Laplacian Theory

Since the planetary cores form in the presence of a great deal of gas Prentice suggests that turbulent drag would result in them having very little spin angular momentum. However, the accreted gas would have an angular momentum slightly exceeding the Keplerian value at the distance of capture and it will tend to be in a direct sense. For this reason, if the gas is to accrete onto the planetary core then some means must be available to remove excess angular momentum.

The suggestion made by Prentice is that the contraction of the primitive hydrogen atmospheres of the major planets occurred by a process similar to that of the Sun's own contraction—that is by the development of supersonic turbulent stress and the shedding of a series of rings. No suggestion is made concerning the source of energy for the turbulent stress and it must be questioned whether under the conditions of acquiring a gaseous envelope such turbulent stress would be present, or even necessary, for the collapse of the atmosphere onto the planetary core.

6.5.3   Satellites from the Capture Theory

The basic mechanism for satellite formation in the Capture Theory was described in section 5.3 and is similar to the suggestion by Jeans. In the Jeans tidal theory, satellite formation was described as a small-scale version of planet formation, in line with Jeans' conviction that the processes should be the same. However, while it was shown that planets could not be drawn out of the Sun to give the Solar System as we know it, so that the Jeans theory became untenable, the same problems were not present for satellite formation. The difference in angular momentum distribution in the two types of system, as illustrated in tables 6.6 and 6.7, shows that to form satellites a small fraction of the total mass contains a somewhat less small fraction of the total angular momentum. Nevertheless some mechanism is required so that the intrinsic orbital angular momentum of satellite material is a few times the intrinsic spin angular momentum of equatorial planetary material.

6.5.3.1   *The induced spin of the tidal bulge*

The relationship between planetary spin and satellite formation for the Capture Theory was dealt with by modelling the behaviour of a tidally-distorted protoplanet in orbit around the Sun (Williams and Woolfson 1983). In figure 6.21 a

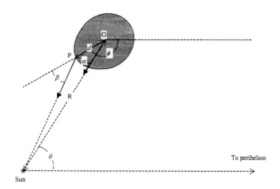

Figure 6.21. A distorted proto-planet in orbit around the Sun.

Table 6.8. Initial orbits of the proto-planets and the deduced average density and radius of the spherical surface containing the retained material.

Planet	Initial mass (kg)	Initial a (10^9 km)	Initial e	$\langle\rho\rangle$ (kg m^{-3})	r_c (10^7 km)
Jupiter	2.00×10^{27}	2.19	0.80	4.55×10^{-5}	2.19
Saturn	5.97×10^{26}	2.79	0.68	5.37×10^{-6}	2.98
Uranus	9.25×10^{25}	5.34	0.69	8.42×10^{-7}	2.97

proto-planet is shown moving towards perihelion with its tidal bulge lagging an angle α behind the radius vector from the Sun. If the proto-planet has a reasonably high central condensation then it may be treated as a Roche model (section 4.4.1) and the form of the critical equipotential for such a model is known (section 4.6.2). When the ratio of masses of the distorting body to distorted body is high, as in this case, then the distance of the tidal tip from the centre of the proto-planet, d, is about 1.4 times the original undistorted radius. It is assumed that the original giant planets, Jupiter, Saturn and Uranus, occupied a critical equipotential—on the basis that anything outside would be lost. From the proto-planet orbits suggested by the Capture Theory the average density $\langle\rho\rangle$ and radius, r_c, of these proto-planets may be found and are shown in table 6.8.

From figure 6.21 it is seen that there is a torque causing a rotation of the tip, P, about the centre, O, of the proto-planet. The acceleration of the spin is given by

$$\ddot{\phi} = \frac{GM_\odot}{d}\left(\frac{1}{r^2}\sin\beta - \frac{1}{R^2}\sin\alpha\right) = \frac{GM_\odot}{dR^2}\sin\alpha\left[\left(\frac{r}{R}\right)^{-3} - 1\right]. \quad (6.40)$$

Using

$$r^2 = R^2 + d^2 - 2Rd\cos\alpha$$

together with the assumption that terms in $(d/R)^2$ can be ignored leads to

$$\ddot{\phi} = \frac{3GM_\odot}{2R^3}\sin 2\alpha + \frac{GM_\odot d \sin\alpha}{R^4}\left(6\cos^2\alpha - \frac{3}{2}\sin^2\alpha\right). \tag{6.41}$$

From the relationship between angles, deduced from figure 6.21,

$$\alpha = \theta + \phi - \pi,$$

and the equation for an ellipse

$$R = \frac{a(1 - e^2)}{1 + e\cos\theta}$$

the independent variable may be changed to θ giving

$$(1 + e\cos\theta)\frac{d^2\phi}{d\theta^2} = -\frac{3}{2}\sin 2(\theta + \phi) + 2e\sin\theta\frac{d\phi}{d\theta}$$
$$+ \frac{3d}{4R}\sin(\theta + \phi)[3 + 5\cos 2(\theta + \phi)]. \tag{6.42}$$

To solve (6.42) numerically requires an initial ϕ and $d\phi/d\theta$ for some initial θ; by multiplying the value of $d\phi/d\theta$ at any point by the corresponding $d\theta/dt$ the value of $d\phi/dt$ may be found.

A large variety of starting conditions and ranges of integration were explored although all with the common characteristic that the perihelion was at the centre of the range. Some results are displayed in figure 6.22. Each curve gives ζ, the final ratio of the angular speed of the tidal tip to the orbital perihelion angular speed, against the range of integration, which is between $-\theta_0$ and θ_0. The plots in each diagram are for different initial tidal lags and the different diagrams correspond to different initial tidal tip spin rates. The results show that the final spin rate of the tidal tip is not very sensitive to initial conditions or range of integration and $\zeta = 1.5$ is a good average figure. Accepting this estimate for ζ the values of $(d\phi/dt)_{\text{final}}$ are: for Jupiter 8.02×10^{-8} s^{-1}, for Saturn 2.66×10^{-8} s^{-1} and for Uranus 1.06×10^{-8} s^{-1}.

6.5.3.2 *Angular momentum in the tidal bulge*

By modelling the tidal bulge as a cone tangential to a sphere (figure 6.23) Williams and Woolfson estimated the mass of the tidal-bulge material as

$$M_b = 0.12\rho_b r_c^3 \tag{6.43}$$

and the induced angular momentum within it as

$$H_b = 0.123\rho_b r_c^5 \dot{\phi}_{\text{final}} \tag{6.44}$$

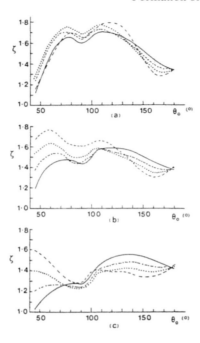

Figure 6.22. The ratio of the final bulge rotation rate to the planetary perihelion rate as a function of the range of integration $(-\theta_0$ to $+\theta_0)$ for: (a) $\phi_0 = 0$; (b) $\phi_0 = -0.5\theta_0$; (c) $\phi_0 = -\theta_0$. The sets of curves are for different initial lags: $\alpha_0 = 0°$, full line; $\alpha_0 = 10°$, chain line; $\alpha_0 = 20°$ dotted line; $\alpha_0 = 30°$ dashed line (Williams and Woolfson 1982).

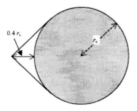

Figure 6.23. A distorted proto-planet modelled as a cone fitted to a sphere.

in which ρ_b is the mean density of the bulge material. In this analysis it has been assumed that there is only one tidal bulge but under the conditions of the Sun–proto-planet interactions there will also be another bulge facing away from the Sun. The ratio of angular momentum in the smaller bulge to that in the main bulge is 0.35, 0.5 and 0.65 for Jupiter, Saturn and Uranus respectively. Accepting these figures, and taking the mean density of bulge material as f times the mean density indicated in table 6.8, gives values of H_b/f where H_b is the total angular momentum in the proto-planet material. The total angular momentum acquired

Table 6.9. Comparison of the angular momentum (AM) induced in the bulges with that expected for planetary spin and the orbits of augmented satellites. All units are kg m^2 s^{-1}.

Planet	H_b/f	Observed spin AM	Orbital AM of satellites	Augmented AM of satellites	Expected total initial AM
Jupiter	3.2×10^{39}	4.4×10^{38}	4.2×10^{36}	4.2×10^{38}	8.6×10^{38}
Saturn	6.2×10^{38}	9.1×10^{37}	9.5×10^{35}	9.5×10^{37}	1.9×10^{38}
Uranus	4.2×10^{37}	1.6×10^{36}	1.3×10^{34}	1.3×10^{36}	3.0×10^{36}

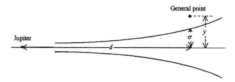

Figure 6.24. The lateral spread of a proto-satellite filament contained by the pinch forces from a planet.

by the proto-planets should be sufficient to explain both their spin and the satellite orbits but, as previously indicated in reference to table 6.7 the satellites are almost certainly the solid residues of gas-plus-dust bodies from 50 to 100 times more massive. Taking this into account the values of H_b/f are compared with the required angular momentum in table 6.9.

The result from observations that the orbital angular momentum of augmented satellites is approximately equal to the spin angular momentum of the parent planet is interesting and suggests that the induced angular momentum was equally shared by the planet and its satellites.

In the process of proto-planet collapse the tidal bulge would stretch into a filament so that some appreciable fraction of the bulge material would be available for the formation of the augmented satellites. The total mass of the augmented Galilean satellites is approximately 4×10^{25} kg but the estimated mass of the bulge material, assuming its density was one-half of that of the proto-planet, is about 3×10^{25} kg. This estimate is a bit low considering that not all the bulge material would go to satellite formation. However, the assumed conical form of the bulge somewhat underestimates its volume and the deficiency of mass is not too serious.

6.5.3.3 The masses of individual satellites

The proto-satellite filament will have a flared shape, as shown in figure 6.24, due to a balance between a pinch effect from the component of the solar gravitational field perpendicular to the length of the filament and a pressure gradient in the same direction. This leads to an expression for the pressure gradient

$$\frac{dP}{dy} = \frac{GMy}{(d^2 + y^2)^{3/2}} \tag{6.45}$$

where M is the planet mass and other quantities appear in figure 6.24. This expression can be integrated and gives a density variation along y:

$$\rho = \rho_0 \exp\left[-\frac{GM\mu d}{GM\mu - k\theta d}\left(1 - \sqrt{\frac{1}{1 + y^2/d^2}}\right)\right], \tag{6.46}$$

in which μ is the mean molecular mass of the filament material of temperature θ and ρ_0 is the density on the axis. The characteristic fall-off distance for the density, where its value is $1/e$ of its peak value, is given by

$$\sigma = \left[\left(\frac{GM\mu d}{GM\mu - k\theta d}\right)^2 - d^2\right]^{1/2}. \tag{6.47}$$

Taking $M = 2 \times 10^{27}$ kg (Jupiter), $\mu = 3.3 \times 10^{-27}$ kg (molecular hydrogen) and $d = 10^{10}$ m the value of σ is 1.26×10^9 m.

If a proto-satellite condensation forms from material between a distance $d+s$ and $d - s$ from the planet centre then, from (4.15)

$$2s = \sqrt{\frac{\pi k\theta_T}{G\mu\rho}},$$

where θ_T is the temperature of the filament material.

Taking the mean density of the filament at distance d as ρ gives an estimate of the mass of the condensation

$$M_c = 2\pi\sigma^2\rho s. \tag{6.48}$$

Eliminating ρ from (4.15) and (6.48) gives

$$s = \frac{\pi^2 k\theta_T \sigma^2}{2G\mu M_c}. \tag{6.49}$$

With $M_c = 10^{25}$ kg, a typical augmented satellite mass, and $\theta_T = 20$ K then (6.49) gives $s = 1.2 \times 10^9$ m with a corresponding density 8.5×10^{-4} kg m^{-3}. The Jeans critical mass is given by (2.22):

$$M_J = \left(\frac{375k^3\theta^3}{4\pi G^3\mu^3\rho}\right)^{1/2} \tag{2.22}$$

and for the values of the various parameters considered here it is 1.16×10^{25} kg. This suggests that the augmented proto-satellite would just not be stable but in view of all the approximations in the analysis it seems probable that stable proto-satellite condensations could actually form in a filament.

It will be noticed that the distance d chosen in this analysis is more than five times the orbital radius of Callisto. This is to allow for the fact that the material in the filament left behind by the collapsing proto-planet will also be moving inwards, although at a slower rate. Thus it is anticipated that newly formed proto-satellites will fall in towards the parent planet and when the orbit has rounded off it will be much closer to the planet than the distance of its formation.

6.6   Successes and remaining problems of modern theories

Here we shall summarize the extent to which the various theories considered have been successful in explaining the formation of the Sun, planets and satellites and also highlight the problems they still face.

6.6.1   The Solar Nebula Theory

This theory is the one on which most work has been done and there is an abundance of approaches to all aspects of the formation of the Solar System. Its main attraction is that there is some evidence for the existence of the conditions it postulates, a dusty disc surrounding a new star, but it has not been successful in showing convincingly that a star could form from a collapsing nebula. The basic requirement is the removal of angular momentum from the collapsing core of the nebula during the whole period of the process of collapse and while material is still moving inwards towards it. It is not enough to propose such a mechanism coming into play *after* the star has substantially collapsed because, without the mechanism, there is so much angular momentum in the system that the star could not collapse and form at all. Given that, somehow or other, a very rapidly spinning star would form then the only type of mechanism that seems plausible at present is one dependent on magnetic coupling between the inner and outer material of the nebula. If some rather extreme parameters are adopted for the ionization state of the material and the strength of the dipole field generated by the forming proto-star then sufficient angular momentum can be transported outwards. However, in view of the evidence for other planetary systems it will be necessary to show that such high fields are routinely generated and not a special phenomenon associated with the Solar System.

The question of the tilt of the solar spin axis was raised in section 6.2.2 and presents a serious problem for monistic theories. One possibility to solve the problem is to perturb the planets by a passing star after the Sun had become a highly condensed body. This could then change the plane of the planetary orbits without seriously affecting the spin axis of the Sun. Such a perturbation would not only change the plane of the planetary orbits but also make the orbits much more eccentric and this is especially true for Neptune which, at present, has an orbit of very low eccentricity. However, if the perturbation by the passing star occurred when the planets were still in the presence of a resisting medium then their orbital planes could be changed and subsequently their orbits could be rounded off by the action of the medium.

In forming planets there are competing requirements which cause difficulties. With a higher nebula density, planets can either form directly by gravitational instability or on a reasonably short time-scale by first producing planetesimals. There is then a considerable problem of disposing of surplus material, the great bulk of which must be eliminated from the Solar System. Any large addition of material to the Sun would reintroduce the angular momentum problem, assuming that it had been previously solved at the Sun formation stage. With a lower nebula density there are serious planet time-of-formation problems. The initial difficulty, first noted by Safronov, was that the outer major planets would take too long to produce—even longer than the age of the Solar System according to most estimates. A newer difficulty arises from the observations which are sometimes regarded as supporting the Solar Nebula Theory—that the lifetime of the disc is less than 10^7 years, which presents no serious difficulty for the formation of the terrestrial planets but does for the planets from Jupiter outwards.

The formation of satellites is just stated to be a small-scale version of the planet-forming process in which satellites would form in a dusty disc surrounding the collapsing proto-planet. In view of the angular momentum distribution in the planet–satellite systems, illustrated in tables 6.6 and 6.7 this mechanism of satellite formation could be plausible, although no detailed analysis of the process is available.

6.6.2   The Accretion Theory

This dualistic theory has the advantage of sidestepping the star-formation process, which is acceptable as long as some plausible mechanism is available for forming stars independently of planets. The capture of a dusty disc by passage through an interstellar cloud is similar in some ways to the Capture-Theory process by which material from a diffuse proto-star is captured by a more condensed star and this has been well established by modelling of ever-increasing complexity. Once the dusty disc is in place then the Accretion Theory faces the same difficulties as the Solar Nebula Theory, as will any theory seeking to assemble planets from highly dispersed material.

6.6.3   The Modern Laplacian Theory

The very first step in the Modern Laplacian Theory removes a great deal of the angular momentum problem, but not all of it, by postulating stellar formation from solid hydrogen grains. The basis of this hypothesis is not generally accepted and there is no direct evidence for temperatures in dense cool clouds that could give hydrogen grain formation. The next critical step is the action of supersonic turbulence to eject material in the form of spikes which changes the moment-of-inertia factor and creates a density inversion leading to ring formation. No successful modelling of this mechanism has been done and it is not intuitively obvious that such a process would occur. It is also necessary to postulate the formation of rings within the orbit of Mercury in order that angular-momentum extraction from the core could be sufficient. Since the masses of all the rings are approximately equal, the total mass in these inner rings is considerable and their disposal becomes a problem. If they are re-absorbed by the Sun then their angular momentum would not be lost.

There must be serious doubts about the stability of the rings in view of the analysis leading to table 6.5 so that even if the condensing solids would eventually fall towards the core of the ring there may not be time for this to happen. The behaviour of condensed particles in the rings is likely to be much more complex than that which has been described. Because of radiation pressure the gas and very small particles in the ring will not be moving with Keplerian speeds and when larger particles assemble they will experience serious drag effects. How this would influence the process of aggregation of solids is uncertain but the possibility exists that it would lead to diffusive spreading of solids rather than their concentration.

Satellite formation is again described as a miniature version of planetary formation. Once proto-planets form then the situation appears to be that described for the Solar Nebula Theory and the need for supersonic turbulence is not obvious.

6.6.4 The Capture Theory

As for the Accretion Theory the Capture Theory separates the processes of star and planet formation, although they are connected through the existence of a forming cluster, necessary for two stars to interact. The spin axis of the Sun and the plane of the planetary orbits (approximately the plane of the Sun–proto-star orbit) were not directly connected but solid material absorbed by the Sun pulled the spin axis towards the normal to the mean plane. Since the Sun had an initial component of angular momentum in some other direction there has to be a residual tilt of the spin axis, albeit a small one.

By having planets formed in a filament the Capture Theory meets the needs previously mentioned for the Solar Nebula Theory in having a sufficiently high density to produce planets directly while having no great surplus of material which must be disposed of. Modelling in various ways all supports the view that the filament would have escaped from the proto-star and that condensations within it would be captured.

The process by which satellites are produced from the collapsing tidally distorted planets has been very crudely modelled. The physical properties of the planetary material that would affect its behaviour have not been included in the model—although the numerical results are supportive of the general idea.

There are, however, some numerical inconsistencies. Table 6.9 suggests values of $f = 3.7$, 3.3 and 14 for Jupiter, Saturn and Uranus respectively. Intuitively it would be expected that the values of f should be fractional since they represent the ratio of the density in the tidal bulge to that of the planet as a whole. One reason for the high value of H_b/f is that it is assumed that the intrinsic angular momentum imparted to the tidal tip is characteristic of that for the whole tidal bulge. This is not so for two reasons. In the first place, as will be seen from (6.42), the angular acceleration depends on d, the distance of the material from the proto-star centre, O. In the second place, if the tidal lag is not too large then the bulge material on the opposite side of the line Sun–O from the tidal tip will be contributing angular momentum in the opposite sense to that of tidal-tip material. Taking these effects into consideration the values of f for Jupiter and Saturn are not so large that they could not be brought to a reasonable value by some minor modification of the model. Nevertheless the large value of f for Uranus suggests that the angular momentum predicted by the model is too high for Uranus and that a small modification of the model would not correct this. Actually Uranus is anomalous in another respect, in the direction of its spin vector, and solar-induced tidal effects could not explain what is observed in any case. A suggestion for the spin axis direction of Uranus, and perhaps a smaller angular momentum content than the present theory indicates, is given in section 7.4.

A requirement for further supporting the Capture Theory is to model in detail the whole basic process of stellar interactions, formation of a filament and condensations within it and the rotational disruption of the proto-planets to give satellites. What has been done is to factorize the whole process into independent scenarios without showing that one stage would inevitably lead to the next. Another requirement is to try to estimate the frequency of planetary systems of one sort or another that would come about in this way. A previous assessment (Woolfson 1979) suggested that this could be as low as one in 10^5 stars having a planetary system. However, this was based on the stars being in an environment of stellar density based on observed clusters. It is now believed that evolving stellar clusters go through an embedded stage (Gaidos 1995) where stellar densities can be as high as 10^5 pc^{-3} for a period of about 10^6 years—which will greatly change the estimates of the frequency of stellar interactions. Indeed, such high stellar densities might give many capture events but also raise the question of whether some planetary systems, if formed in such an environment, could survive the ravages of stellar perturbations.

6.6.5   The Proto-planet Theory

The Proto-planet Theory is unique in straddling the monistic–dualistic divide. While the Sun and the planets are created independently the source and form of material that produces them is the same—proto-planetary blobs. Its great attraction is that starting with few assumed parameters it enables derivation of a number of other parameters that agree tolerably well with the observed characteristics of the Solar System. Nevertheless it is open to the criticism that it does not justify some of the basic assumptions which underpin it.

The formation of the Sun by large numbers of small objects coming from random directions substantially solves the problem of the slow spin of the Sun very neatly but there are, nevertheless, some important reservations. In estimating the angular momentum of the Sun it is assumed that the size of the body on which proto-planets are accumulating is that of the present Sun. This could hardly be so as part of the process of proto-planet evolution requires them to collapse and undergo fission that implies that they have a low initial density. Consequently it would be reasonable to assume that the proto-Sun produced by this process would be a very extended body with much greater angular momentum than that given following (6.4*b*). Another difficulty is that the modelling of star formation by Woolfson (1979) suggests that initial condensations in a cloud are of stellar rather than planetary mass. There is a relationship between the McCrea and Woolfson models here; in the Woolfson model the proto-star is produced by the accumulation of just two condensations, each of stellar mass, and the mechanics of the system ensures that the compressed material so produced has little spin angular momentum.

The McCrea model considers that the forming stellar cluster is broken up into a large number of independent regions within each of which a star, plus possibly planets, will form. Since the star accounts for so little of the angular momentum in a region then it is assumed that the rest must go into a system of planets. This ignores the fact that the regions are coupled together by proto-planets which cross the boundaries between one region and another. A proto-planet coming from one region and joining the star in another region will be contributing to the relative motion of the centre of masses of the two regions. Thus the 'missing' angular momentum does not necessarily have to be present in a system of planets—it could equally well appear in the relative motion of the stars in the cluster.

Another problem with planet formation in the Proto-planet Theory is that it gives orbits in completely random directions and no process is described by which they would settle down into a directly-spinning almost planar system. However, there are possible scenarios involving a disc-like resisting medium developing around each proto-star and collisions eliminating retrograde bodies which could lead to the required outcome, so this criticism is not a strong one.

The mechanism of fission of the initial proto-planets both to remove angular momentum and to produce satellites is attractive but has not been demonstrated in detail. There is also no explanation as to why the whole planet is involved in the major-planet region but only the cores in the case of terrestrial planets.

Chapter 7

Planetary orbits and angular momentum

7.1 The evolution of planetary orbits

The present orbits of the planets are mostly near-circular, the ones furthest from that state being those of Pluto ($e = 0.249$), Mercury ($e = 0.206$) and Mars ($e = 0.093$). Planetary eccentricities do not remain constant and, due to mutual perturbations of the planets, change in a cyclic fashion. When Laplace put forward his nebula theory an important condition he sought to satisfy was that the orbits of his planets should be circular. This would also be an outcome from the Modern Laplacian Theory for which planets originate within rings left behind by the retreating nebula. The Accretion Theory and the Solar Nebula Theory, for which planets are produced in a similar way, would lead to planets in near-circular orbits, but not precisely so. The need for planetesimals to interact with growing planetary cores ensures that non-circular orbits are necessary. On the other hand, interactions between planetesimals and growing cores should not be too violent otherwise abrasion of the cores will take place rather than accretion of planetesimals. A scenario with modest eccentricities of the various involved bodies would seem to be required.

The remaining two theories, the Proto-planet Theory and the Capture Theory, by their very nature, would give planets in very eccentric orbits and some rounding-off mechanism is required to give what is seen today.

7.1.1 Round-off due to tidal effects

By whatever process planets were first produced they would have been extended objects which collapsed on time-scales of tens of years to their present state (section 6.4.2.5). A possible tidal mechanism for reducing the eccentricity of an orbit was proposed by Goldreich (1963). If there is no lag in the direction of the tidal bulge in the secondary body and if the primary body remains spherical then there will be no angular perturbation due to gravitational effects. However, if the orbit is eccentric then the tide will increase and decrease in a periodic fashion and, through dissipation processes, energy will be lost. Loss of energy with constant angular momentum implies that the eccentricity will fall. This can be seen from the relationship

$$e^2 = 1 + \frac{H_{\mathrm{p}}^2 E_{\mathrm{p}}}{G^2 M_\odot^2} \tag{7.1}$$

in which e is the orbital eccentricity and H_{p} and E_{p} are, respectively, the intrinsic angular momentum and intrinsic energy of the planet in its orbit. When E_{p} becomes less, i.e. more negative, then e also becomes less. The final outcome is a body in circular orbit of radius equal to the original semi-latus rectum, which is a measure of the orbital angular momentum.

If we assume that the distorted secondary body takes up the shape of a prolate spheroid then we can make an estimate of the rounding time by this process. The gravitational potential energy of a prolate spheroid of uniform density has been given by Lamb (1932) as

$$V = -\frac{3GM^2}{5R} \left\{ \frac{(y^2+1)^{1/6}}{y} \coth^{-1} \left[\frac{(y^2+1)^{1/2}}{y} \right] \right\} \tag{7.2}$$

where M is the mass of the spheroid, $R = (AB^2)^{1/3}$, $y = (A/B)^2 - 1$ and A and B are the semi-major and semi-minor axes of the spheroid. The values of A and B can be estimated from the solar tidal field at perihelion and aphelion, assuming that the planet occupies a volume equal to that contained within an equipotential surface, and hence the values of R and y can also be estimated. This enables the variation in the gravitational potential energy in one orbit, δV, to be found. If some fraction c (≤ 1) of this is lost in each orbit then Dormand and Woolfson (1974) found that the rate of change of orbital semi-major axis is

$$\frac{da}{dt} = -\frac{ca^{1/2}\delta V}{\pi (GM_\odot)^{1/2} M_{\mathrm{P}}} \tag{7.3}$$

where M_{P} is the mass of the planet. Taking a proto-planet of mass 10^{27} kg, modelled as one-third uniform atmosphere and two-thirds point mass with $c = 1$, Dormand and Woolfson found that a proto-planet of constant radius 10^{10} m in an orbit with $(a, e) = (2.5 \times 10^{12}$ m, $0.8)$ would round off to $e = 0.1$ in 3.2×10^7 years. Since the planetary radius would not be maintained at such a large value for so long it was concluded that a tidal mechanism had no part to play in rounding off the orbits.

7.1.2   Round-off in a resisting medium

All the theories of planet formation we have described lead to their origin within a resisting medium. Assuming that the planets are formed quickly then the observational evidence on discs around young stars suggest that there is a period of a few million years for the medium to act on them. However, it is necessary for the medium to be effective when the planets are in a compact form since they collapse quite quickly to a small radius.

The resisting medium will give rise to a force opposing the motion of the planet and hence to a tangential acceleration, A. From a theory given by Kiang (1962) the rate of change of the intrinsic energy and intrinsic angular momentum of the planet are

$$\frac{dE_P}{dt} = -\frac{A}{W}(v^2 - H_P\omega) \tag{7.4a}$$

and

$$\frac{dH_P}{dt} = -\frac{A}{W}(H_P - r^2\omega) \tag{7.4b}$$

where W is the speed of the planet relative to the medium, v the Kepler velocity, r the solar distance and ω the angular velocity of the medium. These equations assume a two-dimensional scenario so that $\mathbf{W}$ is parallel to $\mathbf{A}$ and $\boldsymbol{\omega}$ is parallel to $\mathbf{H}$. In terms of the orbital elements a and e, Dormand and Woolfson (1974) found from (7.4a) and (7.4b)

$$\frac{da}{dt} = -\frac{2a^2}{\mu}\frac{A}{W}[v^2 - (\mu a)^{1/2}\omega(1 - e^2)^{1/2}] \tag{7.5a}$$

and

$$\frac{de}{dt} = -\frac{(1 - e^2)^{1/2}}{\mu e}\frac{A}{W}[(1 - e^2)^{1/2}(av^2 - \mu) + \omega(a^3\mu)^{1/2}(e^2 - 1 + r^2/a^2)] \tag{7.5b}$$

where $\mu = G(M_\odot + M_P)$.

Dormand and Woolfson considered the case where each particle of the resisting medium was in Keplerian orbit so that

$$W^2 = \dot{r}^2 + (r\dot{\theta} - r\omega)^2 \tag{7.6}$$

with (r, θ) the polar coordinates of the planet. Their computational algorithm for calculating the round-off of the planet took advantage of A being small. They found the average rate of change of the orbital elements by finding the change over one orbit and then dividing by the orbital period. A convenient form for expressing this is

$$\langle \dot{a} \rangle = \frac{1}{\pi}\int_0^\pi \dot{a}(1 - e\cos E)\, dE \tag{7.7a}$$

and

$$\langle \dot{e} \rangle = \frac{1}{\pi}\int_0^\pi \dot{e}(1 - e\cos E)\, dE \tag{7.7b}$$

where E is the eccentric anomaly. By this means, in a Runge–Kutta algorithm they were able to use orbital periods as time steps and so to accelerate the computational process.

One piece of physics that has to go into the calculation is the form of the resistance of the medium, as expressed by A. If it is assumed that any particle of the medium coming within the sphere of influence of the planet describes a hyperbolic path in its gravitational field then the force on the planet can be found by summing the changes in momentum of all the particles. The sphere of influence of a planet is that distance within which motions of small bodies are dominated by the planet's gravitational field rather than by that of the Sun. The sphere of influence is taken to have radius

$$S = r \left(\frac{M_P}{2M_\odot} \right)^{1/3}. \tag{7.8}$$

Dodd and McCrea (1952) considered the resistance problem in relation to a star's motion through the interstellar medium. Adapting their result to the present case

$$A = 2\pi\rho \frac{G^2 M_P}{W^2} \ln \left(1 + \frac{S^2 W^4}{G^2 M_P^2} \right) \tag{7.9}$$

where ρ is the density of the medium.

The effect of accretion of the medium by the planet must also be considered. This gives a resistance law

$$A_a = \pi\rho x^2 W^2 / M_P \tag{7.10}$$

where x is the accretion radius has already been given in (2.41) and in present notation is given by

$$x^2 = R \left(R + \frac{2GM_P}{W^2} \right) \tag{7.11}$$

where R is the radius of the planet. Combining the two types of resistance, allowing for the fact that material within the accretion radius does not contribute to the non-accretion resistance, we find

$$\frac{A}{W} = \frac{\pi\rho}{M_P} \left[x^2 W + \frac{2B}{W^3} \ln \left(\frac{1 + S^2 W^4 / B}{1 + x^2 W^4 / B} \right) \right] \tag{7.12}$$

where $B = (GM_P)^2$.

Since the planet is accreting material its mass is not a constant and a third differential equation is necessary if the evolution of the planet's orbit is to be followed. This is

$$\langle \dot{M_P} \rangle = \rho \int_0^\pi x^2 W (1 - e \cos E) \, dE. \tag{7.13}$$

Dormand and Woolfson (1974) found that the non-accretion mechanism is dominant and neglecting accretion increased the rounding-off times by only 25% or so. Thus most of their calculations ignored accretion and the results they gave

Table 7.1. Rounding times ($e < 0.1$) for the Gaussian and exponential distributions of the medium and for various planetary masses. In each case the thickness of the medium is 10^{12} m and for the Gaussian distributions $r_0 = 10^{12}$ m. The initial planetary $(a, e) = (2.5 \times 10^{12}$ m, $0.8)$. Times are in units of 10^6 years.

	$\alpha\ (10^{-12}$ m)	$\beta\ (10^{12}$ m)	10^{27}	10^{26}	10^{25}	10^{24}	10^{23}
			Mass of planet (kg)				
Gaussian		0.5	0.056	0.39	3.0	24	200
Gaussian		1.0	0.062	0.42	3.2	26	220
Exponential	2.0		0.076	0.48	3.6	29	250
Exponential	1.0		0.069	0.43	3.2	26	220

were upper bounds for the rounding time. They took a mass of medium equal to 10^{28} kg ($5M_{\text{Jupiter}}$) with a uniform thickness of 10^{12} m (~ 6.7 AU). Two models for the radial distribution were taken. The first was a Gaussian type profile with

$$\rho(r) = \rho(0)\exp\{-[(r - r_0)/\beta]^2\} \qquad (7.14a)$$

and $r_0 = 10^{12}$ m, giving a density peak just outside the orbit of Jupiter. Different values of β, that defined the variance of the distribution, were used. The second model was an exponential distribution

$$\rho(r) = \rho(0)\exp(-\alpha r) \qquad (7.14b)$$

with α as a variable parameter.

Table 7.1 shows the rounding time, defined as that for which the eccentricity falls below 0.1, for various distributions of resisting media and various planetary masses. It will be seen that the distribution of the medium, within the limits of these calculations, makes very little difference but rounding times are very dependent on the planetary mass. To a rough approximation the rounding time is inversely proportional to the mass. The evolution of the orbital elements with time is shown in figure 7.1.

For the major planets the round-off times are quite short, well within the observed lifetime of discs, and they are also proportional to the density and hence the mass of the medium. To explain the spin axis of the Sun, assuming that it had a very low initial spin and that most of its angular momentum comes from solid material spiralling inwards, it would need to absorb about one-quarter of a Jupiter mass (section 6.2.2). If the solids are 2% of the total medium and all are in grains large enough to move inwards then this would indicate a medium mass of about $12M_{\text{Jupiter}}$, more than twice that taken by Dormand and Woolfson. Taking the influence of orbital radius and planetary mass into account it is just about possible that the larger terrestrial planets could round off within a few million years but it seems extremely unlikely that Mars and Mercury could do so.

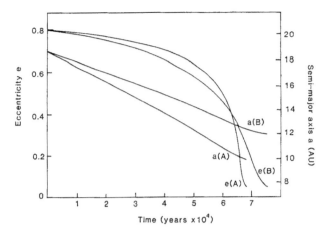

Figure 7.1. Evolution of an orbit in a resisting medium. Mass of medium $5M_{\text{Jupiter}}$, mass of planet 10^{27} kg; $\beta = r_0 = 1/\alpha = 6.67$ AU. (A) is for the Gaussian distribution and (B) is for the exponential distribution.

7.1.3 Bode's law

When von Weizsäcker introduced his vortex theory (section 4.8.1) he found that having five vortices per ring gave something similar to Bode's law. The ratio of the size of a vortex in ring $n + 1$ to that in ring n was the same throughout the system and this led to the ratio of successive orbital radii being the same. This is very roughly what Bode's law gives although the ratio of neighbouring orbital radii for the planets out to Uranus vary from 1.38 (Earth:Venus) to 2.01 (Uranus:Saturn). It is not possible to get a convincing fit of orbital radii over the whole range with a single ratio. A reasonable fit can be obtained by using a ratio 1.86 between Mars and Uranus and 1.57 between Mercury and Mars but this is just playing with numbers and has no particular significance.

The Modern Laplacian Theory (section 6.3.1.1) also leads to a constant ratio between the principal radii of successive rings. This can give a rough match to the actual planetary radii (see table 7.2) with the ratio equal to 1.73. The ratio that comes from the model depends on a turbulence parameter that controls the strength of the supersonic turbulence. The Bode's law relationship exposes another problem of the Modern Laplacian Theory according to which there will be several rings released within the orbit of Mercury. Clearly there cannot be too many; with a ratio of 1.73 between the radius of one ring and the next, the eighth ring within Mercury would be skating around the surface of the present Sun. Most theories of stellar evolution assume that there is a Kelvin–Helmholtz contraction stage, lasting 4×10^7 years for the Sun. This stage begins when the radius of the Sun equals that of Mercury's present orbit and for a short time the Sun would be between 1000 and 100 times as luminous as at present. It is difficult to see how the Mercury ring could form in such an environment and there seems to be even less chance of there having been rings interior to Mercury.

7.1.4   Commensurability of the Jovian satellite system

Although there is no agreement amongst cosmogonists about the significance of Bode's law there is no dispute concerning the validity of the commensurabilities that occur between the periods of some neighbouring planets and within some satellite systems. These have already been mentioned in sections 1.2.3, 1.4.2 and 1.4.3.

The Jovian satellite system is a particularly interesting one. The orbital periods of Io, Europa and Ganymede are almost, but not exactly, in the ratios 1:2:4. However the Laplacian-triplet relationship between the mean motions and the concomitant one between the orbital longitudes, given in section 1.4.2, hold precisely. Although the orbits of the inner Galilean satellites are very nearly circular, they are not precisely so. The commensurability between neighbouring pairs means that their mutual perturbations are a maximum at the same points of their orbits and so are reinforced by resonance effects. This point was appreciated by Peale *et al* (1979) who argued that the non-zero but small eccentricity of Io, the closest satellite to Jupiter, would lead to periodic tidal stressing and hence heating of that satellite. They predicted that, as a consequence, the satellite would show volcanic activity; three days after the paper appeared Voyager I took photographs of Io and their prediction was fully confirmed.

The interesting question is how the Laplacian-triplet condition is established in the first place. It may be related to the actual formation mechanism of the satellites that constrains them to form in a commensurable relationship. Alternatively it could be that they originally have some quite arbitrary periods and an evolutionary process brings them into their present state.

The action of Jupiter on each of the satellites taken singly is to make its orbit circular. This is due to the mechanism analysed by Goldreich (1963) and described in section 7.1.1. The departure of the orbits from exact resonance is what controls the strength of the mutual perturbations and therefore the eccentricities of the orbits. The departure from exact resonance is such that

$$n_1 - 2n_2 = n_2 - 2n_3 = k = 0.74° \text{ day}^{-1} \tag{7.15}$$

where k is the rate at which the longitude of conjunctions moves around Jupiter. If k was smaller then there would be an even larger concentration of mutual perturbations in one part of each orbit and the eccentricity would be higher; this would lead to a much higher production of heat energy in the satellites. The heat energy produced must come from the satellite orbit that should therefore decay but there is no evidence that this is actually happening for the satellites, which have been observed over more than 300 years. The answer seems to be in a mechanism described in section 1.4.6.3 relating to the Earth–Moon system. Io raises a tide on Jupiter but, since the planet is spinning so rapidly, the tidal bulge is dragged forward. The gravitational effect of this on Io is to pull it forward in its direction of motion so increasing both the angular momentum and energy of its orbit. Since the effect is strongest at perijove it has the effect of adding kinetic energy when the kinetic energy is a maximum anyway and this increases the eccentricity of the orbit. A stable equilibrium is set up where the eccentricity due to both the action of Jupiter and the other satellites gives a loss of energy just balanced by the input of energy by Jupiter.

Yoder (1979) described a mechanism by which the near 1:2:4 resonance could have arisen. If the period of Io, P_{Io}, was well below $\frac{1}{2}P_{Europa}$ then Jupiter's tidal influence, acting most strongly on Io, would have pushed it out until P_{Io} was just less than $\frac{1}{2}P_{Europa}$. At this stage the eccentricity of Io's orbit would build up and the resultant rate of loss of energy would prevent any further expansion of Io's orbit relative to that of Europa. However, the tide on Jupiter would be able to drive the coupled satellites, Io and Europa, outwards until P_{Europa} was just less than $\frac{1}{2}P_{Ganymede}$. The three orbits would then be linked as now with the mutual gravitational effects associated with the Lapace triplet system keeping the config-uration stable. In principle the three satellites could move outwards and link up with Callisto but the Solar System will not survive long enough for this ever to happen.

The type of mechanisms described here explain reasonably, if not in detail, how the satellite commensurabilities occur and may also be applied to those that occur in the Saturnian system. However, they cannot be applied to the planetary case that we shall now consider.

7.1.5   Commensurability of planetary orbits

The main near-commensurabilities between planets have been described in sec-tion 1.2.3 together with the 3:2 resonance of several Kuiper-belt objects with Nep-tune, similar to that of Pluto with Neptune. Computational work shows that con-junctions between Neptune and Pluto occur when Pluto is at perihelion. Although the 3:2 ratio of periods is not constant it is the average over long periods of time. The mechanics of the situation ensures that when the periods drift so that Pluto can get closer to Neptune an opposite tendency sets in so that the minimum dis-tance between the two planets increases again.

Another phenomenon that shows commensurability in a negative sense is the formation of the asteroid Kirkwood gaps and the major divisions in Saturn's rings. The basic mechanism here is that the main orbiting body, either Jupiter around the Sun or a close satellite around Saturn *removes* energy from particles in *interior* resonant orbits. Why this might be so for bodies in near-circular orbits with, say, a 2:1 resonance can be illustrated by a general non-mathematical argument. In figure 7.2(*a*) the main body is shown at Q and the minor body, with half the period, is at T. After every complete orbit of Q the bodies will meet at the same position and the overall accumulated perturbation of T by Q is in the direction TX. This means that at the conjunction position, T will be induced to move in a

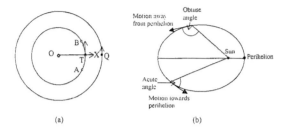

Figure 7.2. (*a*) With a 2:1 resonance, the perturbation on T is always in direction TX. Thus T moves at an obtuse angle to OT. (*b*) When the motion of a planet makes an obtuse angle with the radius vector it is moving away from the perihelion.

direction making an obtuse angle with the orbital radius vector OT. The motion of a body in an elliptical orbit is indicated in figure 7.2(*b*). It is clear that when direction of motion of the orbiting body makes an obtuse angle with the radius vector then it is moving away from perihelion and hence slowing down. Now we return to figure 7.2(*a*) and now imagine that it shows the motion of T relative to Q. During the part AT of the minor body's motion it is gaining energy from Q. During the part TB it is losing energy to Q. However, since it is slowing down it spends longer in going from T to B than in going from A to T so the net effect is a loss of energy. The argument is admittedly a rough one and difficult or impossible to extend to general orbits and complicated resonance ratios but it does illustrate the general principle of what is happening. Thus an asteroid which happened to have one-half of the period of Jupiter would slowly lose energy by moving inwards until it had gone out of resonance, so creating the most prominent Kirkwood gap. The same style of argument can be used to show that, conversely, the main body would *add* energy to a particle in an *exterior* resonance.

There has been considerable work done in studying the behaviour of bodies moving around the Sun in the presence of a resisting medium, which is an initial scenario in any theory of the origin of the Solar System. As we can see in figure 7.1 the semi-major axis of a body reduces with time so that the body is losing orbital energy and the orbit decays. Weidenschilling and Davies (1985) studied resonance trapping of a planetesimals by a planet in the presence of gas drag. If the planetesimal is exterior to the planet then the loss of energy due to gas drag can be balanced by the gain due to the planet and so maintain the planetesimal's orbit without decay. The stability of the orbits and the probability of capture of planetesimals have been investigated by Beaugé and Ferraz-Mello (1993), Malhotra (1993) and Gomes (1995). One possible outcome of this mechanism is that if there was a pre-existing Jupiter then a swarm of planetesimals could be captured in a 2:5 resonance and these could accumulate over time to form Saturn.

Other variants of this basic idea have been studied. The decay of small-particle orbits due to the Poynting–Robertson effect can be substituted for gas drag and this is the mechanism for the capture of a dust ring in close orbit to the Earth (Sidlichovský and Nesvorný 1994). In all the situations mentioned so far an essentially massless particle becomes locked into resonance with a massive perturber and its semi-major axis librates around the commensurate position. Resonance locking of bodies of comparable size has been studied in the case of pairs of satellite orbits evolving under tidal dissipation (Goldreich 1965, Allan 1969, Greenberg 1973).

Melita and Woolfson (1996) numerically investigated a general three-body problem under the action of accretion and a dynamical friction force for different mass ratios and initial conditions. The drag due to the medium was taken as the sum of a dynamical component, (7.9), and an accretion component, (7.10). A constant density of 10^{-11} kg m^{-3} was used for the medium, corresponding to 16 Jupiter masses occupying a spherical volume of radius 50 AU. Calculations in which the masses of the planets changed due to accretion, which required two extra differential equations, showed that, although the masses increased by up to 70%, there was no quantitative difference in the outcome over assuming that the masses remained constant. Since the computations took a long time, covering a simulated time-span of 10^7 years, the economy was made of assuming that the planetary masses remained constant.

The first result obtained by Melita and Woolfson illustrated the characteristics of the mechanism. In this the inner body had the mass, radius and, initially, the present semi-major axis of Jupiter and the outer body the corresponding characteristics of Saturn. In each case the initial eccentricity and inclination of the orbits were 0.1 and 0.06 radians ($\sim$3.4$°$) respectively. Three runs were made, corresponding to different relative initial positions in the orbits. Figure 7.3 shows the changes over time of the semi-major axis, eccentricity and inclination for 'Jupiter' and 'Saturn' and figure 7.4 the evolution of the ratio of periods. The ratio departs from 5:2 and settles down close to 2:1, actually oscillating about 2.02. For both bodies the eccentricities initially fall, as indeed was found by Dormand and Woolfson (1974), but once the resonance is established the eccentricities rise again. The effect of rising eccentricity is to increase the relative velocity of the planet relative to the medium and hence to increase the rate of dissipation of energy. In the case of Saturn this ensures that the *net* loss of energy, including the gain due to perturbation by Jupiter, gives an orbital decay which keeps it in resonance with Jupiter. In the case of Jupiter the rise of eccentricity causes an increase in energy dissipation, seen as an increase in the magnitude of da/dt. It is clearly a complex pattern of adjustment but the ratio of periods is a stable feature of an otherwise dynamically evolving system.

The result of changing the initial eccentricities was also investigated by Melita and Woolfson. Each planet had an initial eccentricity of 0.1, 0.2 or 0.3 and all nine combinations were examined. Figure 7.5 shows the changes of ratio of periods and of the eccentricities of the two planets. Some combinations retain the 5:2 commensurability while others evolve to 2:1. Other numerical experiments by Melita and Woolfson with the Jupiter–Saturn model showed that,

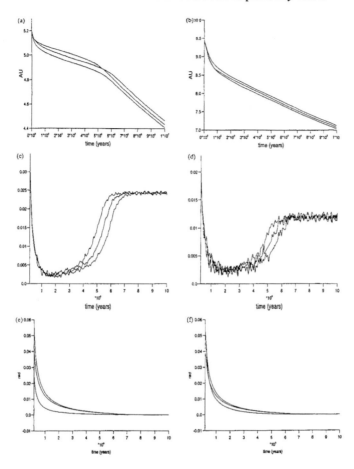

Figure 7.3. The variations of the semi-major axes, eccentricities and inclinations for a Jupiter–Saturn system starting near the 2:5 resonance. Semi-major axis—(*a*) Jupiter, (*b*) Saturn. Eccentricity—(*c*) Jupiter, (*d*) Saturn. Inclination—(*e*) Jupiter, (*f*) Saturn.

starting with the 5:2 commensurability, it was only retained if the internal body is the more massive. It was also found that, depending on the starting conditions, other ratios could be reached, e.g. 3:1.

The result of an examination of the Uranus–Neptune system is shown in figure 7.6. The system started in its present configuration, and ended with a period ratio of 2.01. This result is interesting since the outer planet, Neptune, is slightly more massive than Uranus.

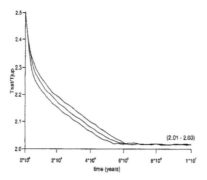

Figure 7.4. The ratio of the periods for a Jupiter–Saturn system starting near the 2:5 resonance. The curves are for different initial relative starting positions.

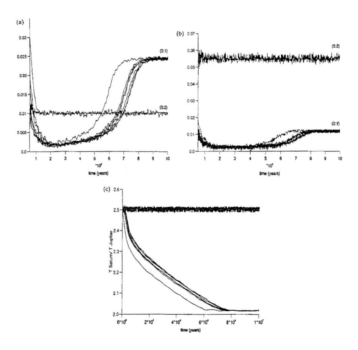

Figure 7.5. The effect of different pairs of initial eccentricities for the Jupiter–Saturn system. The numbers in parentheses represent the resonance to which the system evolved. (a) Eccentricities for Jupiter. (b) Eccentricities for Saturn. (c) The ratio of periods.

7.1.5.1 Relevance of resonance locking to the present Solar System

The resonance ratios that occur between the major planets are all less than the exact integer or half-integer ratios, i.e. Saturn–Jupiter, 2.48; Uranus–Saturn, 2.85;

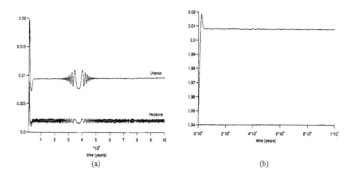

Figure 7.6. Resonance of the Uranus–Neptune system started near the present ratio of periods. Resonance is obtained despite the fact that Neptune is slightly more massive than Uranus. (*a*) Eccentricities. (*b*) Ratio of periods.

and Neptune–Uranus, 1.96. On the other hand all the final ratios found by Melita and Woolfson from their numerical work oscillated about values that were exactly commensurable or slightly higher—2.01, 2.51 or 3.00. To produce results that would be relevant to the major-planet system it would be necessary to consider the whole system of major planets together. The maximum gravitational attraction of Jupiter on Uranus and also on Neptune is greater than that of Saturn, and greater than that of the two outer planets on each other, so considering the system as isolated neighbouring pairs of planets is not really valid. Another deficiency in the Melita and Woolfson simulations was that they *began* with semi-major axes corresponding to the present values rather than ending with those values. Again, the assumption of a uniform time-invariant density over the whole region of the Solar System was not a very realistic one. A final uncertainty is the effect of point-mass interactions after the gaseous medium had disappeared.

  Despite the inadequacies of the resonance-locking analysis it does indicate that an initially non-resonant system evolving in a resisting medium will be induced into a more resonant form. The 'great' resonance between Jupiter and Saturn, the most massive planets, will be least disturbed by the other major planets and the small departure from 5:2 could well be explained by other disturbances of the types previously referred to.

7.2   Initial planetary orbits

The mechanisms of planet formation for the various theories have been described in chapter 6 and this was sometimes intimately bound up with where the planets formed—as for the Modern Laplacian Theory. Here each of the present theories, with the exception of the Modern Laplacian Theory dealt with in section 7.1.3,

Figure 7.7. The dominance regions for planets n and $n + 1$ of lengths αr_n and αr_{n+1}.

will be examined in respect to what they suggest for the sequence of planetary orbits.

7.2.1   The Accretion and Solar Nebula Theories

The Accretion and Solar Nebula Theories both postulate the formation of planets from initially diffuse material through the intermediate process of producing planetesimals. The question of *where* the planets grow has not been explicitly addressed. Safronov's analysis (section 6.4.3.3) indicates that there will be one dominant mass, i.e. planet, forming in *each region* but what constitutes a region is not specified. Here we offer a possible pattern that could give something like the present distribution of planetary orbits.

The radius of the sphere of influence of a body orbiting the Sun is given by (7.8) and is seen to depend both on the distance from the Sun and also on the mass of the body. If we take each of the embryonic planets to have had the same mass then their spheres of influence would have been proportional to their distance from the Sun. The assumption of equal mass is a severe one but is softened somewhat by the one-third power that appears in (7.8). The assumption is now made that the radius over which the planetary embryo is dominant is proportional to its sphere of influence and hence to its distance from the Sun. In figure 7.7 we show the contiguous regions dominated by planets n and $n + 1$. We may write

$$r_{n+1} - r_n = \tfrac{1}{2}\alpha(r_{n+1} + r_n) \tag{7.16}$$

where the dominance region for planet n has width αr_n. This can be transformed to

$$\frac{r_{n+1}}{r_n} = \frac{1 + \tfrac{1}{2}\alpha}{1 - \tfrac{1}{2}\alpha} = \beta. \tag{7.17}$$

Taking $\beta = 1.73$ gives the match with Solar System values shown in table 7.2. Because of orbital decay in the medium in which the planets form and the resonance-locking mechanism it would be more appropriate to have initial orbital radii considerably greater than the present ones. However, perhaps all that can be expected is that a theory should give orbital radii on the right scale and increasing approximately in a geometric progression. It seems quite plausible that, from such a beginning, evolutionary factors will give the present system.

Table 7.2. Equation (7.17) matched to the actual planetary radii, r_{SS}, with $\beta = 1.73$.

	Mercury	Venus	Earth	Mars	Ceres	Jupiter	Saturn	Uranus	Neptune
Equation (7.17)	0.37	0.65	1.1	1.9	3.3	5.8	10.0	17.3	30.0
r_{SS}	0.39	0.72	1.0	1.5	2.8	5.2	9.5	19.2	30.1

7.2.2 The Proto-planet Theory

Although it was not part of the theory as presented by McCrea (1988) it is possible that one proto-planet body could become gravitationally dominant in a particular region surrounding the newly-formed Sun giving rise to something like the pattern in table 7.2. The situation is nothing like as clear as for the accumulation of planetesimals that are on orbits of modest eccentricity when they are accreted. The blobs in McCrea's theory are much larger and traverse the whole system.

The main concern with the Proto-planet Theory is not whether it gives a particular pattern of planetary orbits, corresponding to commensurabilities or Bode's law, but rather whether the overall scale of the system will correspond to what we have in the Solar System. The only reason for considering that planets will form at all with realistic orbits is that so little of the angular momentum in a star-forming region goes into forming the star itself. The assumption is that the only other way that it can be taken up is in planets. That this is so is not at all clear. This matter is discussed in more detail in section 7.3.1.

7.2.3 The Capture Theory

The tidal disruption and capture process postulated by the Capture Theory can be investigated by a very simple model. Where the ratio of primary-to-secondary mass is more than three or so, which would probably apply to the Sun and proto-star, the tip of the bulge of the limiting equipotential surface, as shown in figure 4.12, is distant from the centre of the proto-planet about 1.4 times the original radius. We may take the proto-planet blobs as originating within the tidal bulge which eventually stretches out into a filament.

We now consider a proto-star of mass $\frac{1}{4}M_{\odot}$ and radius, $R = 21$ AU (density 3.8×10^{-9} kg m^{-3}) in a hyperbolic orbit around the Sun with perihelion distance 30 AU and eccentricity, $e = 2$. Starting with the proto-star with true anomaly $-90°$ and the tidal bulge pointing towards the Sun calculations have been made of the path of particles, representing potential proto-planets, at different distances, D, from the proto-planet centre. For a range of values of D/R the particles are captured by the Sun. The result of a calculation with $D/R = 1.36$ is shown in figure 7.8. After its release from the proto-star the particle moves into an elliptical orbit with $a = 99.1$ AU and $e = 0.882$. Its initial motion in an orbit with a period of nearly 1000 years is towards aphelion, which allows ample time for planetary

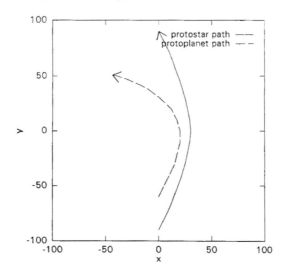

Figure 7.8. Motions of the proto-star and proto-planet for $D/R = 1.36$. The Sun is at the origin.

condensation, as described in section 6.4.2.5.

The orbital round-off process described in section 7.1.2 depends on the distribution of the resisting medium. If it has a very high central condensation then it will tend to round off the orbit to a radius equal to the original perihelion distance. For a rather improbable distribution, with an increase of density with distance from the Sun, the round-off would tend to be towards the aphelion distance. However, for the two distributions giving figure 7.1, where the initial orbit had semi-major axis 18 AU and eccentricity 0.8, the round off is approximately to an orbital radius equal to the geometric mean of the perihelion and aphelion distances, $a(1 - e^2)^{1/2} = 10.8$ AU.

In figure 7.9 the results of calculations for different values of D/R are shown. For D/R less than 1.115 the particle is not captured but is retained by the proto-star. The semi-major axis and eccentricity of the particle orbit as a function of D/R is shown in figures 7.9(a) and (b) and the semi-latus rectum and $a(1 - e^2)^{1/2}$ in figures 7.9(c) and (d). Taking the geometrical mean as a measure of the final rounded-off orbit this is seen to vary between 9 and 120 AU, but different models can easily give results scaled down to correspond more closely to the distances in the Solar System.

If proto-planets came from equally spaced parts of the original tidal bulge then it can be seen from figure 7.9(d) that there would be something like a geometrical progression of orbital radii, which is an acceptable starting condition. It should also be noted that since the resonance-locking process described by Melita and Woolfson requires the planetary orbits to decay the scale of the orbits as indi-

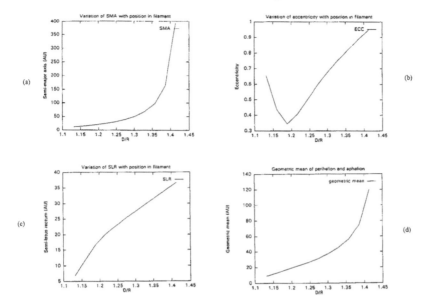

Figure 7.9. Variations with position in filament of: (*a*) the semi-major axis; (*b*) eccentricity; (*c*) semi-latus rectum; and (*d*) geometric mean of the perihelion and aphelion distances.

cated by figure 7.9(*d*) could be consistent with what would give the present orbits of the major planets.

7.3 Angular momentum

Inevitably in dealing with the formation of the Sun, planets and satellites the angular momentum associated with these types of body has been considered, if only implicitly. Here we shall pull together the strands of previous references to angular momentum applying to the Sun, the planets and satellites for the different theories.

7.3.1 Angular momentum and the Proto-planet Theory

This is the one theory that links the process by which planets are made to the process that forms not just the Sun but a whole cluster of stars. If the whole cloud of blobs, i.e. that which is to form a cluster of stars, is considered then the application of the Virial Theorem just to a small part of it, as is done in (6.19), is invalid. The mass of the cloud may be, say, 1000 times that of the Sun and the radius 10 times that assumed in (6.19). This implies that the root-mean-square speed of the blobs is ten times that estimated by McCrea (1988) relative to the centre of mass of the total cloud, or about 100 km s^{-1}. Thus blobs will not be bound into any particular region and the model of isolated and independent regions, which is at the heart of the Proto-planet Theory, will not be valid. Blobs joining a growing proto-star will have come from anywhere in the total cloud and their contribution to the angular momentum of the star will be a little larger than suggested by (6.4b). However, since McCrea took the blobs as joining the proto-Sun at a speed corresponding to the escape speed from the present Sun, over 600 km s^{-1}, this will make little difference. The satisfactory estimate of the spin angular momentum of the Sun is unaffected.

If the original angular momentum of the cloud was zero, which could not actually be so, then the total angular momentum will have to be zero at all subsequent times. The contributions to the total angular momentum are:

(i)   the spins of the stars that form;
(ii) the relative motions of stars; and
(iii) the motions of planets around stars.

We now consider a cloud with mass $1000M_\odot$ and radius 7.8×10^{12} m, which is ten times the value given by McCrea for a single star-forming region. For a star formed from 3000 blobs, the root-mean-square momentum it acquires is

$$p_* = \sqrt{3000}mV = M_*V/\sqrt{3000} \tag{7.18}$$

where m is the mass of a blob, V its speed, assumed to be the same for all blobs, and M_* is the mass of the star. The angular momentum of the star about the centre of mass of the cloud is $p_* r \sin\theta$ where r is the distance of the star from the centre of the cloud and θ is the angle between the radius and momentum vectors. For a system of 1000 stars the root-mean-square of the expectation value of the total angular momentum is

$$H_{\text{stars}} = \sqrt{1000p_*^2\langle r^2 \sin^2\theta\rangle}. \tag{7.19}$$

The average values of $\sin^2\theta$ and r^2 are 0.5 and $0.8(R_{\text{cloud}})^2$ respectively. From this the angular momentum associated with the motion of the stars is 6.7×10^{47} kg m^2 s^{-1}. Now we suppose that each of the 1000 planetary systems formed has its angular momentum exactly anti-parallel to that of the stellar motion. Then each planetary system would have to have angular momentum of magnitude 6.7×10^{44} kg m^2 s^{-1} or about 25 times that of the Solar System.

This approximate calculation is not meant to give applicable results. What it shows is that the angular momentum associated with the stellar motions must be constrained by conservation to be much less than the statistically derived figure of 6.7×10^{47} kg m^2 s^{-1}. Alternatively, on average the individual planetary systems would have to have much more angular momentum than the Solar System. What is clear is that the scale of the potential variation in the angular momentum contained in the stellar motions makes assessment of the probable angular momentum associated with the planetary motions quite impossible. For this reason it is not possible to make any meaningful analysis that would indicate the scale of the planetary orbits. However, that does not necessarily rule out the Proto-planet Theory as a possible way of forming the Solar System.

The way in which satellites are produced, by the rotational instability of a rapidly spinning planet, is likely to give the kind of relationship between the angular momentum of a planet and its satellites that was noted by Williams and Woolfson (1983).

7.3.2 Angular momentum and the Modern Laplacian and Solar Nebula Theories

The angular-momentum problem for these nebula-based theories is not just to produce the Sun spinning reasonably slowly but that of actually producing the Sun at all. If sufficient angular momentum cannot be removed from the material moving inwards then it will simply produce a very extended spinning contracted nebula at the edge of rotational stability.

A simple analysis, the result of which was given in section 6.3.1.2, will confirm this point. The extreme assumption is made that the Sun is formed in its *present condensed state* by material spiralling inwards in the equatorial plane. As it joins the Sun it spreads itself out to maintain the growing Sun as a sphere. These assumptions describe a quite impossible situation but they err in the direction of predicting a lower angular momentum than any realistic physical event could give.

To model the distribution of matter in the Sun we take an analytical form

$$\rho(r) = \rho(0)\cos^n\left(\frac{\pi r}{2R_\odot}\right) \tag{7.20}$$

where r is the distance from the centre of the Sun and $\rho(0)$ the density at the centre. The moment of inertia factor for such a distribution is

$$\alpha = \frac{\int_0^{R_\odot}\cos^n\left(\frac{\pi r}{2R_\odot}\right)r^4\,\mathrm{d}r}{\int_0^{R_\odot}\cos^n\left(\frac{\pi r}{2R_\odot}\right)r^2\,\mathrm{d}r}. \tag{7.21}$$

For $n = 13$ a simple numerical calculation gives $\alpha = 0.055$ which is the accepted value for the Sun. The value of $\rho(0)$ giving the correct mass of the Sun is then 7.526×10^4 kg m^{-3}.

The mass contained within a spherical surface of radius r is

$$M(r) = 4\pi\rho(0)\int_0^r x^2\cos^{13}\left(\frac{\pi x}{2R_\odot}\right)\mathrm{d}x \tag{7.22}$$

and the total angular momentum is

$$H = 4\pi\rho(0)\int_0^{R_\odot}\sqrt{GM(r)r^5}\cos^{13}\left(\frac{\pi r}{2R_\odot}\right)\mathrm{d}r. \tag{7.23}$$

Numerically it is found that $H = 2.2 \times 10^{44}$ kg m^2 s^{-1} which is about 1500 times that of the present Sun. This must be a gross underestimate by at least a factor of 10—although it assumes that the Sun could actually form. Even accepting that magnetic braking, as described in section 6.2.1, can give some reduction over time it is clear that most of the angular momentum must be removed from the material while it is on its way inward to join the Sun. Thus it must not join the growing Sun tangentially; on the contrary it must move almost radially with a very small tangential component. The only effective mechanism for angular momentum removal that has been suggested, magnetic braking, requires an implausibly high early magnetic moment and rate of loss of fully-ionized material. To this must be added the complication of requiring the ionized material to move outwards while at the same time normal material is moving inwards to join the Sun.

The Modern Laplacian Theory is in a somewhat better position than the Solar Nebular Theory in that it removes so much angular momentum initially by the assumption of forming the Sun from solid hydrogen. Nevertheless, in forming the Sun it must still avoid the problem of having material spiralling inwards and the many rings which are postulated to form within the orbit of Mercury are part of this problem.

The Solar Nebula Theory postulates that the process of satellite formation is a small-scale version of the formation of planets but few details are available. Given that a giant proto-planet is formed by an accumulation of planetesimals followed by the acquisition of a gaseous envelope then it is not clear how satellites come into being. If the matter and angular momentum of the present satellites were subsumed into their parents then these planets would spin very little faster and be nowhere near rotational instability. How then would they leave matter behind on their way to becoming compact bodies? If, on the other hand, they once were rotationally unstable and left a disc behind in which satellites formed then how did they lose so much of the angular momentum they once contained?

Prentice has given more details about satellite formation and assumes that supersonic turbulence plays the same role for satellite formation as it does for planet formation. It seems unlikely that the mechanisms postulated for planetary formation would be applicable to the very much smaller system where the energy available, for example from the collapse of proto-planets, would be much smaller in proportion to the total mass than that available in the Sun

7.3.3   Angular momentum and the Capture Theory

In this theory the slow spin of the Sun comes from the mode of formation by the collision of turbulent elements and is therefore not directly linked to the process of forming planets. The angular momentum contained in the planetary orbits is derived from the intrinsic angular momentum of the Sun–proto-star orbit; the perihelion distance of this orbit dictates the scale of the system and hence, *ipso facto*, the angular momentum in the planetary orbits.

Satellites are produced by a completely different mechanism within the filament left behind by a collapsing distorted proto-planet. The intrinsic angular momentum of Io in its orbit is only eight times that of Jupiter's equatorial material. That of Mercury in its orbit is 2000 times that of the Sun's equatorial material. Taken in conjunction with the information in tables 6.6 and 6.7 it makes it more likely than not that the mechanisms should be different—despite Jeans' and Alfvèn's assertions to the contrary (section 6.5). Indeed, any theory that assumed that the basic mechanisms were the same would require an auxiliary theory to explain the differences in the present systems. The Capture-Theory model for satellite formation by Williams and Woolfson (1983) describes satisfactorily the relative angular momentum in the planetary spin and satellite orbits. The basic mechanism is similar to that suggested by Jeans and this was a part of the Jeans tidal model that was never in dispute. Since he had the same mechanism for forming planets they had no way of acquiring enough angular momentum—which was the basis of Russell's damaging criticism of the tidal model.

7.3.4   Angular momentum and the Accretion Theory

As for the Capture Theory, the Accretion Theory avoids the problem of forming the Sun with low angular momentum by separating that event from planetary formation. The parameters of the model, for example as given by Aust and Woolfson (1973), ensure that there is sufficient angular momentum in the captured material so the only problem is that of organizing that material into the form of planets. The theory has little to say on satellites but, if planets formed in eccentric orbits, one could envisage a similar process of formation as described for the Capture Theory.

7.4   The spin axes of the Sun and the planets

The tilt of the solar spin axis to the mean plane of the system, mostly quoted as 7° but actually 6°, is not taken into account by most theories. It is an awkward quantity, too large to consider as a good approximation to zero yet too small to be comfortably a matter of chance. The probability that two random vectors are inclined to each other at an angle β or less is $\frac{1}{2}(1 - \cos \beta)$ or 0.0027 for $\beta = 6°$. So it is incumbent on theories, according to their nature, to explain either why the angle is not exactly zero or why it is so small.

The tilts of the planetary spin axes are very variable and are shown in table 7.3.

Three of the nine tilts, those of Venus, Uranus and Pluto, correspond to retrograde spin. The smallest tilts in a direct sense are those of Mercury, certainly due to solar tidal influence, and of Jupiter. We shall now see what the various theories have to suggest for explaining the tilts of the planetary and solar spin axes.

Table 7.3. The tilt of the planetary spin axes to the normal to their orbital planes.

					Planet				
	Mercury	Venus	Earth	Mars	Jupiter	Saturn	Uranus	Neptune	Pluto
Tilt	$0°$	$178°$	$23°27'$	$25°12'$	$3°7'$	$26°45'$	$97°53'$	$28°48'$	$118°$

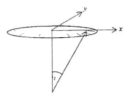

Figure 7.10. The 'cap' defines the limits of the inclinations of planetesimal orbits. It is almost planar and the x- and y-axes are in the approximate plane.

7.4.1 Spin axes and the Solar Nebula Theory

The Solar Nebula Theory defines a very planar system and there is no mechanism within the nebula itself for disturbing that pattern. Given that the Sun contains most of the mass of the original nebula it would certainly have its spin axis normal to the mean plane of the nebula, as defined by its spin. Several different possibilities have been raised for the mass of the disc which surrounded the newly-formed Sun, varying from $0.01 M_\odot$ to $0.1 M_\odot$; more massive discs, up to a solar mass, are no longer seriously considered. Since the disc is required to contain almost all the angular momentum of the original nebula then, as for the Sun, its spin axis must be parallel to that of the original nebula.

Even with the least massive discs it is required that anything from 85% to 98.5% of the mass of the disc must eventually be lost and the question then arises as to whether what is left can be unrepresentative of what is lost in terms of its angular-momentum vector. In that case we are considering not so much the tilt of the solar spin axis but the mean tilt of the planetary orbits. We can explore this possibility with a model, admittedly crude but nevertheless indicative of the probability of such a tilt.

Taking a 'heavy' disc of mass $0.1 M_\odot$ it would contain about 3×10^{27} kg of solid material. If this was all organized into planetesimals of radius 50 km they would each have a mass of 10^{18} kg and there would be 3×10^9 of them. Assuming that the total mass of the terrestrial planets plus the cores of the major planets is about $30 M_\oplus$ then approximately 2×10^8 planetesimals are required. This is many fewer than the total number so we may assume that the angular momentum associated with the planetesimals forming planets or cores is dictated by statistics rather than by conservation laws.

The mutual scattering of planetesimals will give a range of directions of their angular momentum vectors although with an average tilt of zero. The maximum tilt is not likely to be large and from what we know about asteroids is likely to be within about 30°. The tilt vectors will then be constrained within the range of directions shown in figure 7.10. For tilts less than 30° is is a reasonable approximation to take the cap-shaped region as a plane and we take two axes x and y in that plane. The distribution of tilts is modelled in Gaussian form, the x and y directions as

$$p(x) = \frac{1}{\sqrt{2\pi\sigma^2}} \exp\left(-\frac{x^2}{2\sigma^2}\right) \qquad (7.24a)$$

and

$$p(y) = \frac{1}{\sqrt{2\pi\sigma^2}} \exp\left(-\frac{y^2}{2\sigma^2}\right). \qquad (7.24b)$$

The combination of these two distributions gives an axially symmetric Gaussian distribution about the mean direction with standard deviation σ. If n vectors are chosen at random then the average value of x will have a Gaussian distribution with mean zero and standard deviation $\sigma/n^{1/2}$. The same applies in the y direction so the distribution of the tilt of a combination of n vectors is Gaussian with mean zero and standard deviation $\sigma_n = \sigma/n^{1/2}$. With $n = 2 \times 10^8$ and $\sigma = 30°$ (it should be smaller for almost all tilts to be within 30°) this gives $\sigma_n = 0.002°$. It is clear that no random selection of planetesimals can give the required angle of 6°.

Another possibility, suggested by Tremaine (1991), is that after the planetary system formed a star passed close to the Sun, changing the plane of the planetary orbits but without affecting the more compact Sun. This is certainly dynamically possible. The problem with this scenario is that the orbits of the outer planets would certainly have been greatly disturbed and Neptune, in particular, could not have been left with its very small present eccentricity (0.0086). However, if the disturbance to Neptune's orbit had not been too great and the resisting medium was still present then perhaps it could have been rounded off subsequent to the stellar passage.

The tilts of the planetary axes for terrestrial planets or planetary cores do not present any difficulty to theories that involve the accumulation of planetesimals. The sizes of the bodies collecting together would increase with time and the collision of one or a few large bodies in the last stages of accumulation could tilt the axis in almost any direction. However, when the core of a major planet acquired a gaseous envelope then the addition of so much material, with a circulation $\nabla \times v$ corresponding to that of the nebula, would seem inevitably to pull the spin axis towards the normal. The tilt of Jupiter's spin axis is acceptable but tilts of more than 25° are difficult to explain for the major planets in this scenario.

7.4.2 Spin axes and the Modern Laplacian Theory

The problem of the solar spin axis is more challenging for the Modern Laplacian Theory than for any of the other theories under review. The central core must have its spin axis in the same direction as the original nebula and planets are formed within rings that are strictly in the mean plane of the system. One is left with the possibility of a stellar passage after the system has formed—but, as indicated previously, that also raises new problems.

The accumulation of material within a ring to form a planet would have all the same characteristics as planet formation with the Solar Nebula Theory. Again the cores of the major planets, when they formed, could have had a variety of spin directions but the addition of the gaseous component would pull the axis towards the normal.

7.4.3 Spin axes and the Accretion Theory

For the Accretion Theory there would be no special relationship between the direction of the spin axis of the Sun and the mean plane of the captured material. Although it might be expected that the process of capturing material from an interstellar cloud would give a rather turbulent nebula it would soon settle into a more quiescent form. The process by which planets formed would be similar to that for the Solar Nebula Theory, via planetesimals, and the problems mentioned in section 7.4.1 would also apply here.

As stated in section 6.2.2 there is no solar-spin-axis problem for the accretion theory. The addition of solid material to the Sun, drawn in from the surrounding material by the Poynting–Robertson effect, would pull the resultant solar spin vector towards the normal to the mean plane. It is not possible to estimate how much material would have to be added to the Sun to explain what is observed today because there are many possible scenarios. From table 6.1 it is reasonable to suppose that the Sun lost more than one-half of its original angular momentum from magnetic braking in the first few million years. If the material drawn in by the Poynting–Robertson effect accounts for the present component of the spin angular momentum normal to the mean plane then, allowing for subsequent loss, this would imply an addition of about one-half of a Jupiter mass of solid material. Assuming that 2% of the surrounding material was in the form of solids and that most of the solid material joined the Sun (only one-tenth of a Jupiter mass is required for terrestrial planets and cores of major planets) then the mass of the surrounding medium would have been $25M_{\text{Jupiter}}$ or $0.025M_{\odot}$. Since most of the gaseous component would have to be expelled from the system we should also examine how long this might take. Hoyle (1960) estimated that due to solar heating and evaporation a considerable proportion of a gaseous nebula would be lost over a period of 3×10^7 years but he was considering a rather massive nebula. To remove the amount of gas we are considering here would require very little of the solar radiation over a period of a few million years, the observed lifetime of discs. If the nebula mass was uniformly spread out over a spherical volume of radius 50 AU then its mean density would have been $\sim 3 \times 10^{-11}$ kg m^{-3}. The total energy required to remove it from the Sun would be just over 10^{36} J which is the output from the present Sun in about 100 years. Thus if the efficiency of converting solar energy into escape energy of the material was only 0.01% it could be expelled in one million years. There are many uncertainties in this rough analysis, especially concerning the distribution of the material surrounding the Sun but there can be little doubt that the medium can be dispersed over a period corresponding to the likely lifetime of observed discs.

To summarize, the Accretion Theory has no difficulty with explaining either the tilt of the solar spin axis or that of the individual planets.

7.4.4   Spin axes and the Proto-planet Theory

As for the Accretion Theory the Proto-planet Theory would not have any preferred direction of solar spin relative to the planes of the planetary orbits. In this theory background material not forming planets would have constituted a resisting medium the action of which would have been twofold—first, modifying the planes of the planetary orbits and second, the spin axis of the Sun. If planets were formed in both direct and retrograde orbits then their evolutionary patterns in the resisting medium could have been quite different. For planets in retrograde orbits the resistance in a directly rotating medium would have been enhanced and they might have rapidly spiralled in towards the Sun. If the total mass joining the Sun in this way had been more than one or two Jupiter masses then this would have given the Sun too much angular momentum—although in a retrograde sense. The planets in direct orbits would have rounded off with reduced inclinations as described in section 7.1.2. Since the Proto-planet Theory can give planetary orbits with all possible inclinations then the question of what happens to orbits with inclinations not far from 90° arises. Those orbits that are borderline retrograde could be flipped over into a direct sense as the proto-planets capture material from the medium. It is even possible that fairly extreme retrograde orbits could flip over, depending on the relative rates of orbital decay and of capturing material. Because of the totally random nature of the processes in this theory it is difficult to assess whether or not a well-ordered set of planets and a solar spin axis slightly inclined to the normal to the mean plane could be an outcome, but it is possible and cannot be excluded.

The proto-planets formed by condensations in a turbulent medium would be expected to have had a wide range of spin axis directions. The subsequent capture of material, or interaction with the resisting medium, could have given the bias towards direct spins which is actually observed.

7.4.5 Spin axes and the Capture Theory

The scenario for the solar spin axis that was described in relation to the Accretion Theory fully applied to the Capture Theory as well. A solar spin axis making a small angle to the normal to the mean plane is a natural outcome of this theory.

The orbital planes of the proto-planets in the Capture Theory are close to the plane of the Sun–proto-star orbit. The spin of the proto-star, in some general direction, could throw proto-planet material out of that plane but a five or six degree spread in the orbital planes is all that could be expected from this source. In section 6.5.3 the process of satellite formation which involved a tidal interaction between the Sun and the newly-formed proto-planet was described. This would give rise to a planetary spin axis normal to the orbital plane so the observed departures from this condition require explanation.

We have noted in section 7.12 that the orbits round off in a resisting medium but that is not the only thing they do. Since the resisting medium forms a flattened system the gravitational force it exerts on the planet is not centrally directed. This gives a torque (a vector quantity) on the rotating system that is not parallel to the orbit axis and hence gives rise to a precession of the orbit axis. This can be described in terms of the rates of change of the argument of the perihelion, ω, and the longitude of the ascending node, Ω (section 1.2.1) and the changes in these quantities for the model which gave rise to figure 7.1 are shown in figure 7.11. The rates of precession would have been different for the different proto-planets since, as for a gyroscope, it would have depended on the characteristics of the orbit and the imposed torque. Consequently the relative positions of pairs of orbits will vary with time and, from time to time, will give orbits passing close to each other or even crossing. Proto-planets that pass close to each other while they are still extended objects, although compact enough to be well outside the solar Roche limit, will impart to each other considerable spin angular momentum. This will be perpendicular to the plane of the relative motion of the two proto-planets when they are closest. We can see from (1.8) that the tidal effects exerted by one body upon another are proportional to the mass and inversely proportional to the cube of the distance. Thus Jupiter at a distance of 0.1 AU gives a similar tidal effect to the Sun at 1 AU.

We now give an illustration of the way in which tilted planetary spin axes can arise by taking the example of Uranus, the spin axis of which is tilted at $97.9°$ to the normal to its orbit. The orbit of Uranus, soon after its formation and when its radius was 0.25 AU, is taken as having semi-major axis 35.6 AU and eccentricity 0.69 (see table 8.1). The planet is modelled by a cubical distribution of 203 particles within a sphere, each with the same mass, $M_{Uranus}/203$, so that the density is uniform. The gravitational forces on the particles are found numerically and a constant pseudo-pressure force is applied to each particle that opposed a fraction 0.999 of the gravitational force. This means that an isolated model planet would be able to collapse slowly. In fact we are interested in the behaviour of the planet for such a short time that it is acceptable to consider it as being in a state of quasi-

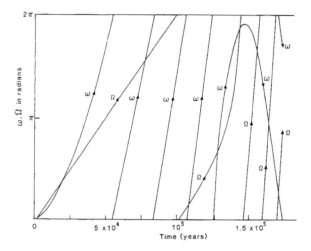

Figure 7.11. Variations of the argument of perihelion, ω, and the longitude of the ascending node, Ω, during a planetary round-off.

equilibrium. The model Uranus interacts with a model point-mass Jupiter on an orbit with $a = 14.8$ AU and $e = 0.826$, similar to the parameters in table 8.1, which passes over Uranus with nearest approach 4.6 Uranus radii at a distance of 21.2 AU from the Sun (true anomaly for Uranus 100°). However, Jupiter passes through the point of closest approach to Uranus's orbit some 9×10^6 s before Uranus passes through the closest-approach point of *its* orbit. It can be seen that there are many parameters that can be varied in this model.

Initially the model Uranus is set spinning about an axis normal to its orbital plane so that the total angular momentum is 3.00×10^{36} kg m^2 s^{-1}. From table 6.9 this is seen to be more than the present spin angular momentum of Uranus but what is expected taking satellite formation into account. Actually the final outcome of the simulation is very insensitive to this initial angular momentum. The interaction between Jupiter and Uranus changes the orbit of Uranus to $a = 37.85$ AU, $e = 0.707$ and $i = 2.3°$, a small change from the original orbit. However, the tilt of the spin axis is at 98.7° to the new orbit and the new magnitude of the angular momentum is 3.00×10^{36} kg m^2 s^{-1}, just what it was originally.

It is not being argued that what is given here indicates what actually occurred early in the evolution of the planetary orbits. The parameters giving this outcome were easily found and it is probable that quite different parameters could have given a similar outcome. What *is* significant is that interactions between the early proto-planets, while their orbits were rounding off, could influence their spin axes without at the same time substantially changing their orbits. In terms of this interaction process all the spin axes seem readily explained, except perhaps that of Venus which will be dealt with in section 8.2 and Pluto which is anomalous in many other respects. The tilts of Jupiter's orbit and the Uranus post-interaction orbit are 4.5° and 2.3° with respect to the original orbit of Uranus. This means that the near co-planarity of the system predicted by the Capture Theory is not changed by the close tidal interactions.

There are no problems in explaining either the spin axis of the Sun or those of the planets in terms of the Capture Theory. The processes described for the Capture Theory, Poynting–Robertson accretion of solid material by the Sun and interactions between early proto-planets in near-co-planar orbits, would apply to any other theory giving a similar initial state.

Chapter 8

A planetary collision

8.1 Interactions between proto-planets

In section 7.4.5 there was a description of an interaction between two proto-planets, one representing Uranus and the other Jupiter, which gave Uranus an axial tilt similar to that observed today. The interaction was very strong in a tidal sense but actually made only minor modifications to the orbits of the involved bodies. The reason that such an interaction had a fairly high, rather than a very low, probability is that the evolving non-co-planar orbits of the proto-planets precessed, thus constantly changing their relative configurations.

Although Uranus is an extreme case there are also substantial tilts of other planetary spin axes, which suggests that tidal interactions were the rule rather than the exception during orbital round-off. This raises the question of the possibility of more extreme interactions that could lead to major orbital changes or even to a direct collision between planets. There now follows a theoretical treatment of this matter as given by Dormand and Woolfson (1977).

8.1.1 Probabilities of interactions leading to escape

In figure 8.1 the relative orbit of a proto-planet of mass M_2 relative to one of mass M_1 is shown. Initially it is assumed that the orbits are co-planar but later that condition will be relaxed. The approach speed of M_2 is V with impact parameter D but due to mutual gravitational effects the closest approach distance is R. The major change in the hyperbolic orbit of M_2 takes place in a region so small that perturbation by the Sun can be ignored. The relative direction of motion of the two bodies is changed by an angle β (figure 8.1) where

$$\tan \frac{1}{2}\beta = \frac{G(M_1 + M_2)}{V^2 D}. \tag{8.1}$$

We now transform the description of the motion so that it is relative to the Sun. This can be followed by reference to figure 8.2(a). The velocities of

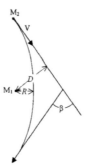

Figure 8.1. The hyperbolic orbit of a planet of mass M_1 relative to one of mass M_2 with impact parameter D and closest approach distance R.

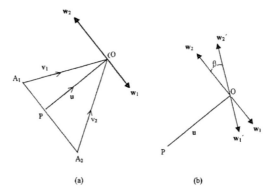

(a)   (b)

Figure 8.2. (a) The velocities of the interacting planets, v_1 and v_2, resolved into the velocity of their centre-of-mass, u, and their velocities relative to the centre-of-mass, w_1 and w_2. (b) Rotation of w_1 and w_2 by an angle β to give w_1' and w_2'.

the bodies relative to the Sun are v_1 and v_2, represented by A_1O and A_2O respectively. The velocity of the centre-of-mass, u, is represented by PO where $A_1P : PA_2 = M_2 : M_1$. The line A_2A_1 represents V, the velocity of M_2 relative to M_1. The vectors w_1 and w_2, equal to A_1P and A_2P, are the velocities of M_1 and M_2 relative to the centre-of-mass of the two bodies. From this we see

$$w_1 = \frac{M_2}{M_1 + M_2} V \qquad (8.2a)$$

$$w_2 = \frac{M_1}{M_1 + M_2} V \qquad (8.2b)$$

$$v_1 = u + w_1 \qquad (8.2c)$$

$$v_2 = u + w_2. \qquad (8.2d)$$

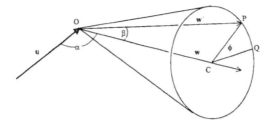

Figure 8.3. Rotation of the velocity w of a planet by an angle β to give w'. The locus of the possible vectors w' lies on the surface of a cone of semi-angle β with w along the axis. The point P lies on a circle and CQ is a radius of the circle in the plane of interaction of the two planets.

After the interaction, which is assumed to take place in such a short time that the bodies will not have moved very far relative to the Sun, the vectors w_1 and w_2 will have rotated through the angle β. The axis of the rotation will be along $D \times V$ (figure 8.1) and for co-planar orbits this is perpendicular to the plane of figure 8.2(a). The rotations of w_1 to w'_2 and w_2 to w'_2 are shown in figure 8.2(b) and the new velocities relative to the Sun are now $u + w'_1$ and $u + w'_2$ which will, in general, be very different from the previous velocities both in magnitude and direction. Of special interest in the evolution of the proto-planets is the possibility that the new orbit would be hyperbolic with respect to the Sun so that the proto-planet would leave the Solar System.

In figure 8.3 the rotation of the vector w to w' is shown with the constraint of co-planar orbits removed. The possible loci of the vector w' lie on the surface of a cone and its position can be described by the angle ϕ in figure 8.3 where CQ is in the plane of the original orbit. The speed of the planet relative to the Sun after the interaction, v', is given by

$$(v')^2 = u^2 + w^2 - 2uw \cos \beta \cos \alpha + 2uw \sin \beta \sin \alpha \cos \phi. \tag{8.3}$$

From (8.3) it is seen that the dependence of v' on ϕ is such that v' is a maximum when $\phi = 0$ when

$$(v')^2 = u^2 + w^2 - 2uw \cos(\alpha + \beta) \tag{8.4}$$

and, for a given u and w, escape will occur when

$$\cos(\alpha + \beta) < \frac{(u^2 + w^2 - v_{esc}^2)}{2uw}. \tag{8.5}$$

To obtain a feeling for what this means we consider an interaction for which $u = 11 \text{ km s}^{-1}$, $w = 6 \text{ km s}^{-1}$ and $\alpha = 120°$ at a distance of 10^9 km from the Sun. In this region $v_{esc} = 16.33 \text{ km s}^{-1}$ and (8.5) gives

$$\cos(\alpha + \beta) < -0.832$$

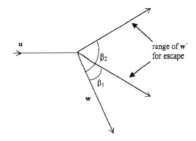

Figure 8.4. The range of directions of w' for which $|u + w'|$ is greater than the escape speed.

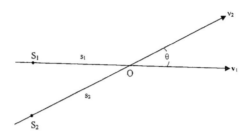

Figure 8.5. Two planets approaching an interaction near point O.

or

$$26° < \beta < 94°.$$

Given that $\phi = 0$ then (8.5) gives a range of impact parameters that would enable escape to take place. The situation is illustrated in figure 8.4. If the original w is rotated to w' anywhere in the range shown then the proto-planet will leave the Solar System.

Since β depends on D we shall now look at the probability of having a particular value of D. It will be assumed that the orbits are very nearly co-planar, so that the probability distribution for the component of D in the orbit of one of the proto-planets, x, is the same as if the orbits were co-planar. The probability distribution of the component perpendicular to the plane, y, will then be treated separately.

The approach of two proto-planets to an interaction in the region of the point of intersection, O, of two co-planar orbits is shown in figure 8.5. If the first body, S_1, is at distance s_1 from O when the second body, S_2, is at distance s_2 from O then the nearest approach of the two bodies, without any interaction, will be

$$x = \frac{(s_2 v_1 - s_1 v_2) \sin \theta}{V} \tag{8.6}$$

where $V = |v_1 - v_2|$ is the approach speed of the bodies. The sign of x indicates which of the bodies first passes through the point O. We now assume that S_1 reaches O first and that $s_1 = 0$. The range of s_2 to give a closest approach between x and $x + dx$ is found from (8.6) as

$$ds_2 = \frac{V}{v_1 \sin \theta} dx. \tag{8.7}$$

The probability that S_2 will be at the range of distances to give a closest separation between x and $x + dx$ is the duration of passing through the distance ds_2 divided by the period of the orbit of S_2, P_2. This gives the probability per orbital period of S_1, the time between successive passages through O. Combining these considerations we find the probability *per unit time* that the closest separation will be between x and $x + dx$ is

$$P(x)\, dx = \frac{V}{v_1 v_2 P_1 P_2 \sin \theta} dx. \tag{8.8}$$

From the symmetry of this result it is clearly independent of which of the two bodies first passes through the point O.

The vertical separation is more difficult to handle with a formal treatment but an approximate treatment gives a result that will overestimate the time-scale for events—which is the correct bias if one is interested in establishing the credibility of such events. If the inclinations of the two orbits are i_1 and i_2, both of which are small, and if the point O is distance r from the Sun then we may write

$$-r(i_1 + i_2) < y < r(i_1 + i_2) \tag{8.9}$$

in which a sign is associated with y. The assumption that the probability density is uniform in the range given by (8.9) will overestimate the probability of large separations. This assumption gives

$$P(y)\, dy = \frac{dy}{2r(i_1 + i_2)} \tag{8.10}$$

where the separation, y, is given a sign.

The probability per unit time that the closest separation will be in a region $dx\, dy$ around the point (x, y) is

$$P(x)P(y)\, dx\, dy = \frac{V}{2r(i_1 + i_2)v_1 v_2 P_1 P_2 \sin \theta} dx\, dy. \tag{8.11}$$

From (8.3), if all other quantities are fixed, then the range of values of ϕ giving escape is found from

$$\cos \phi > \frac{v_{esc}^2 - u^2 - w^2 + 2uw \cos \alpha \cos \beta}{2uw \sin \alpha \sin \beta}. \tag{8.12}$$

If the magnitude of the right-hand side of (8.12) is less than unity then this defines a range of angles ϕ. Since β varies with D then so does ϕ and for a particular D we may write

$$-\phi_D < \phi < \phi_D.$$

For a particular configuration of orbits the x and y of the closest separation will depend on D and ϕ and the area defined by D and ϕ that gives escape, inserted in place of $dx\,dy$ on the right-hand side of (8.11), will give the probability of escape. The area when the impact parameter is between D and $D + dD$ is $2\phi_D D\,dD$ and the total probability of the escape of the planet per unit time is

$$Q = \frac{V}{r(i_1 + i_2)v_1 v_2 P_1 P_2 \sin\theta} \int_{D_2}^{D_1} D\phi_D\,dD \qquad (8.13)$$

where D_1 and D_2 are defined by the limits of β which allow escape when $\phi = 0$.

The assumption in all this is that two approximately defined proto-planet orbits will define the approach angle θ and also v_1 and v_2 but will allow variation of D and ϕ. The value of w inserted in the various formulae defines which planet is being considered as a candidate for escape. The value of Q^{-1} defines a *characteristic escape time*, τ_{esc}, which is such that the probability of escape in time t is given by

$$P_{esc}(t) = 1 - \exp(-t/\tau_{esc}). \qquad (8.14)$$

In section 7.4.5 and in figure 7.11 the way that the proto-planet orbits precess in the resisting medium was described. The precession rates vary from one planet to another and are sufficiently rapid to assume that during round-off of the orbits all possible relative configurations of the two orbits are possible. Of course the orbital parameters are constantly changing but one can find an average characteristic escape time by taking an average value of Q over all possible relative configurations of the planetary orbits.

8.1.2   Probabilities of interactions leading to a collision

The relative motion of two proto-planets, showed in figure 8.1, will be hyperbolic with a closest approach R, taking their gravitational interaction into account. The value of R may be found from

$$D^2 = R^2 + \frac{2G(M_1 + M_2)R}{V^2}. \qquad (8.15)$$

From the proto-planet orbits the approach speed, V, can be found and there will be a collision if $R < a_1 + a_2$ where a_1 and a_2 are the radii of the two planets. This gives a limiting value of D, D_L coming from

$$D_L^2 = (a_1 + a_2)^2 + \frac{2G(M_1 + M_2)(a_1 + a_2)}{V^2}. \qquad (8.16)$$

By substituting the limits D_L and 0 and $\phi_D = \pi$ in (8.13) a characteristic time for collisions can be found for any particular configuration of two orbits and, as done previously for escape, this can be averaged over all configurations.

Table 8.1. Initial characteristics of the planets and their orbits.

Planet	Mass (kg)	Density ($\times 10^3$ kg)	Semi-major axis ($\times 10^9$ km)	Eccentricity	Semi-latus rectum (AU)	Inclination (°)	Rounding time (yr)
Neptune	1.03×10^{26}	1.20	9.34	0.720	30.20	3	2×10^6
Uranus	9.25×10^{25}	1.20	5.34	0.690	18.80	2.5	2×10^6
Saturn	5.97×10^{26}	0.50	2.79	0.680	10.07	1.5	3×10^5
Jupiter	2.00×10^{27}	1.00	2.19	0.800	5.29	2	1×10^5
A	2.00×10^{26}	1.20	1.83	0.874	2.90	1	2×10^6
B	3.15×10^{25}	0.80	1.36	0.908	1.60	1	6×10^6

8.1.3 Numerical calculation of characteristic times

Estimates from various Capture-Theory models, tempered by observational values for the major planets, suggest an initial system with the characteristics shown in table 8.1. The densities of the planets have been set below their present values on the assumption that the original proto-planets would collapse quite quickly at first but then settle towards their final states much more slowly. Interactions are more likely to take place at the slow collapse stage since it is of longer duration. In section 6.4.2.4 it was suggested that orbital inclinations up to about 3° were reasonable for the Capture Theory and the inclinations in the table are randomly chosen within that range.

The rounding-off times given in the final column, to one significant figure, were not all calculated individually but some were inferred from previously calculated results on the assumption that rounding-off times are approximately proportional to the inverse of mass—as indicated in table 7.1.

Numerical computation has been applied to the analyses given in sections 8.1.1 and 8.1.2 to find the characteristic times of escape or collision for particular pairs of planetary orbits. In addition the characteristic time, τ_{ME}, for some major event, either escape or collision, has been found from

$$\frac{1}{\tau_{\mathrm{ME}}} = \frac{1}{(\tau_{\mathrm{esc}})_1} + \frac{1}{(\tau_{\mathrm{esc}})_2} + \frac{1}{\tau_{\mathrm{col}}}. \tag{8.17}$$

The results are displayed in table 8.2.

The significance of the times for major interactions can only be judged in relation to the rounding-off times. For example, the characteristic time for some major interaction involving Jupiter and Saturn is only 2.11×10^5 years but then we find from table 8.1 that the Jupiter orbit rounds off in 10^5 years and Saturn will have considerably rounded-off in that time. Nevertheless the overall picture suggests that some major event, but one cannot say which, is likely to have occurred in the early Solar System. If we take, say, 10^4 years as a time during which the orbital characteristics will not greatly change then we can estimate the probability

Table 8.2. Characteristic times for major interactions in the early Solar System.

Planet 1	Planet 2	$(\tau_{esc})_1$ (yr)	$(\tau_{esc})_2$ (yr)	τ_{col} (yr)	τ_{ME} (yr)
B	A	2.41×10^6	1.79×10^9	3.33×10^7	2.24×10^6
B	Jupiter	1.08×10^5	$\propto$	1.31×10^7	1.07×10^5
B	Saturn	2.53×10^6	$\propto$	4.12×10^7	2.38×10^6
A	Jupiter	9.04×10^4	$\propto$	5.91×10^6	8.90×10^4
A	Saturn	2.56×10^6	3.39×10^8	2.81×10^7	2.33×10^6
A	Uranus	$\propto$	1.83×10^8	4.65×10^8	1.31×10^8
Jupiter	Saturn	4.53×10^8	2.15×10^5	1.39×10^8	2.11×10^5
Jupiter	Uranus	$\propto$	3.41×10^5	1.11×10^8	3.41×10^6
Jupiter	Neptune	$\propto$	9.47×10^5	3.27×10^8	9.44×10^5
Saturn	Uranus	$\propto$	4.40×10^6	3.14×10^8	4.32×10^6
Saturn	Neptune	$\propto$	1.78×10^7	1.15×10^9	1.75×10^7
Uranus	Neptune	3.42×10^9	2.19×10^9	5.24×10^9	1.06×10^9

that at least one major event will have occurred in that period. This is

$$P_{\geq 1 \text{ event}} = 1 - \exp\left(-10^4 \sum_{\text{all pairs}} \frac{1}{\tau_{ME}} \right) \tag{8.18}$$

where the summation is over all pairs of planets and the characteristic times are in years. Inserting numbers into (8.18) we find that the probability of one or more major events in 10^4 years is 0.25. After 3×10^4 years the probability would be 0.57, but then the assumption that the system had not greatly changed from its original state would be less valid.

Despite the uncertainties and approximations in the analyses, most of which have veered in the direction of underestimating the probabilities of interactions, it is plausible to consider that evolutionary processes involving interactions played an important role in the early Solar System.

8.2   The Earth and Venus

Nebula-based theories of planetary origin assume that the early Sun was more luminous than now and that the terrestrial planets formed from non-volatile material which could survive close to the Sun. The Proto-planet Theory also assumes that there is some distinction between the major and terrestrial planets. Rotational instability of the major planets occurred in the planet as a whole while for the terrestrial planets it only occurred in the non-gaseous core. The process for producing planets by the Capture Theory, as described in section 6.4.2, is by condensation in a tidal filament. The initial motion of the condensations is away from the Sun so placing them in a cooler environment and giving them time to condense before being subjected to large tidal forces at perihelion. One of the conditions stated in section 6.4.2 for a planet to be produced is that it must be sufficiently condensed by the time of its perihelion passage to survive. It will be seen from the progression of semi-major axes listed in table 8.1 that this condition is more difficult to satisfy for the planets which will round-off closer to the Sun. For example, the initial period of Neptune is 490 years with a perihelion of 17 AU while for planet B the corresponding quantities are 27 years and 0.83 AU. On the basis of very early modelling of planetary collapse Dormand and Woolfson (1971) concluded that the capture mechanism could not explain the formation of the terrestrial planets and the conclusions from table 8.1 support this view. Indeed part of the reason for only having six planets in the initial system is that it is impossible for planets to survive further in.

Although the terrestrial planets constitute only a tiny part of the mass of the system they cannot be ignored and it is necessary to explain their existence.

8.2.1 A planetary collision; general considerations

We have already deduced from (8.18) that the probability of some major event in the early Solar System is quite high. It is not possible to deduce what, if anything, did happen although taking rounding-off times and characteristic interaction times together it can be deduced that some events are more likely than others. However, what we can do is to look at the Solar System as it is today and to deduce what possible event could explain what we have. If a single event is capable of explaining a large number of features of the system then this will increase the plausibility of the hypothesis. In this spirit we now examine the consequences of a collision between planets A and B. Planet A would eventually have rounded-off in the region of the asteroid belt where there is no planet and planet B would have rounded-off close to the orbit of Mars but is far too massive to be the direct source of that planet. What we require of our collision is not only that it removes what is *not* in the Solar System at present but also provides what *is* in the system.

Collisions at high speed—hypervelocity impacts—have been studied both experimentally and theoretically because of interest in weaponry, in assessing hazards from micrometeorites to spacecraft and in understanding impact features on planets, the Moon and other satellites. Very basic work in this field was reported in a paper by Gault and Heitowit (1963) who presented a theoretical analysis of impacts between aluminium projectiles and basalt targets at velocities of about 6.25 km s^{-1}. Their analyses were based on experiments carried out at the Ames Research Center in California.

The impact of the projectile sends shock waves travelling into both bodies. By the time that the shock waves have reached the back surface of the projectile the energy of the system is in four forms:

(i)   kinetic energy of the projectile which is still moving;
(ii) internal energy of the projectile mostly in the form of shock compression;

(iii) kinetic energy of target material; and

(iv) internal energy of target material, again mostly as shock compression.

When the shock wave has reached the end of the projectile a rarefaction wave travels backwards restoring some kinetic energy but also producing heat. Material is ejected, initially at low angles to the surface of the target but at speeds up to three times that of the projectile. Material ejected at a later time is at greater angles to the surface but travels more slowly. The final destination of the energy originally in the projectile is approximately—heat 31%, ejected material 49% and crushing (producing surface energy) 19% with the small residue going into spallation.

It is not at all certain how to extrapolate these results into the regime of a planetary collision in which there are many additional features:

(i)   For bodies of comparable size there is no distinction between projectile and target.

(ii) The impact velocities and hence the shock compressions will be much higher than in any laboratory experiment.

(iii) The ejecta may be massive and their dispersion greatly influenced by gravitational forces.

(iv) Gravitational energy will be important for planetary bodies, especially in influencing their approach speeds.

Here we shall repeat the description of a planetary collision given by Dormand and Woolfson (1977) but later, in chapter 11, we shall amend the description taking other factors into account.

8.2.2 A collision between planets A and B

Consider the head-on collision of two planets with masses M_1 and M_2 and radii a_1 and a_2. The kinetic energy at the moment of impact is

$$E > \frac{GM_1M_2}{a_1 + a_2}. \qquad (8.19)$$

The right-hand side is the minimum possible energy but in practice the planets will always approach each other with finite velocities when they are far apart and the actual kinetic energy will normally be considerably greater than the right-hand-side value. If some fraction of this energy, say ϕ, ends up as heat in about twice the lesser mass (taken as M_2) then the thermal energy per unit mass is

$$E_T = \frac{\phi GM_1}{2(a_1 + a_2)}. \qquad (8.20)$$

If the density ρ is the same for both planets then

$$E_T = \frac{4\pi}{3} \phi \rho^{1/3} \frac{GM_1}{2(M_1^{1/3} + M_2^{1/3})}. \qquad (8.21)$$

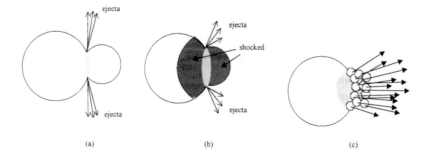

Figure 8.6. The head-on collision of two planets. (*a*) High velocity ejecta soon after impact. (*b*) The smaller planet is completely in a shocked state. The material in the interface region is vaporized. (*c*) Expansion of the vaporized material pushes the planets outwards and breaks up the smaller one.

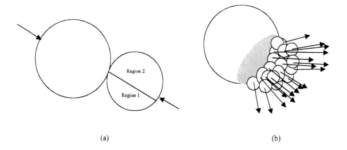

Figure 8.7. A collision of two planets with an offset. (*a*) At time of contact. (*b*) At a stage equivalent to figure 8.6(*c*), the fragments of the smaller planet form two streams.

For the masses given in table 8.1, with $\rho = 10^3$ kg m^{-3} and $\phi = 0.31$, suggested by laboratory experiments, this gives $E_T = 4.0 \times 10^7$ J kg^{-1}. This is likely to be a underestimate because of the finite approach speed of the planets at large distance and also the larger collision speeds give greater shock compression with more of the shock energy converting into heat on relaxation. The main conclusion from the estimated E_T is that the impact will cause vaporization of material and hence that the impact will be explosive in nature. The vaporized material will expand, so thrusting the planets apart and adding to the effective elasticity of the collision.

Figure 8.6 gives a schematic representation of a head-on collision between two planets, different in size with the smaller one taking the role of a projectile. In figure 8.6(*a*), shortly after the impact occurs, high-speed ejected material is thrown out making a small angle with the larger planet (target) surface. In figure 8.6(*b*) the smaller planet is in a totally shocked state and vaporized material is trapped between the two bodies. Finally, in figure 8.6(*c*), the explosive ex-

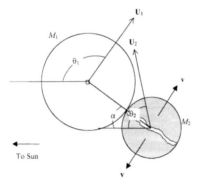

Figure 8.8. A planetary collision with the smaller planet breaking up into two equal fragments. The parameters of the collision are given in table 8.3.

pansion of the vapour pushes the two planets apart; the smaller planet is broken up while the bulk of the larger planet remains intact. This pattern of behaviour is consistent with what has been observed in small-scale laboratory experiments. If the collision is oblique rather than head-on, which in a centre-of-mass frame just appears as offsets of the motion of the centres of the two bodies with parallel motions, then extra shearing forces come into play (figure 8.7). Numerical modelling by Dormand and Woolfson (1977) showed that this divided the smaller body into two regions (figure 8.7(a)) and the projectile debris was lost in two slightly-diverging streams (figure 8.7(b)). The relative speeds of the ejected fragments in each stream were not high enough to overcome their mutual gravitational attractions so they would eventually combine to form two separate bodies.

A computer analysis was made of this type of collision, in particular to follow the progress of the two fragments of the smaller planet. The simplified model is shown in figure 8.8. The speeds of the bodies, U_1 and U_2, are relative to the Sun at the moment of impact and these make angles θ_1 and θ_2 with the Sun direction. The line-of-centres of the bodies at impact make an angle α with the Sun direction. If the bodies are not broken up and the coefficient of restitution is ε then the energy not taken up in the rebound motion would be

$$E_P = \frac{M_1 M_2}{2(M_1 + M_2)}(1 - \varepsilon^2)[(\boldsymbol{U}_1 - \boldsymbol{U}_2) \cdot \hat{\boldsymbol{n}}]^2, \tag{8.22}$$

where $\hat{\boldsymbol{n}}$ is the unit vector along the line-of-centres. The break-up of the smaller planet into two parts, assumed equal n mass, is simulated by adding velocities $\boldsymbol{v}$ and $-\boldsymbol{v}$ to the fragments perpendicular to $\hat{\boldsymbol{n}}$. This velocity was chosen to take up a fraction 0.7 of the available energy EP so that

$$\tfrac{1}{2}M_2 v^2 = 0.7 E_P. \tag{8.23}$$

With $\varepsilon = 0.75$, about 87% of the original kinetic energy then appears as kinetic

Table 8.3. Pre- and post-collision orbital parameters: a = semi-major axis, e = eccentricity, p = semi-latus rectum.

Collision parameters

$M_1 = 2.00 \times 10^{26}$ kg ($33.5 M_\oplus$) $U_1 = 32.2$ km s^{-1} $\theta_1 = 100.1°$
$M_2 = 3.15 \times 10^{25}$ kg ($5.3 M_\oplus$) $U_2 = 35.8$ km s^{-1} $\theta_2 = 49.7°$
Distance from Sun = 2.37×10^8 km; closest approach of centres, $s = 6 \times 10^4$ km
$\alpha = 60°$ $\varepsilon = 0.75$

Pre-collision orbital parameters

Larger planet			Smaller planet		
a_1 (km)	e_1	p_1 (AU)	a_2 (km)	e_2	p_2 (AU)
2.63×10^9	0.91	2.93	1.51×10^9	0.92	1.53

Post-collision orbital parameters

Larger planet		Fragment 1			Fragment 2		
a_1 (km)	e_1	a_2 (km)	e_2	p_2 (AU)	a_2 (km)	e_2	p_2 (AU)
Hyperbolic	1.002	1.06×10^8	0.79	0.70	1.59×10^8	0.64	1.06

energy after the collision, some of it having been in the form of heat energy before the compressed vaporized material re-expanded.

Table 8.3 gives the collision parameters just as the collision occurred and by integrating backwards and forwards in time from the collision it is possible to deduce the orbital parameters of the planets before the collision and of planet 1 and the two fragments of planet 2 after the collision.

The orbital parameters in the simulated collision were chosen so that the *semi-latera recta* of the fragments correspond closely to the orbital radii of the Earth and Venus. If the round-off of orbits was to some other value, for example, close to the geometric mean of the perihelion and *semi-latus rectum* distances, as suggested in section 7.2.3, then an adjustment of the orbital parameters could have given this. What is shown by the calculation is that *the larger planet can be expelled from the Solar System and that two fragments of the smaller planet can go into orbits in the terrestrial region which, after orbital round-off, would closely correspond to the present orbits of Venus and the Earth.*

This calculation does not necessarily portray the precise details of an event in the evolution of the Solar System. A wide range of initial conditions can give an outcome consistent with present observations. Again, the choice of the masses of the two planets, in ratio 6.6:1 was selected by Dormand and Woolfson (1974) to illustrate how much reserve of power was available to expel the larger planet. Clearly with a ratio, say 3:1, the eccentricity of the hyperbolic orbit of the escaping planet could be much larger than that shown in table 8.3.

It has previously been noted that the primary Capture-Theory mechanism for producing planets in highly elliptic orbits makes it unlikely that such a planet could survive to round-off in the terrestrial region. The form of the elliptical orbits and the way they evolve makes a major interaction, including a collision, a probable event. The Capture Theory proposes that the two most massive terrestrial planets, the Earth and Venus, are the non-volatile residues of fragments of a planet which would have rounded-off in the region of Mars, had it survived.

Later, in chapter 11, a model different in detail, although still involving a planetary collision, will be suggested which also explains other important features of the Solar System.

Chapter 9

The Moon

9.1   The origin of the Earth–Moon system

The Earth–Moon system has long been recognized as a notable feature of the Solar System, the explanation of the origin of which may be intimately linked with that for the whole system. The Moon is the fifth most massive satellite in the Solar System, having one-half the mass of Ganymede and more than three times the mass of Triton, and it is intermediate between Io and Europa in both mass and density. Thus, in itself, it is not extraordinary in any way but its large size relative to the Earth suggests that there is something special about its origin. This can be seen both in terms of the ratio of masses and in the ratio of the angular momentum of the planet in its spin to that of the angular momentum of the satellite in its orbit. This is illustrated in table 9.1 for the Earth–Moon system and also for the pairs of bodies Jupiter–Callisto and Saturn–Titan.

It is usually assumed that the way that the Moon came into association with the Earth differs from that which gave satellites to the other planets and that some special event was involved. We now look at, and comment on, the ways that have been suggested within the modern era of astronomy.

9.1.1   The fission hypothesis

In 1878 George Darwin suggested that early in its existence the Earth had been subjected to a periodically fluctuating solar tidal force which happened to coincide in frequency with the natural frequency of Earth oscillations. The large amplitude of the oscillations so built up led to instability and part of the Earth was ejected thus giving rise to the Moon. This fitted in with analyses of the Moon's motion showing that it was gradually retreating from the Earth due to tidal effects so that in the past it had been much closer to the Earth. The mechanics of this situation is shown in figure 9.1 where the tides on the Earth are dragged forward by the Earth's spin because the period of the Earth's spin is less than that of the Moon in its orbit. The main bulk of the Earth gives a centrally directed force but the

251

Table 9.1. The ratios of masses and of angular momenta for three planet–satellite combinations. M_P/M_S is the ratio of the masses and H_P/H_S the ratio of the spin angular momentum of the planet to the orbital angular momentum of the satellite.

Planet	Satellite	M_P/M_S	H_P/H_S
Jupiter	Callisto	18 730	404
Saturn	Titan	4 015	140
Earth	Moon	81	1.9

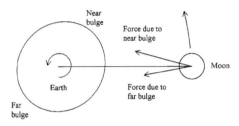

Figure 9.1. Tidal bulges on the Earth give a force on the Moon increasing its orbital angular momentum.

two tidal bulges give non-central forces. The bulge facing the Moon is dominant by virtue of its closeness so that there is a net torque acting in the sense which increases the angular momentum of the Moon's orbit.

Since the Earth–Moon system should approximately conserve angular momentum (tidal effects due to the Sun make the conservation only approximate) Darwin was able to work backwards to deduce that the Moon originated some 6000 miles from the surface of the Earth. The Earth–Moon combination then rotated as a rigid system so that the month and the day were both equal to 5 hr 36 min. It is not necessary to be concerned about the details of the interaction or to follow the process backwards to deduce the form of the initial state—it is only necessary to assume that the Earth–Moon system is isolated and that angular momentum has been conserved. The present angular momentum is given by

$$H = \alpha_\oplus M_\oplus R_\oplus^2 \omega_\oplus + \frac{M_\oplus M_m}{M_\oplus + M_m} r^2 \omega_o + \alpha_m M_m R_m^2 \omega_o \tag{9.1}$$

in which α represents the moment-of-inertia factor, M mass, R body radius, ω angular speed, r the Earth–Moon distance and subscripts $\oplus$, m and o represent the Earth, Moon and orbit respectively. The first term on the right-hand side of (9.1) is the angular momentum in the Earth's spin, the second term is the angular momentum of the Earth–Moon system orbiting around the centre-of-mass and the final term is the angular momentum associated with the Moon's spin.

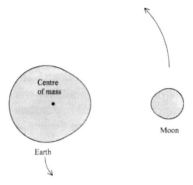

Figure 9.2. The early Earth–Moon system according to Darwin. The spin of both bodies is locked to the orbital rate as both bodies orbit the centre-of-mass.

Taking both values of α equal to 0.4, which is good enough for the present purpose, and other values as observed the total angular momentum of the system is 3.57×10^{34} kg m^2 s^{-1}. For the initial state, represented by figure 9.2, it is convenient to express the angular momentum in terms of ω_I, the initial angular speed, without reference to the distance between the two bodies. Hence ω_I comes from the solution of

$$(\alpha_\oplus M_\oplus R_\oplus^2 + \alpha_m M_m R_m^2)\omega_I + \frac{G^{2/3} M_\oplus M_m}{(M_\oplus + M_m)^{1/3}\omega_I^{1/3}} = H \qquad (9.2)$$

where the final term on the right-hand side is the orbital angular momentum of the two bodies around the centre-of-mass. The value of ω_I satisfying this equation is equivalent to an orbital period of 20 780 s, or 5 hr 46 min. This agrees reasonably well with Darwin's estimate of 5 hr 36 min and corresponds to a centre-to-centre separation of the two bodies of 16 425 km or about 2.6 Earth radii. This is just outside the Roche limit. Darwin also estimated that the separation of the two bodies had to occur at least 54 million years ago. If the Moon with all its angular momentum were to be absorbed into the Earth then the spin period of about 5 hours would not lead to any instability. Lord Kelvin suggested that a resonance could have occurred between the semidiurnal period of the Sun's tidal effect, which would have been $2\frac{1}{2}$ hours, and the free oscillation period of a fluid Earth. The period of free oscillation of a sphere with the present density of the Earth is about $1\frac{1}{2}$ hours but since the period varies as $\rho^{-1/2}$ it could have been longer if the Earth had passed through a less-dense stage. However, it was pointed out by Jeffreys (1930) that internal dissipation within the Earth would prevent large amplitude vibrations from building up so that Darwin's idea was not really tenable.

The theory was popular for many years and, in particular, it was suggested that the effect of removal of the Moon's material was the formation of the Pacific Ocean basin, regarded as a scar on the Earth's surface. There were many attacks on the theory. For example, Nolan (1885) pointed out that just after leaving the Earth the potential lunar material would be within the Roche limit and hence be broken up and spread out into a ring around the Earth. This would negate all ideas about tidal effects between a nearby Moon and the Earth. Darwin countered by saying that tidal effects would operate on the small bodies to push them outwards and that once they were outside the Roche limit they would recombine. However, it seems unlikely that a ring of bodies would lead to the same kind of tidal effects as would be produced by a single body unless the ring was very lumpy.

For the first two decades or so of the 20th century Darwin's fission theory had both supporters and detractors but by 1930 support for it had ceased.

The Proto-planet Theory puts forward an alternative and more plausible model based on fission (section 6.5.1). The original proto-planets disposed of surplus angular momentum by a fission mechanism which separated them into two bodies with a mass ratio about 8:1. There is a valid theory by Lyttleton (1960) describing this fission process. In the inner system both fragments were retained and the pairs Earth–Mars and Venus–Mercury are presumed to have originated in this way. For the original proto-planets in the terrestrial region the fission is presumed to have operated only within the solid core and presumably the Moon would be regarded as an outcome of the fission of the Earth–Mars proto-planet.

9.1.2 Co-accretion of the Earth and the Moon

The idea that the Earth and the Moon were produced separately but in the same region and in association was first put forward by Roche (1873). This was based on the Laplace model where the evolving collapsing proto-Earth left behind a ring within which the Moon condensed.

Another co-accretion hypothesis was put forward by Ruskol (1960), based on the concept that planets and satellites were formed by the accumulation of small solid bodies. The starting point for the model is that the Earth has substantially formed and is still growing due to the presence of a surrounding swarm of small bodies with individual dimensions between 10–100 km and total mass between 0.01 and $0.1 M_\oplus$. These bodies would be rotating in the same sense as the Earth and be spread out to a distance of about 100 Earth radii with the concentration increasing towards the Earth. Inelastic collisions of the bodies would lead to some being deflected towards the Earth thus adding to its mass. Other collisions could lead to capture, i.e. amalgamation of the bodies (section 6.4.3.3), and if 1% or more of collisions lead to this outcome then the accumulation of a large body, the Moon, would be possible. Ruskol suggested that the newly formed Moon was initially in orbit at between 5 and 10 Earth radii.

An obvious problem with this model is that there seems to be no obvious reason why the material forming the Earth should be any different from that which produced the Moon. Which of the bodies a planetesimal joined would have been the result of chance and would not have been systematically related to what type of material it was. The mean density of the Moon is 3340 kg m^{-3} and the uncompressed density of the Earth is about 4600 kg m^{-3}. This difference seems to rule against the Ruskol idea. Ruskol argued that the Earth's core could be silicates in a high-density form but this idea has proved to be quite untenable.

Ruskol herself accepted that this model would not be applicable to the formation of the satellites of the major planets, but since it is generally accepted that the Earth–Moon system is a special case this in itself is not regarded as a serious criticism.

9.1.3 Capture of the Moon from a heliocentric orbit

From time to time the idea has been raised that the Moon was produced separately from the Earth somewhere within the Solar System. Both bodies were in separate heliocentric orbits but a chance close passage led to capture of the Moon.

Apart from the problem of explaining how a body with the size and composition of the Moon could be produced in isolation there are also considerable dynamical problems with this hypothesis. For two bodies in isolation the total energy of the two-body system must remain constant. If the two bodies have finite speed when at large (effectively infinite) separation then the total energy of the system is positive and must always be so. If they approach each other then, inevitably, they must separate again.

There are two mechanisms that modify this conclusion in considering the Earth and the Moon within the Solar System. The first is that a close passage could produce tidal effects which would transform mechanical energy into heat which would then be dissipated. In effect, if the Earth and the Moon approach each other closely then the two-body approximation is no longer valid. Through tidal distortion the assumption that all the mass of the bodies acts as though it was concentrated at their centres breaks down and essentially the system becomes a many-bodied one. The second mechanism is that there are other bodies present with which energy can be exchanged although, for practical purposes, only the Sun needs to be considered.

We now consider a close encounter giving tidal dissipation. If the Earth and the Moon were on similar paths then their relative speed at large distance could have been quite small. With a closest approach distance of r_{ME} the total energy of the system would be positive with the magnitude of the kinetic energy of the motion slightly greater than the potential energy which would have been

$$\Omega = -\frac{G M_\oplus M_m}{r_{ME}} \tag{9.3}$$

in which M_m is the mass of the Moon. If the kinetic energy is $-(1 + \varepsilon)\Omega$ and the interaction removes energy $f\Omega$ from the system then it will end up with total energy

$$E_f = (\varepsilon - f)\frac{G M_\oplus M_m}{r_{ME}}. \tag{9.4}$$

With $f > \varepsilon$ the two bodies would form a bound system. If the removal of energy depends on tidal dissipation due to a close passage outside the Roche limit then f would actually be quite small. This puts very heavy demands on the relative orbits of the Earth and the Moon to make ε sufficiently small to give capture. A very close interaction within the Roche limit, or indeed a collision, would remove much more energy but would also lead to a complete or partial break-up of the Moon and also severe damage to the Earth. The Moon debris, left in orbit around the Earth, could then perhaps reassemble as envisaged by Ruskol.

Another constraint which can be imposed is that if the Moon does go into orbit it should not go so far from the Earth that it can be removed by the tidal pull of the Sun. If the Moon is at a distance d from the Earth and directly between the Earth and the Sun then the condition that it is within the sphere of influence of the Earth is

$$d < r_\oplus \left(\frac{M_\oplus}{2M_\odot} \right)^{1/3} \tag{9.5}$$

where $r_\oplus$ is 1 AU, which gives $d < 1.7 \times 10^9$ m, about 4.5 times the present radius of the lunar orbit. An interaction which just satisfied (9.5) would give an elliptical orbit with small perigee so that approximately, for non-disruption of the orbit, $a = \frac{1}{2}d < 8.5 \times 10^8$ m. The energy of the orbit is given by

$$E = -\frac{GM_\oplus M_m}{2a} \tag{9.6}$$

so that from (9.5) and the condition on a for stability of the orbit

$$a = \frac{r_{ME}}{2(f - \varepsilon)} < 8.5 \times 10^8 \text{ m}. \tag{9.7}$$

Taking $r_{ME} = 1.6 \times 10^7$ m, the closest distance for non-disruption of the Moon, requires $f > 0.0095 + \varepsilon$, which is a very demanding requirement.

Actually if the original Moon orbit does not bring it between the Earth and the Sun when it is at perigee for some time so that several orbits have taken place then the removal of energy from the system can be much greater. Nevertheless, the need to find approach orbits of the Moon and Earth which would make ε small plus the need to have an interaction which would make f large makes the capture mechanism very unlikely, although not necessarily impossible.

Possible mechanisms for capture of the Moon by the Earth have been explored in detail by Goldreich (1966) and Singer (1970).

9.1.4   The single impact theory

In the middle of the 1970s a new idea was put forward which seemed to answer most of the difficulties of previous theories—which do not include the collision hypothesis described in section 9.1.5 with which it was almost contemporaneous. This idea, described by Hartmann and Davis (1975) and Cameron and Ward

(1976), involved a sideswipe impact on the Earth by a large impactor with a mass about 0.1 that of the Earth itself. One of the problems it seems to solve is the difference in the spin rates of Venus and the Earth which otherwise have quite similar characteristics. With the model that the terrestrial planets were formed by the random accretion of planetesimals then the expectation is that the final spin rate should be small, the reasoning being similar to that given by McCrea to explain the slow spin of the Sun (section 6.2). The slow spin of Venus, and even its retrograde sense, are quite compatible with this model but the comparatively rapid spin of the Earth is not. If the impactor struck the Earth, originally spinning slowly, tangentially and gave up a fraction β of its momentum to the Earth then

$$\omega = \frac{\beta m v}{\alpha M_\oplus R_\oplus} \tag{9.8}$$

where m and v are the mass and velocity of the impactor on impact, α the moment-of-inertia factor of the Earth about its spin axis and $M_\oplus$ and $R_\oplus$ the mass and radius of the Earth. For an approach speed of the impactor at large distance from the Earth that is not too large, v will be little more than the escape speed from the Earth, 11 km s^{-1}. The moment-of-inertia factor for the Earth is about 0.33 so that (9.8) gives $\beta m/M_\oplus = 0.014$. Since $R_\oplus$ and v might have been bigger at the time of the collision than the values assumed here and since the collision need not have been precisely tangential this result is only indicative.

This hypothesis has been widely accepted by cosmogonists so a fairly detailed account of it is given here. The first detailed study of the mechanism of the impact was made by Benz *et al* (1986) using the technique of smoothed-particle hydrodynamics (Appendix III). A feature of their calculation was to represent properly the equation of state of the material of the two bodies over a wide range of densities and temperatures. Clearly for low temperatures and high densities the material would behave as a solid. Conversely for very high temperatures and very low densities it would behave as a gas. In going from one extreme state to the other some intermediate equation of state is required. In the first 1986 calculation, for simplicity, both impacting bodies were taken as consisting wholly of one material, granite. The forms of the equation of state are:

(i) For a condensed state when either the density $\rho > \rho_0$, some reference density, or when the material is cold so that the intrinsic internal energy $u < u_s$ then the pressure is given by

$$P_g = \left(a + \frac{b}{\frac{u}{u_0 \eta^2} + 1}\right) u\rho + A\mu + B\mu^2 \tag{9.9}$$

where $\eta = \rho/\rho_0$ and $\mu = \eta - 1$. The first term on the right-hand side represents pressure due to thermal expansion while the remaining terms corresponds to non-linear elasticity. If the material is cold but $\rho < \rho_0$ then the pressure becomes negative. This represents a state of tension and has the desirable property that it prevents the material expanding like a gas.

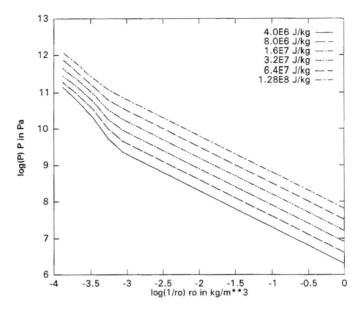

Figure 9.3. The equation of state used by Benz *et al* (1986) for granite.

(ii) For hot low-density states, where both $\rho < \rho_0$ and $u > u_s'$, the equation of state used is

$$P_s = au\rho + \left\{ \frac{bu\rho}{\frac{u}{u_0\eta^2} + 1} + A\mu \exp[-\alpha(\eta^{-1} - 1)] \right\} \exp[-\beta(\eta^{-1} - 1)^2].$$

(9.10)

For very low density only the first term is effective, which gives the equation of state for a perfect gas.

(iii) For intermediate situations where the density is high, so that $\rho > \rho_0$, and also $u_s < u < u_s'$ then a smooth transition from one equation of state to the other is obtained from

$$P_i = \frac{(u - u_s)P_g + (u_s' - u)P_s}{u_s' - u_s}.$$

(9.11)

For granite the constants used in (9.9), (9.10) and (9.11) are: $a = 0.5$, $b = 1.3$, $u_0 = 1.6 \times 10^7$ J kg^{-1}, $u_s = 3.5 \times 10^6$ J kg^{-1}, $u_s' = 1.8 \times 10^7$ J kg^{-1}, $\rho_0 = 2700$ kg m^{-3}, $A = B = 1.8 \times 10^{10}$ J m^{-3} and $\alpha = \beta = 5$. The form of these curves is shown in figure 9.3 for various values of u.

In a later paper Benz *et al* (1987) added an iron core to both the Earth and the impactor. The sequence of events for a particular impact are shown in figure 9.4. In figure 9.4(*a*) the tidally-distorted impactor strikes the Earth a glancing blow.

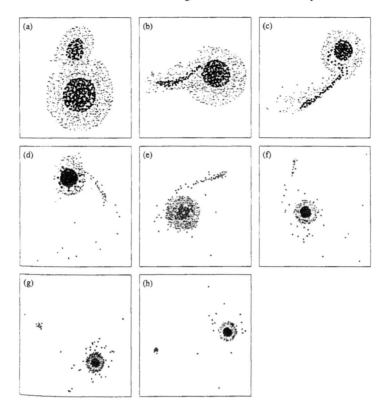

Figure 9.4. Stages in the formation of the Moon by the single-impact process. Light dots represent silicate (granite) and dark dots are iron. The scale of the individual stages is progressively reduced to show the formation of the Moon.

In the next three frames it is seen first to spread out and then for much of it, including virtually all the iron component, to be captured by the Earth. Some mantle material from the impactor does, however, collect together and go into orbit around the Earth.

The parameters of the collision need to be within fairly tight limits to obtain the result shown in figure 9.2. If the collision speed is too low or the collision too head-on then all the impactor is absorbed by the Earth and no Moon is formed. On the other hand, if the collision is too fast then vaporization of impactor material leads to its complete loss. What the model does do is to provide an initially molten Moon with little iron, a lack of volatile materials and also a more-rapidly spinning Earth—all of which are consistent with observation.

9.1.4.1   Problems with, and features of, the single impact theory

Although in its gross features the impact theory seems very satisfactory it still presents some problems, although they may not be serious ones. The impactor has a mass very similar to that of Mars and it is certainly tempting to think of it as a Mars-like body. If so then its density would be about 3940 kg m^{-3} and assuming that this was mostly mantle material with density 3000 kg m^{-3} plus an iron core of density 7900 kg m^{-3} then the proportion of silicate to iron by volume would be around 0.81:0.19. Assuming the same types of component for the Moon then the proportion of silicate to iron by volume would be 0.93:0.07. For all of the Moon material to come from the impactor requires 14.3% of the latter's silicate and 4.6% of its iron to go into Moon formation. The model impact seems to indicate that all the Moon's material comes from mantle material, some of it from the Earth. It seems unlikely that the mantle could have contained enough metallic iron to give rise to the size of core that is now inferred for the Moon. Actually the 'Moons' produced by modelling tend to be more massive than the actual Moon. If the iron sinks to the centre and excess silicate is later lost in some way then this problem is reduced in severity.

Another consideration is that this is a mechanism applied to the Moon to explain its relationship with the Earth and to none of the other large satellites of the Solar System. There are, in fact, seven substantial rocky bodies in the Solar System (excluding those with a large ice content): the four terrestrial planets, the Moon and also Io and Europa, the innermost Galilean satellites of Jupiter. Actually Europa has an icy surface but the ice layer is thin and only has a marginal effect on its mean density. Assuming that the ice layer is 50 km thick brings the estimate of the density of the remainder to 3100 kg m^{-3} rather than 2990 kg m^{-3} for the whole satellite. In considering the mean densities of all these bodies one must take account of compression effects due to self-gravity but this is only significant for Venus and the Earth. Their uncompressed densities are estimated as 4400 and 4600 kg m^{-3}, respectively, although with considerable uncertainty— 5% or more. A plot of density against log(mass), as in figure 9.5, shows a good general relationship for six of the seven bodies. The Moon fits comfortably into the sequence, the exception being not the Moon but Mercury.

The evidence from numerical modelling does not suggest the formation of the Moon with a satisfactory iron content by an impact scenario. However, if an impact event as proposed *did* create the Moon as it is then it is a curious coincidence that it fits so well into the density–mass relationship suggested by five other bodies. It has never been suggested that impacts could have been involved in the formation of the Galilean and other large satellites.

The relationship in figure 9.5 begs the question of why it is that three of the terrestrial planets should fit into the same density–mass pattern as three large satellites, unless there are some common features about the way these planets and satellites were produced. Nevertheless the sandwiching of the Moon between Io and Europa does suggest that there may well be some similarity in the mecha-

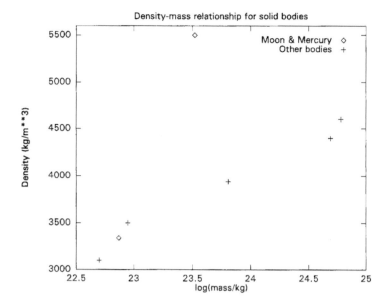

Figure 9.5. The density–mass relationship for seven rocky bodies—Earth, Venus, Mars, Mercury, Io, Moon and Europa.

nisms of formation of the inner Galilean satellites and the Moon.

A final feature of the single-impact theory is that the initial orbit of the Moon clump, seen in figure 9.4(*h*), brings it within the Roche-limit distance from the Earth. This would lead to disruption of the body into several smaller clumps. Tidal effects are taken to have operated on these individual clumps as they do on the Moon today so driving them outwards. Once they had retreated to beyond the Roche limit then it is possible, although perhaps not inevitable, they would once again recombine into a single body. This is just the mechanism suggested by Darwin in answer to Nolan's criticism of the fission hypothesis.

9.1.5 The Earth–Moon system from a planetary collision

The various mechanisms of satellite formation according to different cosmogonic theories were described in section 6.5. There are no detailed analyses or models for nebula-based theories and the Proto-planet Theory explains satellites as droplets in separating fragments of an original proto-planet. The Capture Theory presents a more detailed model in which the tidal effects of the Sun on a proto-planet adds angular momentum in a tidal bulge region. The tidal bulge evolves into a filamentary form and proto-satellites condense within the filament. This gave good quantitative explanations for the characteristics of the satellite systems of the major planets, Jupiter, Saturn and Uranus. Neptune was not included in this account because, being the most distant of the major planets, solar tidal effects would have been much weaker and no large satellites would be expected. In fact Neptune does possess a fairly large satellite, Triton, with a mass one-third that of the Moon and with an anomalous retrograde orbit. This will be discussed further in section 10.4.

The proto-planets involved in the collision described in chapter 8 would have had perihelia much smaller than those of the surviving proto-planets. Consequently they would have been subjected to much stronger tidal effects and hence be expected to have numbers of large satellites as massive as, or even more massive than, those of Jupiter.

The destination of a satellite of one of the colliding planets can be investigated by numerical modelling of the collision. There are only three possible outcomes:

(i)   The satellite is retained or captured by planet A and leaves the Solar System.
(ii) The satellite goes into an independent heliocentric orbit that could be hyperbolic.
(iii) The satellite is retained or captured by one of the fragments of planet B.

Many numerical simulations involving a satellite of the colliding planets were carried out by Dormand and Woolfson (1977). In one series of calculations a satellite of planet A was taken in a circular orbit with radius r. The simulation was begun at a time 5×10^6 s before the collision, at which stage the separation of the planets was about 0.1 AU so that perturbation of the satellite by planet B would have been negligible. The parameters for the planetary motions are given in table 8.3. The outcome for the satellite depends critically on where it is placed in its orbit at the beginning of the calculation. This was defined by the angle, ϕ, between the satellite–planet line and the line from the Sun to the collision region. Most initial positions of the satellite led to it either being retained by planet A or it being released into a heliocentric orbit. However, this was not always the case. For example, with $r = 3 \times 10^5$ km and $\phi = 1.37$ rad the Earth fragment of planet B captured the satellite into a stable low-energy orbit with semi-major axis 5.2×10^5 km and eccentricity 0.74. In other trials semi-major axes down to 2.3×10^5 km and eccentricities down to 0.41 were found.

This is not the only way for the Earth fragment to be left with a satellite; this could also come about from the retention of a satellite of planet B. Since this scenario clearly offers fewer problems than that of capture of a satellite from planet A it was not numerically modelled. Later we shall look at some further consequences of this hypothesis for producing the Earth–Moon system in relation to other properties of the Moon.

For this hypothesis the Moon was produced in the same way as other satellites so that there is no problem in understanding why it fits between Io and Europa in both mass and density.

9.2 The chemistry of the Earth and the Moon and formation of the Moon

The extent to which the Earth and the Moon are chemically similar or distinct has been the subject of a great deal of discussion and investigation. Clearly the overall compositions of the two bodies are different, because of their different densities which can only be partially explained by compression effects. It is generally believed, and it is almost certainly true, that the Earth has a substantial iron core while the Moon's core is much smaller relative to its size. The size of the core or the relative quantities of other components of similar type is not a crucial criterion for judging the similarity or otherwise of the two bodies. Of much more importance is the relative abundance of elements or compounds in situations where it might be expected that similarities should occur or where unusual compounds or compositions suggest particular conditions of formation. When the first samples were brought back from the Moon by United States and Soviet Union spacecraft the subject of lunar cosmochemistry moved from speculation to certainty.

The Apollo missions to the Moon found no solid bedrock but a surface covered with loose pulverized rock, called *regolith* or sometimes *lunar soil*, to a depth of several metres. Regolith is formed by meteorite bombardment and apart from the 1–2% of it that is actual meteorite material it has the composition of the surface on which it rests. The measured age of regolith material is in the range 4.4–4.6×10^9 years which seems to indicate that it is older than the rocks on which it rests. This paradox has not been convincingly resolved although there are suggestions for mechanisms that could disturb the proportions of radiogenic ^{87}Rb and its daughter product ^{87}Sr on which the age estimates are made. One suggestion involves the presence of KREEP, a component of lunar soil rich in potassium (K), rare-earth elements (REE) and phosphorus (P). It is mainly plagioclase, silicates containing sodium, potassium and calcium with aluminium replacing some of the silicon. Apart from the KREEP component it also contains more rubidium, thorium and uranium than is found in other lunar rocks. It cannot be a common lunar component otherwise it would lead to extensive melting of the lunar interior. Most KREEP material is found near Mare Imbrium and it probably has its origin at a depth of 25–50 km and was brought to the surface by the collision that produced the mare basin. However, to explain the larger measured age of regolith requires a relative *increase* in the strontium component. Another suggestion is that impact heating preferentially removed rubidium that is more volatile than strontium.

Maria material is basalt with a composition similar to that from terrestrial volcanic eruptions. The time of eruption of mare material has been dated between 3.16 and 3.96×10^9 years ago but volcanic material produced earlier may have been covered up so volcanism back to the origin of the Moon, 4.6×10^9 years ago, cannot be ruled out. About 100 minerals have been identified on the Moon, compared with about 2000 on Earth, but three lunar minerals were previously unknown. These were *pyroxferroite*, an iron-rich pyroxene, *armalocite* (subsequently discovered on Earth) a Fe–Ti–Mg silicate similar to ilmenite and *tran-*

quillityite, an Fe–Ti silicate enriched in zirconium, yttrium and uranium.

The Moon is deficient in oxygen as compared with the Earth. A consequence of this is that maria material contains particulate iron and iron only in the ferrous form FeO whereas the ferric form Fe_2O_3 also occurs on Earth. Compared with the Earth it is also deficient in *siderophile elements* (those with an affinity for metals, such as gold, nickel and platinum), *chalcophile elements* (those with an affinity for sulphur, such as lead) and volatile elements, as illustrated in figure 1.13. A volatile substance which was thought to be absent from the Moon is water. Hydrated minerals have not been found and certainly no free water was found by the Apollo missions. However, in March 1998 it was reported that water deposits had been discovered near both lunar poles by remote sensing using a neutron spectrometer mounted on the orbiter, Lunar Prospector. It is believed that this water was deposited by comets falling into deep craters near the Moon's poles. Since the Sun's rays cannot penetrate into these craters the water is protected from evaporation. Given this view of the likely source of the water, not too much should be read into the presence of the water in relation to the Moon's origin.

Another source of information is the isotopic composition of lunar materials. Lunar material has a similar oxygen isotopic composition to terrestrial material, in contrast to many meteorites that show a significantly different composition. Observations of isotopic anomalies in meteorites are sometimes taken to suggest that bodies formed in different parts of the Solar System would necessarily have different isotopic signatures. This matter is fully discussed in chapter 11 but for now we can say that if some alternative explanation for meteorite isotopic anomalies is available then similarity of isotopic composition does not necessarily imply a common source or region of origin for different bodies. If all the bodies in the Solar System have common source material, be it a nebula or a proto-star, then similar compositions would be expected with predictable modifications due to local conditions of formation.

The fission theory has difficulty in explaining why lunar surface material is richer in iron, some in metallic form, than is the mantle of the Earth, the presumed source of the Moon. The moment of inertia factor for the Moon is 0.391 which allows for an iron core up to 382 km in radius (Stock and Woolfson 1983a). If iron had somehow settled towards the centre of the Moon then *less* iron would be expected near the lunar surface, not more.

The reduced volatile inventory of the Moon compared with the Earth suggests that if it had been produced from terrestrial material, or from the same pool as terrestrial material, then the whole surface was exposed to a temperature of at least 1700 K. Over a sustained period this could have evaporated off much of the more volatile elements and also modified surface rocks to non-hydrated forms. This may have more to say about the process by which the Moon formed than how the Earth and the Moon came to be associated. Such a condition, with a high temperature of long duration, is consistent with the extensive volcanism that produced the basalt deposits in the mare basins.

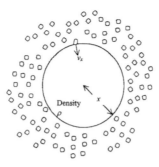

Figure 9.6. A schematic view of the Moon forming by fast accretion.

9.2.1 Possible models of Moon formation

There are three scenarios for Moon formation that are capable of explaining the near-surface high-temperature regime in the early Moon. The single-impact theory is one of these and the computational modelling shows that the whole body would have been in a molten state when it first assembled. If the body had stayed whole then it would have cooled from the outside with the formation of a crust but with molten material below the surface leading to extensive volcanism. If, on the other hand, the body was disrupted by tidal effects then it was the individual portions that would have formed a crust and cooled from the outside. When they recombined there would have been a great deal of heat generated, with a considerable mixing of material so a hot origin for the whole body would again have been an outcome although with a lower initial temperature.

The other two scenarios are somewhat related. One is that the Moon was produced over a long period as a result of gradual accumulation so that the released gravitational energy could be radiated away to keep the body cool. Then at some time, which could have been well after it had formed, a period of bombardment with large projectiles heated the surface material to a considerable depth. A solid surface would then form over the melt but with molten material below the surface giving volcanism. The related scenario is that the Moon formed quickly from the collapse of a proto-satellite condensation, as proposed by the Capture Theory model and described by Williams and Woolfson (1983). This model (figure 9.6) would give an initial state intermediate between the other two. There would certainly be more retained heat than for the slow accumulation but then only the outer parts of the Moon would have been molten rather than the whole Moon as for the single-impact theory. A description of the evolution of the intermediate model will serve as a basis for all three models.

We consider a simple model in which material is falling uniformly onto the growing lunar core. The Moon is forming as a homogeneous body of density ρ and all kinetic energy of in-fall, at the escape speed corresponding to the core, is transformed into heat energy at the point of collision. If the growing core has

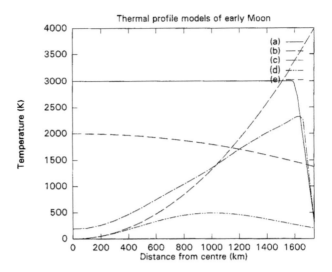

Figure 9.7. Thermal profile models of the early Moon and the solidus curve: (*a*) expected profile from single-impact model; (*b*) fast accretion without shocks or cooling; (*c*) fast accretion with shocks and cooling; (*d*) slow accretion; and (*e*) the solidus curve.

attained a radius x then the intrinsic kinetic energy of colliding material is

$$\tfrac{1}{2}v_x^2 = \tfrac{4}{3}\pi G\rho x^2.\tag{9.12}$$

If the specific heat capacity of the material is C then, assuming no change of state is involved, this will give a temperature at distance x from the centre of the Moon

$$\theta_x = \frac{4\pi G\rho x^2}{3C} + \theta_0\tag{9.13}$$

where θ_0 is the original temperature of the in-falling material. This temperature profile is shown in figure 9.7. Schematic profiles are also shown for the other two possible formation mechanisms.

In practice the initial temperature profile will differ from that given by (9.13). A projectile landing on the surface will send shock waves into the interior of the forming Moon, falling in intensity with distance from the point of impact, and when these decay then heat is generated. In this way internal regions are heated more than is suggested by (9.13). Again, the assumption that the accretion process is so rapid that there is no cooling cannot be strictly true; there will be constant radiation from surface material so that outer regions will be at a lower temperature than is indicated by (9.13). These factors were considered by Toksöz and Soloman (1973) who developed modified thermal profiles for bodies of different sizes accumulating with various time-scales. The general form of a modified thermal profile is indicated in figure 9.7. Another modification of the profile would be a flattening at a temperature corresponding to melting where energy would be absorbed as latent heat. Between a pure solid region and one of pure liquid there would have been an intermediate zone of variable partial melting. Assuming the rapid accumulation of the Moon from a proto-satellite condensation then (9.13) indicates that it would have become molten from about 1100 km from the centre. For a modified initial profile this would have been at a somewhat greater distance. Due to convection in molten material in the outer regions, cooling would have been greatly enhanced and shortly after formation there would have been a solid region 100–200 km thick at a temperature just below the melting point. Between the solid Moon, closer than 1100 km from the centre, and the outer solid shell there would be a molten region more than 400 km thick.

A detailed model describing the thermal evolution of the Moon is given in section 9.3.6.

9.3 The physical structure of the Moon

In 1959 the Soviet Lunik spacecraft transmitted to Earth the first pictures of the rear side of the Moon. These were a surprise in that they showed a hemisphere quite different in appearance from that which faces the Earth. There was a dominance of highland regions with only small mare features. Figure 9.8 is a photograph of the Moon showing part of the near side (bottom right) and part of the far side (top left); the difference in general appearance of the two sides is evident.

Since mare basins were formed by large projectiles, the first idea for explaining the difference of the two sides was that, in some way, the far side had been shielded from very large impacts. From the extensive cratering on the far side it had clearly not been shielded from small impacts—which meant that the shielding had only existed at earlier times when larger projectiles were abundant. This idea had to be abandoned when altimeter measurements from lunar orbiters showed that there *are* large basins on the far side. That side had been impacted by large projectiles but the large impacts had not led to extensive volcanism. The obvious conclusion from this is that molten material had been further below the surface or, in other words, that the crust is thicker on the far side than it is on the near side. This is now generally accepted and seismic evidence is consistent with a near-side upper-crust thickness of 60 km and a far-side thickness of 100 km. This may give an explanation of the difference in appearance of the two sides of the Moon but then raises the question of how that difference came about. The general belief seems to be that it is due to the way that the Moon is tidally locked with one face towards the Earth. The Moon is slightly pear-shaped with the pointed end towards the Earth and the distortion is more than would be expected from tidal influences from the present Earth–Moon configuration. However, we know that tidal effects are causing recession of the Moon from the Earth so that the Moon has been closer to the Earth in past times. Could the effect of having the Moon in close orbit give rise to a thicker crust on the far side? To answer that question

Figure 9.8. A view of the Moon from space. The upper right area, with mare basins, is that seen from Earth. The remaining area, showing only highland regions, is on the far side from Earth (NASA photograph).

confidently would require a proper analysis of a multi-layered Moon orbiting the Earth but a rather cruder analysis suggests that it would not have the required effect. Consider a two-layered Moon tidally locked to the Earth and quite close to it. If the central region has a much higher density than the outer region then the body will not exactly correspond to a Roche model (section 4.6.2) but will resemble one to some extent. From the equipotential contours shown in figure 4.12 it is evident that the central high-density region will take on an approximately spherical form while the surface of the body will occupy one of the egg-shaped contours further out. This suggests that the low-density region will be *thicker* facing the Earth not thinner as is required by the observations. This argument can only be suggestive but it is doubtful that a more accurate analysis would point the conclusion in the opposite direction.

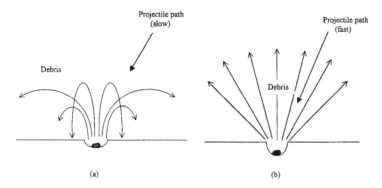

Figure 9.9. (*a*) A slow projectile shares energy with surface material and is accreted. (*b*) A fast projectile shares energy with surface material much of which escapes to give abrasion.

9.3.1 Hemispherical asymmetry by bombardment

We now look at an alternative explanation based on the Moon as a satellite tidally locked to a planet undergoing a collision. From table 8.3 the impacting speed of the planets would have been at least 50 km s^{-1}, taking account of their mutual attraction. Theoretical and experimental studies of hypervelocity impacts (Gault and Heitowit 1963) shows that the first ejecta from a collision is at speeds up to three times the collision speed—or 150 km s^{-1} for the planetary collision. If the Moon was at a distance of, say, 500 000 km from the impact then the first debris would arrive within one hour. Following debris would arrive later, have lower speed and would come from the collision site with a greater angular spread. The effect of colliding objects on a solid astronomical body is shown in figure 9.9. If the speed of impact is little more than the escape speed then the object shares its energy with the surface material so that both it and the surface material have less than escape speed. The result is that the impacting object is accreted. However, if the speed of the impacting object is much greater than the escape speed then sharing its energy gives a mass larger than that of the object itself with more than the escape speed so that material is abraded from the surface. If the escape speed from the body being struck is much smaller than the speed of most abraded material then the ratio of the mass of lost surface material to that of impacting material, ε, will be virtually constant.

We now take the total mass of debris from the collision as m_D spread out over 2π steradians (50% of complete angular spread). For the Moon with radius R_m at orbital distance r the loss of surface material would be

$$m_L = \frac{\varepsilon m_D R_m^2}{2r^2}. \tag{9.14}$$

As an illustration we take $\varepsilon = 200$, $m_D = 6 \times 10^{24}$ kg (mass of Earth) and $r = 5 \times 10^5$ km. This gives the mass of material lost as 7.27×10^{21} kg which is

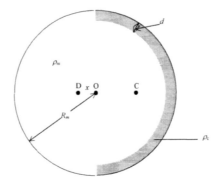

Figure 9.10. The abraded Moon with the shaded region (thickness exaggerated for clarity) removed in one hemisphere. The centre-of-mass of the hemispherical shell is at C and that of the residual Moon at D.

equivalent to a 135 km thick layer from one hemisphere of the Moon with a mean density 2800 kg m^{-3}. Given the possibility of partial shielding from the debris this calculation just illustrates that with one particular set of feasible values the loss of a few tens of kilometres of surface material from one side of the Moon is possible. Other sets of reasonable values give similar results.

Bodies greater than a certain size settle into a configuration of minimum energy which, in general, would be a sphere. Smaller asteroids are often of irregular shape as are some of the smaller planetary satellites. Asteroids and satellites larger than about 300 km in diameter are spherical (Hughes and Cole 1995), or nearly so, since the strength of material of the body can resist the stresses of an irregular shape to some extent. If the Moon had been a perfect sphere and a 40 km thick shell of material had been moved from one hemisphere, the centre-of-mass of the remainder would have moved in a direction away from the centre of the abraded region. This situation is illustrated in figure 9.10. The shell removed has thickness d ($\ll R_m$, the radius of the Moon) and density ρ_c. The thin shell has its centre-of-mass at C and it is easily shown that $CO = \frac{1}{2}R_m$. If the centre-of-mass of the residue of the Moon is at D, distant x from O then

$$(\tfrac{4}{3}\pi R_m^3 \rho_m - 2\pi R_m^2 \, d\rho_c)x = 2\pi R_m^2 \, d\rho_c \times \tfrac{1}{2}R_m \qquad (9.15)$$

where the density of the residue, ρ_c, is taken as that of the Moon as a whole. With $d = 40$ km and $\rho_c = 2800$ kg m^{-3} equation (9.15) gives $x_{com} = 25.9$ km defining the position of the centre-of-mass. The position of the centre of figure—the centre-of-mass if the body had uniform density—is obtained by putting $\rho_c = \rho_m$ in (9.15) and this gives $x_{cof} = 31.1$ km. This indicates that the centre-of-mass is displaced from the centre of figure by 5.2 km towards the evacuated region, that is towards the Earth. The actual displacement as deduced from observations is about 2.5 km, a predictable difference from the calculated value given that

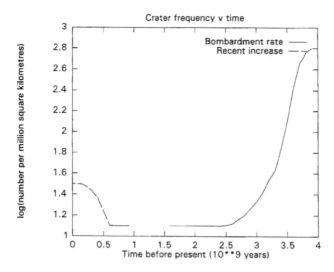

Figure 9.11. The variation of the number of craters greater than 1 km in diameter per million km² with age. The recent increase is of unknown origin.

there would be some rearrangement of the material after abrasion to go towards a minimum-energy configuration.

The tidal coupling of the Moon to the original parent planet gave it a particular shape and distribution of material while the shape and distribution of material ensured that the tidal coupling persisted. When the Moon was captured by the Earth that shape and distribution was disturbed by the abrasion of one face but it appears that what remained of those features was enough to give the same face presented to Earth as had once been presented to the original parent planet.

9.3.2   A collision history of the Moon

Observations of collision features on the Moon and other solid bodies in the Solar System suggest that there was an early period of very large projectiles but that as time progressed so both the size and numbers of projectiles decreased. The deduced production rate of craters more than 1 km in diameter on the Moon is shown in figure 9.11.

A planetary collision, as postulated in chapter 8, requires planets in elliptical orbits. According to the orbital rounding-off times in table 8.1 this means that a collision would have had to occur within a few million years of the formation of the planets and the satellites. This will give time for the Moon to produce a thick solid crust but it also suggests that the planets themselves may have been largely fluid at that time. The collision will have produced a large quantity of debris varying from asteroid size down to quite small bodies. As time progressed the number of large objects would decline because they would have been swept up by major planets, collided with larger solid bodies leaving major impact scars or be fragmented by collisions with each other. Smaller bodies will also decrease in number through all these effects but they will also be reinforced by the break-up of the larger bodies. Over the course of several hundred million years there will have been occasional impacts with the Moon able to excavate large basins. These would have been the source of the mare basins we see today. The initial excavation would have been very deep; modelling and small-scale experiments suggest that the ratio diameter:depth would have been about 5:1. This initial configuration would not have been stable and hydrostatic pressure would have elevated the floor of the basin until the depth was reduced to 10–20 km, at which stage the strength of the material could resist further elevation.

Below the floor of the basins the crust would be cracked and heavily damaged giving channels leading to molten material. Spasmodic eruptions of this material would have given the mare basins, covered with layers of basalt, that are seen today. On the far side of the Moon, with molten material at greater depth, the magma could not penetrate to the surface on the same scale giving fewer and smaller mare features on that side. We must now see what mechanisms were acting to give the spasmodic eruptions during the early active stage of the Moon's existence.

9.3.3   Mascons

An unexpected property of the Moon that was detected by the motions of space-craft was the presence of *mascons*. These are concentrations of mass at the centres of mare basins that enhance the local gravitational field. A spacecraft close to the surface of the Moon speeds up as it approaches a mascon region and slows down again as it departs. The first idea about mascons was that they were due to the buried remains of large iron bodies which had excavated the basins. It was soon realized that this could not be so. When the basins were formed there was molten material not far from the surface and any large dense body would simply have sunk towards the centre of the Moon. In any case careful measurement of the ex-cess field showed that the extra mass is probably in a disc-like form and appears to be due to the basaltic material filling the basin.

Since the lunar basalts are considerably denser than the highland material then it might seem obvious that the formation of a mare basin that involves the replacement of the latter material by the former must inevitably lead to the mas-con effect. This is not so. The principle of *isostacy*, illustrated in figure 9.12, indicates that material welling up from below should fill the excavated basin to a height such that the total pressure at some lower *compensation level* should be everywhere the same. Neglecting the small variation of the acceleration due to gravity with height near the surface the total mass of in-filling basalt would then equal the mass of displaced material forming the basin. The gravitational field above the surface of the basin should not be very different from that elsewhere;

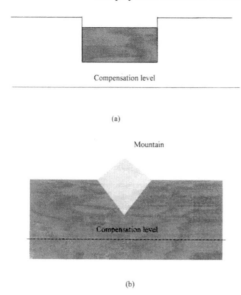

(a)

(b)

Figure 9.12. Idealized representations of the principle of isostacy. (*a*) A crater filled with dense basalt such that the pressure is uniform at the compensation level. (*b*) A mountain of low-density material with a 'root' such that the pressure is uniform at the compensation level.

indeed a naive view might be that since the basalt is at a lower level, and therefore further from the spacecraft, the field it experiences should be less rather than greater.

A solution of this problem was suggested by Dormand and Woolfson (1989) in terms of overfill of the mare basins above the level required by isostacy. Cooling curve calculations by Dormand and Woolfson showed that the main cooling was occurring in the surface regions while the interior of the Moon was subjected to radioactive heating and becoming hotter. The outer solid region, which was gradually cooling, was subjected to forces tending to produce contraction, while the inner material was, on average, heating up due to radioactivity and so was trying to expand. Initially this put solid surface material in a state of tension. Another effect was that material in the interior was subjected to suprahydrostatic pressure, i.e. pressure greater than that just due to the weight of the material above. When this pressure was relieved by material flowing through the cracks below the basins the level it reached was controlled not by isostacy but by the internal pressure. Once this internal excess pressure had been relieved the solidified ejected material would have slowly slumped towards isostatic equilibrium at a rate dependent on the thickness and strength of the lunar crust. If the crust was strong enough then slumping may have been insignificant. Successive bouts of volcanism would repeat the process but rather more sluggishly as the crust thickened and

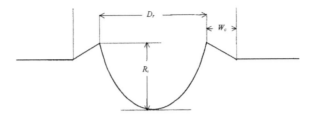

Figure 9.13. The ideal geometry of a fresh crater (Mullis 1993). The complete morphology is defined in terms of the observed dimensions D_r and the free parameter R_i given relationships (9.16).

became mechanically stronger. The end of the process would be a mass excess supported by a strong crust. Eventually the average temperature of the Moon's interior would have fallen and the solid crust, having a lesser volume to encompass, would experience compression forces.

9.3.4   Mascons and basalts in mare basins

One way of estimating the depth of the basalt in a mare basin is to use the *ghost crater technique*. From a study of unfilled craters their stratigraphy is well known. For a particular diameter of crater it is possible to specify within fairly narrow limits both the depth and the height of the crater rim above the original level. If the crater is subsequently flooded to a depth which still shows the crater rim then from the height of the rim above the surface the total thickness of the basalt deposit can be estimated. This technique was used by De Hon and Waskom (1976) and De Hon (1979) to estimate the depth of basalt and therefore the volumes in various mare basins. However, this method breaks down for depths over 1500 m as craters large enough to enable the technique to be used will not be available. Thus estimates of depths in the middle of larger mare basins are not available but the estimated volume from what could be estimated was about 3×10^6 km^3.

A number of workers considered using the gravitational measurements from mascons as a way of judging the depth of material in the larger basins. However, the body of basalt associated with a basin will be in two parts: first, an isostatic root consisting of dense mantle material embedded in the less-dense crust; and then the basalt infill of the residual crater. Mullis (1991, 1992) suggested an approach in which the depth of excavation into the target surface, which defines the geometry of the mantle root and the residual crater depression, is taken as a free parameter. A model of a fresh impact crater is shown in figure 9.13. Pike (1967) showed that small craters are almost half hemi-elliptical in cross-section with semi-major and semi-minor axes

$$a_c = \frac{D_r}{\sqrt{3}} \qquad (9.16a)$$

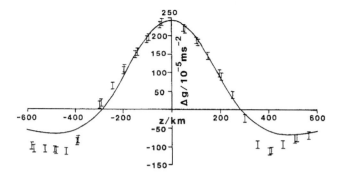

Figure 9.14. The calculated gravitational anomaly over Mare Serenitatis compared with experimental points from the Explorer high-altitude gravity experiment (Mullis 1992).

and

$$b_c = 2R_i \tag{9.16b}$$

where D_r is the diameter of the rim crest and R_i is the total internal depth. Another relationship from Pike (1967) gives the exterior ridge width as

$$W_c = \frac{D_r}{5}. \tag{9.16c}$$

Given that D_r can be estimated from observation of a large mare basin the single free parameter R_i plus relationships (9.16) completely define the morphology of the original excavation. Floor uplift will follow the initial excavation to bring the crater to isostatic equilibrium and it is assumed that this is by the rise of mantle material to fill the crater to the necessary height. Since mantle material is denser than crustal material this leaves a depression which is then filled by basalt coming from below. The crust was taken as a double layer, the upper one of thickness 20 km and density 2700 kg m^{-3} and the lower one of thickness 50 km and density 2900 kg m^{-3}. The mantle material was taken with a density of 3350 kg m^{-3} and the extruded basalt with density 3400 kg m^{-3}. Mullis (1991, 1992) found model mare basins to match the measured graviational anomalies for Imbrium, Serenitatis, Crisium, Humorum, Nectaris and Orientale. Figure 9.14 shows the match he obtained for Mare Serenitatis. The basalt fill for Serenitatis, which gave rise to the mascon since no slumping was built into the model, was 4.9 km. This result and those for the other mare are shown in table 9.2.

The shape of the gravity anomaly is very sensitive to the excavation depth. The depth of the basalt is determined by the maximum gravitational anomaly while the depth of excavation is obtained by getting a 'best fit' of the shape of the anomaly over the whole region. Mullis (1992) estimated the total volume of basalts in the six basins he investigated as about 2.0×10^6 km^3. Based on this work the total volume of surface basalt deposits will be considerably greater than that estimated from ghost craters and will be in the range $4-5 \times 10^6$ km^3.

Table 9.2. Estimated excavation depths, maximum depth of basalt flooding and total volume of extruded basalt for the six large mascon basins. There are two separate estimates of D_r for Nectaris.

Crater dimensions				Basalt fill	
Basin	D_r (km)	R_i (km)	R_i/D_r	Depth (km)	Volume (km^3)
Crisium	630	58	0.0921	3.5	0.20
Humorum	560	44	0.0785	4.2	0.27
Imbrium	970	68	0.0701	5.2	0.76
Nectaris	600	44	0.0733	4.1	0.31
Nectaris	870	72	0.0828	3.7	0.33
Orientale	900	68	0.0755	1.8	0.80
Serenitatis	800	62	0.0755	4.9	0.55

9.3.5   Volcanism and the evolution of the Moon

A detailed theoretical and computational analysis of the Dormand and Woolfson model of volcanism and mascon formation has been given by Mullis (1993). Mullis considered a five-layer model of the Moon based on one given by Bills and Ferrari (1977). This consisted of

- the upper crust, thickness 20 km and density 2703 kg m^{-3};

- the lower crust, thickness 50 km and density 2901 kg m^{-3};

- the upper mantle, thickness 230 km and density 3377 kg m^{-3};

- the lower mantle, thickness 500 km and density 3403 kg m^{-3}; and

- material below 800 km deep.

The density of the material below 800 km is constrained by the overall density of the Moon, 3340 kg m^{-3}.

The pressure at any distance from the centre is subject to the equation

$$\frac{dP}{dr} = g(r)\rho(r) \tag{9.17a}$$

where P is the pressure at radius r and $\rho(r)$ is the local density. The local gravitational field is given by

$$g(r) = -\frac{GM(r)}{r^2} \tag{9.17b}$$

where $M(r)$ is the contained mass out to the distance r.

Figure 9.15. A schematic view of the five layers of the Moon as given by Bills and Ferrari (1977). Magma wells up to fill a basin through a crack from the upper-mantle layer.

Figure 9.15 illustrates the five layers near the surface and a column of magma filling a somewhat idealized crack from some depth to the surface. In general, at zero pressure the density of silicate melts is about 10% less than that of the solid. However, these melts are compressed under pressure and Mullis estimated that at a depth of 375 km the magma would have the same density as the solid. The pressure due to the solids at this depth is about 18.15 kbar and that of the column of magma is only 16 bar less. Taking viscosity and the restraints offered by narrow cracks into account the excess pressure of 16 bar would be insufficient to drive magma through the crack on to the surface.

In order to computationally investigate the thermal evolution of the Moon, Mullis first considered some observational and theoretical considerations that would constrain his model. The first of these came from the dating of lunar basalt that suggests that the main period of volcanism was between 3.9 and 3.2×10^9 years ago. The next was evidence from surface features of the Moon that would indicate whether it had been in a state of expansion or contraction during the period of thermal evolution. Expansion would result in tensional features in the crust, such as rift valleys while contraction would give folding and thrust faults. In considering all the evidence put forward Mullis concluded that contraction, corresponding to a decrease in radius up to 5 km would be reasonable. Such extension features that are seen on the Moon, in particular the cracks called *rills*, can be readily explained by local stresses such as those due to the settling of lunar basalts.

Another constraint, suggested by the work of De Hon (1979) and also by his own work (Mullis 1991, 1992), is that the total volume of mare basalts is 4×10^6 km^3 with an uncertainty of order 25%.

9.3.6 Calculations of thermal evolution

The heat conduction equation for spherical symmetry is

$$\frac{\partial T}{\partial t} = \frac{1}{\rho C_p} \left\{ \frac{1}{r^2} \frac{\partial}{\partial r} \left(r^2 \kappa \frac{\partial T}{\partial r} \right) + H(r,t) \right\}$$ (9.18)

where ρ, C_p and κ are the density, specific heat capacity at constant pressure and thermal conductivity of the mantle material and $H(r,t)$ the rate of heat generation per unit volume as a function of distance from the centre and time.

The change of state from liquid to solid in this model has been simulated by specifying two temperatures, both functions of pressure. The *solidus temperature* is that below which the material is solid. Taking a model pyroxene as lunar material Mullis used a parameterized form of the solidus curve suggested by Mizutani *et al* (1972) for the Moon

$$T_{sol} = 2000 - 2.08 \times 10^{-10} r^{-2}$$ (9.19)

giving the temperature in K if r is in metres. The *liquidus curve* is set at all depths by

$$T_{liq} = T_{sol} + 200 \text{ K}$$ (9.20)

with a specific heat in the melting region given by

$$C_p' = C_p + \frac{L}{T_{liq} - T_{sol}}$$ (9.21)

where L is the latent heat of fusion. It was assumed that the fraction of material melted in the interval $T_{sol} \le T \le T_{sol}$ is

$$p = \frac{T - T_{sol}}{T_{liq} - T_{sol}}.$$ (9.22)

Under the conditions within the early Moon there would have been two contributions to thermal conductivity—*lattice conductivity* and *radiative conductivity* so that

$$\kappa(T) = \kappa_l(T) + \kappa_r(T).$$ (9.23)

These were expressed as functions of T and the resulting expressions gave reasonably good agreement with measured conductivities for a number of common minerals. To allow for the porosity of the near surface lunar material Mullis followed a suggestion of Binder and Lange (1980) in taking the surface conductivity as 20% of the value given by the function for $k(T)$ rising linearly to 100% at a depth of 20 km.

Once partial melting was taking place, with p defined by (9.22) for $T_{liq} \ge T \ge T_{sol}$ and $p = 1$ for $T > T_{liq}$, then normal fluid convection would be taking place. This was simulated as an enhanced conduction so that

$$\kappa' = \kappa \exp(\chi p)$$ (9.24)

with χ as an adjustable parameter. Convection can also occur in solid material close to the melting point and was again simulated as enhanced conductivity. For the temperature range $0.7T_{sol} \leq T \leq T_{sol}$ the effective conductivity was taken as

$$\kappa' = \kappa(T)\left(1 + A\frac{T - 0.7T_{sol}}{0.3T_{sol}}\right) \tag{9.25}$$

with A as another adjustable parameter.

Heating of the Moon was taken to be by radioactivity due to ^{238}U, ^{232}Th and ^{40}K although other forms of heating might also have been present—for example, heating by short-lived ^{26}Al, electromagnetic induction heating (see section 9.4.2.2) or heating through tidal effects. The concentrations of the three relevant radioactive elements are so high in the crust that it is clear that they cannot be characteristic of the Moon as a whole, otherwise the heat flow through the surface would be many times that observed—typically 16–22 W m^{-2}. On solidification of magma the solid preferentially takes up the large high valency cations U^{4+} and Th^{4+}. This has been modelled by assuming that at each time step newly formed solid has η times the radioactive content of the concentration in the melt at that time.

Due to thermal expansion and contraction effects the radius of the Moon changes during the simulation. During a time interval Δt the change of radius is given by

$$\Delta R(\Delta t) = \frac{1}{R^2}\int_0^R \alpha'(T)\Delta T(r, \Delta t)r^2\,dr \tag{9.26}$$

where $\alpha'(T)$ is the effective coefficient of expansion at temperature T and $\Delta T(r, \Delta t)$ is the change of temperature at radius r during the time interval Δt. The thermal expansion coefficient, α, was taken as 2.5×10^{-5} K^{-1} and independent of pressure an allowance was made for melting by taking between the solidus and liquidus temperatures

$$\alpha' = \alpha + \frac{\gamma}{T_{liq} - T_{sol}} \tag{9.27}$$

where γ is the fractional change of volume on melting.

Finally the stresses could be calculated throughout the body. For the change of radial stress

$$\Delta\sigma_r(r, \Delta t) = \frac{2\alpha'E}{3(1-\nu)}\left\{\frac{1}{R^3}\int_0^R \Delta T(r', \Delta t)r'^2\,dr' \right. $$
$$\left. - \frac{1}{r^3}\int_0^r \Delta T(r', \Delta t)r'^2\,dr'\right\} \tag{9.28a}$$

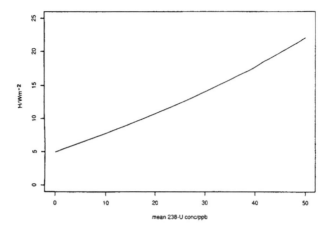

Figure 9.16. Present surface heat flow as a function of the present ^{238}U concentration.

where E is Young's modulus and ν is Poisson's ratio for lunar mantle material. The change in tangential stress is given by

$$
\Delta\sigma_t(r, \Delta t) = \frac{\alpha' E}{3(1-\nu)} \left\{ \frac{2}{R^3} \int_0^R \Delta T(r', \Delta t) r'^2 \, \mathrm{d}r' \right.
$$
$$
\left. + \frac{1}{r^3} \int_0^r \Delta T(r', \Delta t) r'^2 \, \mathrm{d}r' - \Delta T(r, \Delta t) \right\}. \quad (9.28b)
$$

The model had the constraint that it had to give the correct present heat flow through the surface and lithosphere thickness as ascertained from seismic results. Estimates of heat flow through the surface vary from 12–22 W m^{-2} but the upper end of the range is more typical. Results from the computational model as shown in figure 9.16 show that, as expected, the surface flux is dependent on the radioactive content expressed here in terms of the present ^{238}U concentration in parts per billion (ppb). The other radioactive components are defined by a present K/U ratio of 2000 and a ^{232}Th/U ratio of 3.8. At the low end of ^{238}U concentration the influence of the residual heat of formation is evident.

Seismic velocity profiles of the lunar interior (Nakamura *et al* 1982) indicate that the lithosphere is solid to depths of 800–1000 km. Moonquake activity seems to be confined to the region between 800 and 1000 km deep and suggests that this may be where subsolidus convection is taking place. With subsolidus convection included in the model, surface heat-flow values of 12, 16, 19 and 22 × 10^{-3} W m^{-2} are consistent with a 1000 km solid lithosphere with values of A in (9.25) of 1.4, 2.6, 4.3 and 6.0 respectively. Without subsolidus convection thermal conductivities of 1.2, 1.5, 1.75 and 2.0 times that of the mineral dunite are required.

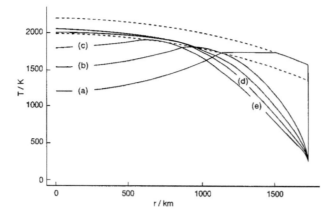

Figure 9.17. Sequence of thermal profiles in the Moon: (*a*) at its formation; (*b*) after 10^9 years; (*c*) after 2×10^9 years; (*d*) after 3×10^9 years; and (*e*) at present (Mullis 1993).

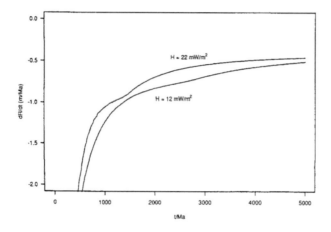

Figure 9.18. Rate of change of the radius of the Moon for two different heat flows as a function of time.

A sequence of thermal profiles for one model is shown in figure 9.17. The initial temperature at the centre is 1200 K corresponding to heavy transport of shock-wave energy into the Moon as it was accreting. The solid crust region is just under 200 km thick with, below that, over 400 km of partial melting. At the present time the model indicates a solid lithosphere down to 1000 km with a partially molten zone below that all the way to the centre. The present bulk ^{238}U concentration is indicated as 34 ppb giving a heat flow of 16 W m^{-2}. The conductivity is taken as that of polycrystalline dunite together with subsolidus convection.

Curves for the rate of change of the lunar radius are shown in figure 9.18 for two extreme values of surface heat flow. They both indicate a total contraction of between 4 and 5 km during the lifetime of the Moon. The volume of basalt generated may be estimated from the model. The rate of accumulation of excess pressure in the magmasphere is found from

$$\frac{dP}{dt} = \frac{1}{r_u^3 - r_1^3} \int_{r_1}^{r_u} \frac{d\sigma_r(r,t)}{dt} r^2 \, dr \qquad (9.29)$$

where r_1 and r_u are the lower and upper radii of the magmasphere. The rate of extruding basalt to relieve this excess pressure is

$$\frac{dV}{dt} = \frac{4\pi(r_u^3 - r_1^3)}{3K} \frac{dP}{dt} \qquad (9.30)$$

where K, the bulk modulus, is taken as 10^{11} Pa. The total volume of basalt emitted is thus

$$V = \int_{t_2}^{t_1} \frac{dV}{dt} \, dt \qquad (9.31)$$

where t_1 and t_2 correspond to 3.2 and 3.9×10^9 years before the present, the dated period of volcanism. Various reasonable sets of parameters give a total volume of 5×10^6 km^3 which is at the upper end of what is estimated from the analysis of mascons (section 9.3.4).

9.4   Lunar magnetism

Spacecraft measurements have revealed that, at some distance from its surface, the Moon has no discernible external magnetic field—which was as expected of a small body spinning very slowly. Closer to the surface weak fields of up to a few hundred nT are measured which vary in both strength and direction on a kilometre scale. However, the first impression, that the Moon was uninteresting from a magnetic point of view, was soon dispelled once magnetometers were placed on its surface.

Even if a body has no intrinsic magnetic field of its own a great deal can be learned about it by studying how it behaves in an external field. Unfortunately the presence of the Sun makes such studies difficult because the field due to the Sun changes very quickly and erratically. However, for some part of every month the Moon passes into the shadow of the Earth and sits in the part of the Earth's field called the *geomagnetic tail*, where the field is comparatively stable at 10^{-8} T. Magnetometer readings on the surface then give information about the average permeability of lunar material. Ferromagnetic materials and, to a lesser extent, paramagnetic materials are magnetized in the same direction as the applied external field and so augment the field. On the other hand, diamagnetic materials are magnetized in the opposite sense to the applied field and hence deduct from

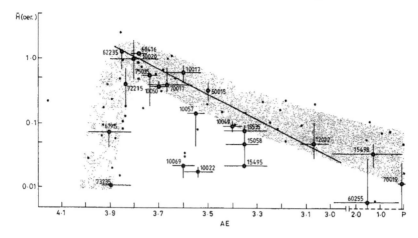

Figure 9.19. The lunar surface field as a function of time. The observations without error bars are subject to large error. The horizontal scale is in units of 10^9 years before the present and the vertical scale in units of 10^{-4} T (Runcorn 1988).

it. Surface magnetometer measurements indicate an overall lunar permeability in the range 1.012 ± 0.010 so that it is paramagnetic or perhaps weakly ferromagnetic. One can learn something about the iron content of the Moon from this observation. Below the Curie point, 1043 K, iron is ferromagnetic but above that temperature it is paramagnetic. From the overall permeability the Moon must contain at least 0.8% by mass of ferromagnetic iron and this must then be within 200–300 km of the surface to be below the Curie point. On the other hand, the maximum metallic iron content is 4.8%, by mass assuming that it is all above the Curie Temperature and therefore paramagnetic

Because of the low state of oxidization of lunar material the lunar basalt contains particles of free iron. When a ferromagnetic material is at a high temperature, above the *Curie point* it cannot be permanently magnetized. If the ferromagnetic material cools then, as soon as it reaches the Curie point temperature, it is permanently magnetized with an intensity characteristic of the ambient magnetic field. If the external field is subsequently removed it retains its state of magnetization—that is referred to as its natural remnant magnetization (NRM). The ambient field at the time of cooling through the Curie point can be estimated from laboratory experiments with the sample. The sample is magnetized in a known field and then its intensity of magnetization is measured. Since for weak fields the intensity of magnetization is proportional to the magnetizing field it is possible to estimate the magnetizing field that produced the NRM.

When the deduced magnetizing fields are plotted against age for lunar basalt samples the result is as shown in figure 9.19. The general trend seems to show that at the beginning of the dated period of volcanism, 3.9×10^9 years ago, the surface field was greater than 10^{-4} T and then fell with time to a few per cent of that value by 3.2×10^9 years ago. Subsequently it has fallen to the effectively-zero value today. The early fields are unexpectedly strong, being about twice that of the present terrestrial surface field.

There is a great deal of scatter in the points contributing to the graph, some due to inevitable experimental error but some seeming to indicate that basalt deposits of the same age were magnetized in fields of different strength. There is also a downturn in the estimated field strength prior to 3.9×10^9 years ago and the significance of this is not understood. Once a rock has acquired NRM it could lose it by being reheated. When it subsequently cooled it would then record the NRM corresponding to a later field. The Curie point temperature for iron is well below the melting points both of iron and the host basalt so basalt can lose its original magnetization but still remain as a closed system as far as radioactive dating is concerned. While it is difficult to specify scenarios which would lead to this re-heating and re-cooling while still leaving the basalt accessible to collection it seems sensible to accept the highest field corresponding to any age as the most significant. This still leaves open the question of the significance of the downturn prior to 3.9×10^9 years ago.

9.4.1   A dynamo theory

It is accepted that planetary magnetic fields are due to the action of a dynamo mechanism acting within an internal conducting fluid. Since there is evidence of strong fields at the Moon's surface in the past then this suggests the possibility of a dynamo within the Moon when it was younger and had a more fluid interior. Runcorn (1975) strongly proposed this idea. He argued that the only alternative was that the Moon was magnetized by an external field. The pattern of surface NRM produced by an internal dipole source which gradually decayed would be such as to give no external dipole field, which is what is observed. On the other hand if the Moon had been magnetized by a uniform external field then the pattern of NRM *would* give an external field—which is contrary to what is observed.

From a mapping of lunar magnetic anomalies by Hood *et al* (1978), measured by three-component magnetometers, Runcorn (1988) deduced the position of the N-magnetic pole that would correspond to each anomaly. He found that the positions correspond to three distinct groups corresponding to three axes heavily inclined to the Moon's spin axis. From this Runcorn concluded that the surface of the Moon had reoriented itself due to collisions by lunar satellites. A satellite with a decaying orbit would have arrived at a low angle to the surface and through a bouncing effect would have created collinear impact features. In addition the creation of the impact basins would have changed the moment-of-inertia tensor of the Moon and so caused the polar axis to wander (see section 10.2.5).

There are a number of problems with this hypothesis. One is that it would require collisions with the Moon some 800 million years after its formation and this is greater than the estimated lifetimes of lunar satellites if such bodies existed. Another is the need for a large conducting core in the Moon that, according to Runcorn, would be an iron core of radius about 500 km. However, based on the estimate of the moment-of-inertia factor for the Moon, 0.3905 ± 0.0023 (Ferrari *et al* 1980) plus its mean density it was shown by Stock and Woolfson (1983a) that the most likely radius of an iron core is 382 km and that this estimate is comparatively insensitive to other parameters of the model used, e.g. bulk modulus, thermal coefficient of expansion and the thickness and density of the crust. The result *was* sensitive to the density of the core and a FeS (troilite) core of radius about 570 km would be possible. Such a core is extremely unlikely. A further difficulty is that most models for the thermal evolution of the Moon, such as portrayed in figure 9.17, do not suggest a molten core for the first 1.5×10^9 years. The dating of magnetized basalt requires the field to be present within 700 million years of the Moon's formation.

9.4.2   The induction model of lunar magnetism

Stock and Woolfson (1983b) suggested another model that gives a field of internal origin, but one induced in it by an external field. The model is linked to the occurrence of volcanism and we now look at the distribution of the maria which is a manifestation of that volcanism. In the first two numerical columns of table 9.3 the positions of the centres of the largest mare basins are shown.

They are all within $30°$ of the equator with the exception of Imbrium and the partly-filled basin giving Mare Australe. In general it is not possible to identify the sources of the magma which filled the basin but the centre of the basin, corresponding to the deepest part of the original excavation, is the most likely region. Molten lunar rocks have a very low viscosity, roughly the same as engine oil, so they flow readily and do not usually leave discernible flow fronts (Guest *et al* 1979). In the case of Imbrium, flow fronts *can* be seen; the flows were sufficiently viscous to indicate a source region near the crater Euler which is much closer to the equator than the basin centre. The centre of the Australe basin is well to the south but the lava flows are concentrated in the north. Table 9.2 shows a range of likely source latitudes. In the dynamical evolution of its orbit around the Earth, with much stronger tidal effects than now, the spin axis of the Moon could have been modified. All the source regions are within $29°$ of a great circle inclined at $6°$ to the equator, if the most northerly possible source region is taken for Australe.

This evidence seems to suggest that volcanism has an equatorial bias since, even when basins are well away from the equator, the sources are then displaced from the centre towards the equator. A reason for this is now suggested. If the Moon is a captured body, as proposed in section 9.1.5, then it would begin its association with the Earth in a highly eccentric orbit. Singer (1968) has described the evolution of the lunar orbit starting from this state. The eccentricity of the orbit would subject the Moon to a periodic stress. Hysteresis effects would dump thermal energy into the Moon, this coming from a loss of orbital energy. The

Table 9.3. The positions of the centres of mare basins or of likely source positions. The figures in the final column give 'latitudes' with respect to an 'equator' tilted at 6° to the actual equator.

Mare	Latitude (°)	Longitude (°)	Modified Latitude (°)
Imbrium (centre)	40N	22W	(38N)
Imbrium (source)	23N	29W	20N
Serenitatis	26N	18E	28N
Tranquillitatis	10N	28E	13N
Crisium	17N	58E	22N
Marginus	18N	82E	24N
Smythii	1N	88E	7N
Foecunditatis	3S	48E	1N
Nectaris	15S	33E	12S
Nubium	19S	16W	21S
Humorum	20S	42W	24S
Procellarum	5S	48W	9S
Orientalis	19S	170W	20S
Moscoviense	25N	145E	28N
Australe (centre)	45S	90E	(39S)
Australe (range of likely sources)	35–42S	90E	29–36S

result would be evolution to a nearly circular orbit with radius close to the original perigee distance. At this stage of rapid dynamical evolution the spin axis of the Moon would tend to become parallel to its orbital axis. The heat energy would be generated mainly in the equatorial region where the tidal stresses are greatest and since the orbit and spin would not be synchronous this energy would be spread uniformly around the equator. Thus it would be expected that in the equatorial region molten material would be both hotter and closer to the surface. After the initial rounding-off stage a slower evolutionary process would begin, which is still acting, causing the Moon to recede slowly from the Earth while at the same time the eccentricity of its orbit gradually increases (section 9.1.1).

Silicates are semiconducting materials, which means that their conductivities, σ, increase with temperature. The variation with temperature of some common minerals is shown in figure 9.20. According to figure 9.17 the temperature of the molten material at the time of interest would have been about 1750 K but this does not take account of tidal heating which may have brought the temperature nearer to 2000 K and the location closer to the surface near the equator. It is clear that the molten material around the equator constituted a conducting ring, the conductivity of which would have been several times higher than that of material away from the equator.

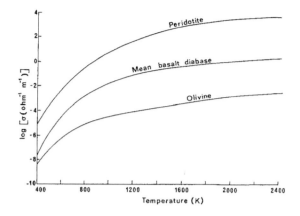

Figure 9.20. The conductivity of some common minerals as a function of temperature (Stock and Woolfson 1983b).

9.4.2.1 The characteristics of the early Sun

A potentially strong source of early magnetic field was the Sun which could have had a magnetic dipole moment much greater than at present. Borra *et al* (1982) have reviewed theories of stellar magnetic fields. The idea that the Sun would reach the main sequence with a field 1000 times the present value, 8×10^{22} T m^3, and with a decay time of 10^8–10^9 years is well within the range of speculation. Freeman (1978) has considered the primordial solar magnetic field from several points of view and concludes that a range from 10^{24}–5×10^{26} T m^3 covers the likely values.

Due to its interaction with the solar wind the dipole nature of the Sun's field is only maintained very close in. Further out where the field is weaker it becomes coupled to the wind particles, which constitute a plasma, and the fall-off in field is somewhere between r^{-1} and r^{-2} rather than the r^{-3} characteristic of a dipole. The distance at which the dipole form breaks down is governed by the relative strength of the field, described by its pressure (energy density) and the pressure due to the flow of solar wind particles. If the dipole has moment ψ then the distance R where the two pressures are equal is found from

$$\frac{\psi^2}{2\mu_0 R^6} = \frac{\dot{M}v}{4\pi R^2}$$

or

$$R = \left(\frac{2\pi\psi^2}{\mu_0 \dot{M}v} \right)^{1/4} \tag{9.32}$$

where μ_0 is the permeability of a vacuum, $\dot{M}$ is the rate of loss of solar wind material and v its speed. For the present Sun with $\dot{M} = 4 \times 10^8$ kg s^{-1} and

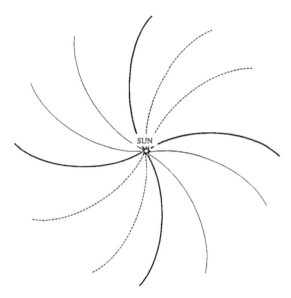

Figure 9.21. An idealized representation of the sector structure of the Sun. The full and dashed field lines are towards and away from the Sun, respectively. There are substantial components of field perpendicular to the plane of the diagram.

$v = 400$ km s^{-1} we find $r = 3.8 \times 10^6$ km or just over five solar radii. An earlier Sun would not only have had a larger dipole moment but would also have been more active with a much stronger solar wind. The estimated outflow from T-Tauri stars is 10^8 times as much as at present; a flow some 10^6 times greater than at present with a dipole moment 1000 times greater would give the same distance of breakdown of the dipole field.

A feature of the magnetic field of the present Sun, and presumably of the Sun in past times, is its sector structure, illustrated in figure 9.21. Viewed down the solar spin axis the magnetic field lines closest to the viewer alternately point towards and away from the Sun in different sectors. There are substantial components of field perpendicular to the plane of the figure and these will point alternately inwards and outwards from the plane. This sector structure spins with the Sun so at a fixed position in the Sun's equatorial plane there would be a field perpendicular to the plane alternating upwards and downwards.

For the field due to the Sun in the plane of the Earth's orbit we may take the field normal to the plane in the form

$$B(R, \theta, t) = \psi \left(\frac{1}{R^3} + \frac{1}{(kR_\odot)^{3-p} R^p} \right) \sin[2(\theta - \omega t)] \qquad (9.33)$$

where the field strength is defined at a point with polar coordinates (R, θ) at time t. The radial part of (9.33) is due to Freeman (1978). At distances from the Sun much smaller than $kR_\odot$ the field has a dipole character. At distances much greater than $kR_\odot$ it varies as R^{-p}.

9.4.2.2 The generation of currents in the molten ring

We consider here an isolated conducting ring of mean radius a and with a circular cross-section of radius b ($\ll a$) perpendicular to a uniform time-varying field $B_0(t)$. The equation for the current, i, generated in the ring is

$$L\frac{di}{dt} + Ri = -\pi a^2 \frac{dB_0}{dt} \tag{9.34}$$

where L is the self-inductance of the ring and R its resistance. For $B_0(t) = (B_0)_{max} \sin \omega t$ the steady-state solution is

$$i = -\frac{\pi a^2 (B_0)_{max}\omega}{(R^2 + \omega^2 L^2)^{1/2}} \sin(\omega t + \varepsilon) \tag{9.35}$$

where $\tan \varepsilon = R/(\omega L)$.

The resistance and inductance of the ring are both frequency dependent since, for high frequencies the skin effect becomes important. The skin depth is given by

$$\delta = \left(\frac{2}{\mu_0 \sigma \omega}\right)^{1/2}. \tag{9.36}$$

To a first approximation currents are restricted to a distance δ from the surface of the conductor. This leads to

$$R = \begin{cases} \dfrac{2a}{b^2 \sigma} & (\mu_0 \sigma \omega b^2 \leq 2) \\[2ex] \dfrac{\mu_0 a \omega}{(2\mu_0 \sigma \omega b^2)^{1/2} - 1} & (\mu_0 \sigma \omega b^2 > 2). \end{cases} \tag{9.37a, b}$$

The self-inductance of the ring is of the form

$$L = \mu_0 a \ln\left(\frac{\eta a}{b}\right) \tag{9.38}$$

where η varies from 1.39 for low frequencies to 1.08 for high frequencies.

When $\sigma \omega$ is sufficiently small the ring is resistance dominated and the maximum current is given by

$$i_{max} = \tfrac{1}{2}\pi ab^2 \sigma \omega (B_0)_{max}. \tag{9.39a}$$

At the other extreme when $\sigma \omega$ is very large the ring is inductance dominated and

$$i_{max} = \frac{\pi a (B_0)_{max}}{\mu_0 \ln(\eta a/b)} \tag{9.39b}$$

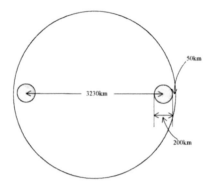

Figure 9.22. A cross-section of the Moon showing an equatorial conducting ring.

which is independent of σ and ω.

The maximum induced field at a point external to the volume of the ring will, if $a \gg b$ and the point is not too far from the ring, be given by

$$B_i(r) = \frac{\mu_0 i_{\max}}{2\pi r} \tag{9.40}$$

where r is the distance from the centre of the circular cross-section. Equation (9.40) assumes that the ring approximates to an infinite rod.

When the induced field adds to the external field then the field amplification factor is

$$\phi = 1 + \frac{B_i}{B_0}. \tag{9.41}$$

A final quantity of interest is the mean power dissipated in the ring which is

$$\langle P \rangle = \tfrac{1}{2} i_{\max}^2 R. \tag{9.42}$$

The numerical description of the inductance effect will now depart from that given by Stock and Woolfson (1983a) to take account of the later work by Mullis (1993) on conditions inside the Moon. It will also take account of tidal effects giving molten material both hotter and closer to the surface in the equatorial region. Although silicate material has some conductivity at any elevated temperature, to a first approximation we may take the situation as having an equatorial ring just below the equator with zero conductivity elsewhere. The configuration we take, corresponding to an early period when the magnetic field is very high, is as shown in figure 9.22. The mean radius of the ring is 1615 km, and the circular cross-section has radius 100 km. The distance of the surface to the nearest part of the conducting ring is 50 km. Other values taken are: $\psi = 8 \times 10^{25}$ T m^3, some 1000 times the present value and within the acceptable range; $\sigma = 6\ \Omega^{-1}\ \mathrm{m}^{-1}$ as suggested by figure 9.20; $\omega = 2.9 \times 10^{-5}\ \mathrm{s}^{-1}$ corresponding to an early solar spin period of 5 days, again within the range of expectation. The skin depth, δ,

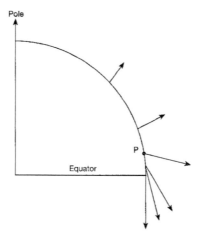

Figure 9.23. The variation of the strength and direction of the lunar surface field with latitude according to the induction model (Stock and Woolfson 1983b).

equals 96 km which is about the same as the radius of the cross-section of the conducting ring. The resistance of the ring, given by (9.37a), is 5.4×10^{-5} Ω. From (9.38) $L = 5.0$ H and $L\omega = 1.45 \times 10^{-4}$ Ω. Since the ring is inductance dominated we find from (9.39) and (9.40)

$$B_{\mathrm{i}}(r)_{\mathrm{max}} = \frac{a}{2d\ln(\eta a/b)}(B_0)_{\mathrm{max}} \qquad (9.43)$$

where d ($=100$ km) is the distance from the centre of cross-section of the ring to the surface. This means that, from (9.41), induction in the ring gives an amplification factor $\phi = 3.8$. Stock and Woolfson (1983b) carried out a somewhat more complicated analysis taking account of the fact that the lateral spread of the conducting region was up to 30° from the equator and slightly larger amplification factors were found.

The present form of the solar magnetic field, as described by (9.33), is best fitted with $k = 1.75$ and $p = 1.8$. If the Earth approached to within 0.5 AU of the Sun in its initial evolving orbit then the maximum field it would have experienced is 0.26×10^{-5} T and conducting ring amplification could have increased that to 10^{-4} T which is what the observations require. The field would have fluctuated both in strength and direction at any point on the Moon's surface. There would also have been different effects at different points on the Moon's surface at any one time and figure 9.23 shows the variation as a function of latitude found by Stock and Woolfson (1983b) taking into account an extended band of conduction around the equator.

A parameter on which these results are very sensitively dependent is the magnitude of the solar dipole moment, ψ. It has been taken as 1000 times the

Table 9.4. The effect of partial ionization of escaping material on the field at the Earth–Moon perihelion.

Degree of ionization	Relative field
1.000	1.000
0.500	0.841
0.100	0.562
0.010	0.316
0.001	0.178

present value and the calculated lunar surface field will be proportional to the assumed value for this quantity. Through their common dependence on the early solar dipole moment the mechanism for producing high fields on the lunar surface is linked to that described in section 6.3.1 for magnetic braking of an early rapidly-spinning Sun. Table 6.1 shows that if the early dipole moment was 8×10^{25} T m^3 and the rate of loss of *ionized* material was at 2×10^{15} kg s^{-1} ($3.2 \times 10^{-8} M_\odot$ year^{-1}) for 10^6 years then an early Sun would retain only 0.0001 of its initial angular momentum. However, the material from T-Tauri stars shows strong neutral hydrogen spectral lines and hence cannot be as strongly ionized as the present solar wind. If one accepts that the escaping material is not strongly ionized then this also has implications for lunar magnetization for then the form of the solar field will not be as in (9.33) with the parameters used here. The solar field will be less distorted and the approximate dipole structure will persist further out which will weaken the field in the perihelion region of the Earth–Moon orbit.

It is possible to estimate the reduction in field if it is assumed that the scaling of the field depends on the distance at which the magnetic pressure equals the dynamic pressure of the flow of ionized material. This distance, R_p, is given by a modification of (9.32)

$$R_p = \left(\frac{2\pi\psi^2}{\mu_0 \delta \dot{M} v} \right)^{1/4} \tag{9.44}$$

where δ is the proportion of the outflow which consists of charged particles. From this it is seen that R_p is proportional to $\delta^{-1/4}$ and scaling the field will replace k in (9.33) which has been taken as 1.75, by $k/\delta^{1/4}$. The effect of this on the magnetic field at perihelion is shown in table 9.4. It is seen that even if only 1% of the escaping material is ionized the field is only reduced by a factor 0.3.

On the other hand, the analysis which led to table 6.1 shows that the effect of partial ionization is much more severe. With a solar magnetic dipole moment of 8×10^{25} T m^3 and a mass loss of 2×10^{15} kg s^{-1} the reduction to 1% ionization changes the proportion of angular momentum retained from 0.0001 to 0.3925. This has serious implications for those cosmogonic theories that require a very large loss of solar angular momentum.

The maximum magnetizing field would have declined with time as the Sun became less active and as the Earth's orbit evolved with increasing perihelion distance. This decline would have had a characteristic time of tens of millions of years. The Earth's eccentric orbit would have introduced another variation, this time with a period of the order of a few years. Finally, with a period of the order of days, there would be rapid fluctuations of field at any point on the Moon's surface due to the spin of the Sun. Both the strength and direction of magnetization of neighbouring basalt deposits could differ depending on the state of the field as they cooled through the Curie point. This is entirely consistent with observations.

The amplification due to ring currents may be regarded as of only marginal interest; increasing the postulated solar dipole field by a factor of 4 would have the same effect and this could not be ruled out. However, if the model suggested for the thermal evolution of the Moon is valid then there is no doubt that the current would flow as described and that some modification of the magnetic environment would follow.

Another effect of the generation of currents would have been ohmic heating within the conducting ring. This would have been a maximum of about 4×10^{10} W and much smaller than that on a time-averaged basis. This rate of energy generation would have been trivial compared with that being generated by tidal effects.

9.5   Summary

The two theories for the origin of the Moon that seem most plausible are the giant-impact theory and capture following a planetary collision. The former theory makes the Moon a unique object in the Solar System in terms of its mode of formation and how it became associated with the Earth whereas the latter theory makes it unique only in the way it became associated with the Earth.

The giant-impact theory has not been developed to the point where it explains anything other than the existence of the Moon. By contrast the collision plus capture model does offer an explanation for the hemispherical asymmetry of the Moon in terms of bombardment at the time of the planetary collision. The magnetic history of the Moon also fits in well with the capture hypothesis but it seems that it could equally well fit in with a giant-impact origin.

Chapter 10

Smaller planets and irregular satellites

10.1   Introduction

With the exception of the Proto-planet Theory all the modern theories would lead to an initially ordered Solar System. Nevertheless it is quite possible that an initially disordered system would evolve into an ordered one in some way or other. Bodies in highly chaotic orbits that could potentially cross each other would inevitably, in the course of time, lead to collisions so that dominant bodies would tend to come about in particular regions. Again, as was described in section 7.1.4, the mechanism of resonance locking due to planetary interactions in a resisting medium gives rise to systematic relationships, that were not originally present, between planetary orbits.

Accretion-based theories ascribe the differences in giant and terrestrial planets to the region of their formation with the assumption that either at the time of their formation, or shortly afterwards, the Sun passed through a highly luminous phase. This either prevented volatile materials from becoming part of planets in the terrestrial region or else removed such material after the planet had formed. The high density of Mercury, the closest planet to the Sun, is then explained by it being formed mainly from dense non-volatile materials. Other terrestrial planets are also formed mainly from silicates and iron. However, the relative sizes of the terrestrial planets, especially the small size of Mars in relation to its position, need to be explained.

Through the primary capture mechanism the Capture Theory only gives major planets. If planets began to form further in than about the region of Mars they would not have survived the first perihelion passage of their elliptical orbits. While the two largest terrestrial planets can be explained in terms of the products of a planetary collision this still leaves the two smallest terrestrial planets unexplained.

The formation of regular satellites for both accretion-based theories and the Capture Theory is intimately linked with the condensation of the planets. The accretion theories do not require a detailed separate mechanism for the formation

294

of regular satellites since it is taken as resembling that of planetary formation on a smaller scale. The Capture Theory has a different mechanism for the formation of regular satellites that involves the disruption of a collapsing proto-planet under the distorting influence of the tidal field of the Sun.

For completeness all theories should attempt to explain the characteristics of small planets and of the presence of non-regular satellites and here we shall consider some of the ideas which have been advanced.

10.2   Mars

Mars, with about one-ninth of the mass of the Earth, seems anomalous in relation to the general pattern of the Solar System. In Jeans' original tidal theory he envisaged a cigar-shaped filament being drawn from the Sun and thus he explained that the most massive planets were those in the middle of the system. From this model it might be expected that a fairly massive planet would exist in the region of the asteroid belt, an obvious gap in the system given numerical support by the Titius-Bode law, and that a planet in the region of Mars should be intermediate in mass between Jupiter and the Earth. This rather simplistic view needs to take into account that the asteroid-belt region is also the division between the major and the terrestrial planets so that perhaps it is unreasonable to expect a smooth variation of planetary characteristics as one passes through that region.

Mars is easily observed from Earth and knowledge about it has greatly increased through information from visiting spacecraft. Its spin period and axial tilt are very similar to those of the Earth and so it shows seasonal variations of a similar kind. The eccentricity of its orbit, 0.093, gives a 19% variation in its distance from the Sun which also heavily influences its climate through the Martian year.

Mars shows several extinct volcanoes, including *Olympus Mons* that towers to a height of 27 km. These must have been active for a considerable portion of the period from the formation of the Solar System. The planet shows hemispherical asymmetry with one-half, mostly in the north, being a smooth, lightly-cratered region while the other half is heavily cratered and resembles the lunar highlands. The division between these two regions runs roughly at 35° to the equator and the regions are separated by a scarp some 2 km in height. On average the northern plains are depressed by about 4 km relative to the southern highlands. The usual theories advanced to explain the asymmetry involve some internal process, such as mantle convection. Wise *et al* (1979) described a process of mantle overturn that would have removed the lower part of the crust on one hemisphere. The lower northern region is then due to isostatic adjustment with volcanism giving rise to the smooth surface.

Mars shows an offset between the centre-of-mass and centre-of-figure, a feature also present in the Moon. For Mars it is about 2.5 km (Arvidson *et al* 1980) and is approximately perpendicular to the plane separating the two hemispheres, directed towards the smooth hemisphere. It is very likely that the hemispherical asymmetry and the COM–COF offset are associated with a common cause.

Also seen on the surface are many channel systems that resemble dried-up river beds and are taken to indicate that, early in its history, liquid water existed on the planet. The residue of that water appears to be permanently locked up in the polar caps. The seasonal variation of the polar caps is due to the sublimation and deposition of carbon dioxide that is the dominant component of the Martian atmosphere.

10.2.1 Mars according to accretion theories

Accretion theories of planetary formation (section 6.4.3) involve the aggregation of planetesimals over a fairly long time-scale to form a planet. This process involves a delicate balance of effects. When the planetesimal aggregations are small the escape speed from them is much less than their orbital speed. For two bodies to amalgamate it is necessary that they should come together at little more than their mutual escape speed so that only little energy loss is required to take place through the collision. This means that their orbits are almost the same so that one body gradually overtakes the other. On the other hand, there is a requirement that planets should form on reasonable time-scales and for this reason a slightly turbulent environment is favourable since this brings bodies together more quickly. However, if the turbulence is too severe then amalgamation will not be possible. To explain the absence of a planet in the asteroid-belt region accretion theories generally assume that Jupiter formed quickly. Perturbations by Jupiter then stirred up the planetesimals in the asteroid belt to the extent that they were unable to assemble into a planet. The small size of Mars is then ascribed to the same effect, except that before the Mars region was cleared of planetesimals a small body was able to form.

The early formation of Jupiter offers solutions to a number of problems in planet formation, not just to explain the absence of a planet in the asteroid region and the small size of Mars but also the formation of Saturn in a 5:2 resonance (section 7.1.4). There are many difficulties, apparently insuperable, with theoretical models for the rapid formation of Jupiter, for example the *runaway growth* model suggested by Stewart and Wetherill (1988, section 6.4.3.3). Nevertheless so persuasive are the theoretical advantages of an early formation of Jupiter that Wetherill (1989) has stated 'because Jupiter very likely did form rapidly, there is probably some way to overcome these difficulties'.

10.2.2 Mars according to the planet-collision hypothesis

In describing the origin of the Moon in terms of the planet-collision hypothesis in section 9.1.5 it was mentioned that the colliding planets would have been subjected to large solar tidal forces as they condensed. For this reason the colliding planets should each have had several massive satellites, probably even larger and more massive than those of Jupiter.

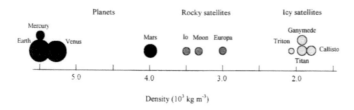

Figure 10.1. The densities of some small solid solar-system bodies.

The numerical experiments on the destination of satellites of the colliding planets, referred to in section 9.1.5, gave as a possible outcome the release of the satellite into a separate heliocentric orbit. It was suggested by Dormand and Woolfson (1977) that Mars had been a satellite of one of the colliding planets. Since Mars is some four times as massive as Ganymede, the most massive satellite in the Solar System, this is a speculation that needs some justification. Figure 10.1 shows the densities of a selection of the larger solid bodies in the Solar System labelled according to the conventional descriptions as planets, rocky satellites or icy satellites. From this picture it is clearly not implausible to suggest that Mars should belong in the category of rocky satellites—especially if one takes into account that its uncompressed density would be in the range 3700–3800 kg m^{-3}. It is also worth noting that while Mars is four times as massive as Ganymede it is seven times less massive than Venus. These numerical comparisons are suggestive rather than conclusive but are worth making to counteract the natural resistance to suggestions for revising well-established categorization.

There are obvious similarities in some of the major surface features of Mars and the Moon, in particular the hemispherical asymmetry and the associated COM–COF offset. However, the volcanism on Mars seems to have been on a larger scale than on the Moon and, rather than the filling in of basins by basalt, the whole hemisphere has been flooded to a considerable depth and can be described as volcanic plains. This can be understood if the abrasion, which removed 25–40 km of the exposed crust of the Moon, removed a similar amount of the exposed crust of Mars. Since Mars is a much larger body than the Moon the accretion energy would have been greater, the mean internal temperature would have been higher and the rate of cooling, relative to its mass, lower. Consequently the abrasion would have penetrated down to, or very close to, the molten mantle and volcanism would have occurred spontaneously and without the need for subsequent basin formation to enable the magma to escape. We now examine the features of the Martian surface in more detail to see how these might be explained in terms of a satellite origin in the presence of a planetary collision.

10.2.3   The Martian crust

It is generally believed that the early atmosphere of Mars was much denser than now and that water existed on the planet in some quantity. Not only is the residue of that water locked into the polar caps but sinuous channels on the surface have all the features expected of dried up river beds.

Predictions of the original amount of water vary greatly. Owen and Bieman (1976), McElroy *et al* (1977) and Anders and Owen (1977) have estimated it to have been the equivalent of a layer over the whole surface of thickness between 10–160 m. At another extreme Allen (1979) has suggested a thick sheet of ice of a thickness of about 1 km.

Connell and Woolfson (1983) considered two extreme situations, one with the equivalent of a 160 m layer of ice and the other with the equivalent of a 10 km thick layer, and also an intermediate situation with a 1 km thick layer. The actual form of the water does not greatly influence the conclusions to be drawn. A very plausible model is that of having a crust with an icy surface and with ice-or-water impregnated silicates at a lower level where the ice may melt under pressure.

The escape speed from Mars is approximately 5 km s^{-1} so that the debris from the nearby planetary collision, arriving with a speed of about 150 km s^{-1} would, in principle, be able to remove up to 900 times its own mass. The removal of a large thickness of crust, and perhaps even mantle material, by abrasion would lead to a readjustment of the surface material in the immediate vicinity and also rearrangement of material in other regions although parts most distant from the abraded hemisphere would have been little affected. In figure 10.2 there is shown a schematic before-and-after picture of an abraded region. The readjustment is designed to restore isostacy so that the pressure at the level of compensation, taken at some arbitrary undisturbed level, is brought back to its original value. From this we have

$$w\rho_i + x\rho_e + y\rho_s + z\rho_m = M \tag{10.1}$$

where M is the mass per unit area above the compensation level, the water is taken to be in the form of ice and the densities of ice, extra material, silicate and mantle material are ρ_i, ρ_e, ρ_s and ρ_m. The indicated densities of ice and silicate are as normally found and the density of mantle material, 3467 kg m^{-3}, is taken from Goettel (1980). The average density of the extra material, consisting of magma, debris and impact melt, is taken as that for the mantle so any analysis must consider the combined quantity $x + z$. Given that the northern plain is lower than the southern highlands by an amount d

$$w + x + y + z = h = 100 \text{ km} - d. \tag{10.2}$$

From (10.1) and (10.2)

$$y = [M - \rho_m h - w(\rho_i - \rho_m)]/(\rho_s - \rho_m) \tag{10.3a}$$

and

$$x + z = h - w - y. \tag{10.3b}$$

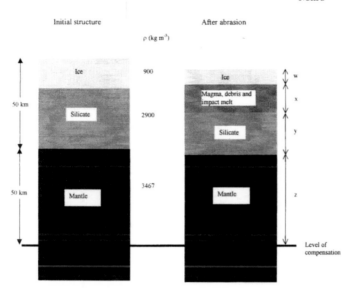

Figure 10.2. The structure of the outer regions of the Martian crust before and after abrasion by ejecta from the planetary collision.

For any assumed final depth of ice, w, it is possible to find the remaining depth of the silicate crust, y, and the sum of the thickness of extra material plus mantle down to the compensation level. Figure 10.3 shows the results of these calculations for the original ice thickness of (a) 10 km, (b) 1 km and (c) 160 m with $d = 4$ km. For the thickest ice layer no more than 45% can be lost otherwise no combination of silicate and other material could restrict the depression of the northern plains to 4 km. At the other extreme about one-half of the silicate layer must be lost no matter how much ice is removed.

To explain these results we consider the point P in figure 10.3(b). A fraction 0.6 of the ice has been lost and the silicate layer has been reduced from 49 to 27.4 km, a loss of 44% of the original material. The thickness of mantle plus added material is 72.3 km; since the original mantle thickness was 50 km this corresponds to a thickness of 22.3 km of added material.

Martian topography can be reasonably explained in terms of the intermediate model. The volcanism that gave much of the added material would have lasted a considerable time—beyond the time when large impacts frequently occurred thus explaining the comparatively few craters in the northern plains. The heat generated by volcanism would have melted and vaporized most of the residual ice creating a water-rich climate for as long as Mars could retain a water-rich atmosphere. The remains of this period are seen in the sinuous channels and collapsed regions thought to be due to the release of water by ice-impregnated surface silicates.

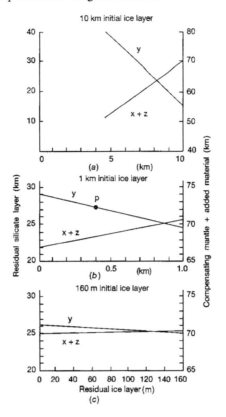

Figure 10.3. The final thickness of the silicate layer (y) and combined thickness of extra material and mantle down to the compensation level ($x + z$) to give a 4 km depression of the northern plain. The equivalent initial ice layer has thickness: (*a*) 10 km; (*b*) 1 km; and (*c*) 160 m.

10.2.4   The COM–COF offset

Figure 10.4 shows a schematic cross-section of Mars with a modified crust in the northern plain region. The northern plains do not quite occupy a complete hemisphere so that the angle α is taken as 80°. The density of the unmodified crust is taken as ρ_1 and that of the thinner modified crust ρ_2. Both the thickness and the density of the modified crust depend on how the estimate $x + z$ is partitioned. The quantity x makes a contribution to the crust, together with the thickness of silicate y, and together these give an estimate of ρ_2. The other densities used in the analysis are that of the mantle, $\rho_m = 3467$ kg m^{-3}, and of the whole planet, $\rho_p = 3940$ kg m^{-3}. The depression of the northern plains is $r_1 - r_4$ and the thickness of crust is $r_1 - r_2$ in the southern hemisphere and $r_4 - r_3$ in the northern hemisphere.

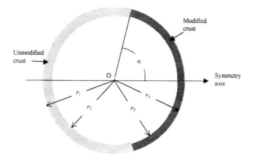

Figure 10.4. A model for Mars after bombardment and crust modification.

The total moment of mass is found to be

$$T = \frac{\pi \sin^2 \alpha}{8}[(r_4-r_3)(r_4+r_3)^3\rho_2+(r_3-r_2)(r_3+r_2)^3\rho_m-(r_1-r_2)(r_1+r_2)^3\rho_1]$$

(10.4)

with the total mass

$$M = \frac{4\pi r_2^3\rho_P}{3} + \frac{2\pi}{3}(1-\cos\alpha)[(r_4^3 - r_3^3)\rho_2 + (r_3^3 - r_2^3)\rho_m]$$
$$+ \frac{2\pi}{3}(1+\cos\alpha)(r_1^3 - r_2^3).$$

(10.5)

The distance of the centre-of-mass along OX is

$$X_{COM} = T/M.$$

(10.6a)

The corresponding distance for the centre-of-figure is found from T' and M' corresponding to T and M calculated from (10.4) and (10.5) with all densities made equal. This gives

$$X_{COF} = T'/M'.$$

(10.6b)

The offset, $X_{OF} = X_{COM} - X_{COF}$, was calculated by Connell and Woolfson (1983) for many different configurations illustrated in figure 10.3 and with various partitions of $x + z$. The value of X_{OF} was very insensitive to the model and was always in the range 2.81–2.86 km, in reasonable agreement with the quoted figure of 2.5 km. This is because for all the models the value of X_{COF} is the same, -2.91 km, since it depends only on the bounding figure. However, the condition for isostatic equilibrium makes the value of X_{COM} small in value, between -0.05 and -0.10 km so that the COM–COF offset has a very small variation.

The model used for Mars in this analysis assumed that all the interfaces were spherical surfaces. Although Mars is slightly pear-shaped this should not affect the validity of the analysis since the offset estimate depends primarily on the difference of thickness and density of the crust on the two sides.

10.2.5 Polar wander on Mars

The relationship of the spin axis of the Moon to the hemispherical asymmetry is consistent with its origin as an abraded satellite. As a regular satellite of its original parent planet its spin axis would have been normal to its orbital plane so that the plane of asymmetry would have contained the spin axis. The association with the original planet would have slightly distorted the Moon so that its spin was synchronous with its orbit and so that it presented one face towards the planet. Assuming that the abrasion did not change the distribution of matter in the Moon too greatly then it would have achieved synchronization of spin and orbital motion with the same damaged face presented towards the Earth as that which previously faced the original parent planet.

If Mars had also been a regular satellite of one of the colliding planets then it too should have had its spin axis contained within the plane of asymmetry. It is actually at 55° to that plane and this requires explanation. A plausible possibility is that there has been polar wander so that the surface features have moved relative to the spin axis. This is not a new idea. Runcorn (1980) suggested polar wander on the Moon to explain some magnetic observations and Murray and Malin (1973) suggested that this could have happened on Mars. It was suggested as an explanation for the laminated appearance of the south polar region and also as an explanation of the near-equatorial position of Olympus Mons.

The mechanism of polar wander requires fairly rigid surface material to move over molten material close to the surface. The surface material will tend to move coherently in regions but there would be local strains giving both compression and tensional features. On Mars it would have been a mechanism similar to plate tectonics but not exactly following the terrestrial pattern with subduction and the creation of new surface material. There is some evidence for such processes to have occurred on Mars (Guest *et al* 1979).

A theorem by Lamy and Burns (1972) states that a rotating body with internal energy dissipation will eventually settle down with its spin axis along the principal axis of maximum moment of inertia. That this should be so is easily seen. A body spinning about an axis has kinetic energy

$$E = \frac{H^2}{2I} \tag{10.7}$$

where H is the angular momentum and I is the moment of inertia. If the body is isolated then H is constant so that the only way that it can decrease its energy is by increasing I. For spinning astronomical bodies this condition is achieved automatically by the flattening of the body along the spin axis. In such a case the inertia tensor will be of the form

$$I = \begin{bmatrix} I_{xx} & 0 & 0 \\ 0 & I_{yy} & 0 \\ 0 & 0 & I_{zz} \end{bmatrix} \tag{10.8}$$

and if the body has axial symmetry and is spinning about the z-axis then I_{zz} is larger than I_{xx} and $I_{xx} = I_{yy}$.

A circular crater of depth h centred on the z-axis and subtending an angle 2α at the centre of the planet of radius R has components of its moment of inertia tensor

$$I_{xx} = I_{yy} = \tfrac{1}{3}\pi h R^4 \rho(4 - 3\cos\alpha - \cos^3\alpha)$$

$$I_{zz} = \tfrac{1}{3}\pi h R^4 \rho(4 - 6\cos\alpha + 2\cos^3\alpha) \qquad (10.9)$$

and all other components zero. If this crater is now centred at the point with latitude and longitude (λ, ϕ) then the components of the tensor are modified by the matrix

$$Q = \begin{pmatrix} \sin\lambda\cos\phi & -\sin\phi & \cos\lambda\cos\phi \\ \sin\lambda\sin\phi & \cos\phi & \cos\lambda\sin\phi \\ -\cos\lambda & 0 & \sin\lambda \end{pmatrix}. \qquad (10.10)$$

A circular, or nearly circular, basin or crater will give components of the inertia tensor with h negative in (10.9) and roughly circular highland region components from (10.9) with h positive. If an astronomical body contains many features then the components of the inertial tensor may be found for each feature individually and the components added to give the overall inertia tensor. By a standard diagonalization process the directions of the principal axes can then be found.

For Mars the major surface features have been taken as in table 10.1. The northern plain is simulated as a large crater and the Tharsis and Elysium regions correspond to bulges in the Martian crust. Hellas is a large and deep basin in the southern hemisphere with an average depression equal to that of the northern plains. The assumption is made that these features are departures from a crust which would be a uniform spherical shell. The principal axis of maximum moment of inertia will then correspond to that of the modelled features alone. Using the data in table 10.1 shows that the principal axis of maximum moment of inertia is inclined at $11.9°$ to the spin axis. The possibility of two random vectors lining up to within this angle is 0.02 so the near-alignment is probably significant. Considering how crudely the surface features have been modelled the result is quite consistent with the idea of polar wander. Another factor to be considered is that some surface features may post-date the period of polar wander after the lithosphere had ceased to be mobile. It is also possible that lithosphere mobility may not have lasted sufficiently long for complete reorientation of the surface to have taken place.

10.3 A general description of Mercury

Photographs of Mercury, taken from spacecraft, show a surface looking superficially very similar to that of the far side of the Moon (figure 10.5). The highland regions are covered with craters, with rays coming from some of them, although the density of craters is less than that in the lunar highlands. Between the craters

Table 10.1. Modelling the major features of Martian topography.

Feature	Mean height (km)	Semi-angular size (°)	Longitude (°)	Latitude (°)
Northern plain	−4	80	150	58
Tharsis uplift	4	50	95	−16
Tharsis supplement	5	15	107	−3
Elysium plain	5	17	210	25
Olympus Mons	12	5	133	18
Hellas	−4	15	291	−44
Argyre plain	2	7	42	−52

Figure 10.5. Part of the surface of Mercury constructed from a mosaic of photographs taken by Mariner 10. The left-hand edge shows part of the Caloris Basin and the concentric rings of mountains are clearly seen.

there are smooth areas and there are also extensive lava plains, similar in some ways to mare features on the Moon. There is one large impact feature, the Caloris basin, so called because at every other perihelion passage it faces the Sun. It has some similarities with the Orientale feature of the Moon in that it is surrounded by concentric rings of mountains up to 2 km high. On the opposite side of the planet is a region where the surface is rippled in a curious way and this is thought to be due to the meeting of shock waves that travelled round the planet after the Caloris impact.

The combination of its mass, 3.33×10^{23} kg, and its radius, 2439 km, gives it a density 5480 kg m^{-3} that is intermediate between that of Venus and that of the Earth. Taking into account that it is much less massive than the large terrestrial planets and there is much less compression due to internal pressure then its intrinsic density is the largest of any of the terrestrial planets. The orbit has a high eccentricity, 0.2056, and the spin period is exactly two-thirds of the orbital period so that there is *spin–orbit coupling*. A deduction from this is that there must be an asymmetric distribution of mass in Mercury so that there is a distinct COM–COF offset, as is observed in the Moon and Mars. At perihelion the offset points either directly towards Mercury or directly away from it. In either case there will be a balance of forces so that there is no tendency for Mercury to rotate away from this condition.

A feature of Mercury's surface, that is absent from the Moon, is the presence of long and high scarps that are clearly due to compression forces on surface material as the planet cooled and shrank. Individual scarps, which can be up to 500 km long and 2 km high, run through various types of terrain and were clearly superimposed after some of the main surface features had developed. It has been estimated that the radius of Mercury could have contracted by cooling as much as 3 km.

It is clear from figure 9.5 that Mercury is anomalous in terms of its density in relation to its mass and we now consider various suggestions that have been made to explain this.

10.3.1   Mercury and accretion theories

A straightforward process of planetesimal accretion would not be expected to produce a body with the composition of Mercury. However, the unique status of Mercury as the planet closest to the Sun offers a possible explanation. Lewis (1972) used a profile of density, temperature and pressure in the early solar nebula suggested by Cameron (1969) to argue that silicates were only partially retained in the inner Solar System but that iron had fully condensed. Mercury was thus produced with a large iron component. This naturally raised the question of why Venus has a lower density than the Earth. By an ingenious construct of arguments, based on different degrees of retention of sulphur and oxygen in the form of FeS and FeO and of hydrous silicates, the pattern of the densities of Mercury, Venus and the Earth was explained.

As mentioned in section 5.4 Cameron (1978) moved away from the position of a very hot solar nebula required by the Lewis model for the formation of Mercury since it presented insuperable problems for planet formation. Instead he postulated that Mercury was formed in the cool solar nebula as a much larger body with a non-anomalous density. Subsequently the Sun went through a T-Tauri stage in its development and greatly increased its luminosity. The high temperatures generated in the vicinity of Mercury then evaporated off a large amount of the outer crust and mantle material leaving the planet with an abnormal proportion of the iron that was in the core and shielded from evaporation.

The single impact theory of lunar formation (section 9.1.4) is consistent with calculations carried out by Wetherill (1986) concerning planetesimal accretion.

He showed that at the stage when one dominant body is forming in a region by the accumulation of planetesimals the second largest body would have a mass about one order of magnitude less. Thus to have a collision between, say, the Earth and a Mars-mass body is a natural outcome of the accretion process. A glancing blow was necessary to produce the Moon but other modes of collision are possible. It has been suggested that an early Mercury, of normal density for its size, was struck by a large projectile in such a way that most of its mantle was removed.

The collision hypothesis is the most plausible of those that have been suggested in relation to accretion theories, especially as it links up with the single-impact process for Moon formation.

10.3.2 Mercury and the Capture Theory

The Capture Theory does not predict the formation of planets in the terrestrial region coming from the proto-star filament. Instead the Earth, Venus, Mars and the Moon have been explained as the products of a collision between early planets. We now examine how Mercury can be explained in terms of the same scenario.

One possibility, mentioned by Dormand and Woolfson (1989), is that it was a high-density fragment of one of the colliding planets. Modelling since that time shows that the iron cores of the colliding planets stay together as coherent units, although they may stretch and greatly distort during the encounter period. Although the possibility of a high-density fragment becoming separated cannot be discounted it is now considered less likely.

A second possibility is that Mercury is the heavily abraded residue of a satellite of one of the colliding planets. This would imply that it was in an orbit of small radius and hence close to the collision and that a large proportion of its crust and mantle material was removed. In some ways this is similar to the impact suggestion at the end of the previous section except that the material is being removed by a rain of smaller projectiles rather than by a single large one.

We now consider the original Mercury as a body consisting of the present Mercury embedded in a thick silicate shell of density ρ_s that was lost in the collision. If the volume of the lost material was α times the present volume of Mercury then the initial density was

$$\rho_I = \frac{\rho_{Merc} + \alpha \rho_s}{1 + \alpha} \qquad (10.11a)$$

and

$$\alpha = \frac{\rho_{Merc} - \rho_I}{\rho_I - \rho_s}. \qquad (10.11b)$$

We may anticipate that to achieve ρ_{Merc}, the present density of Mercury, a great deal of material had to be lost so that the original mass of Mercury was of the same order as that of Mars. That being so we take $\rho_I = 3940 \text{ kg m}^{-3}$, the density of Mars, and in table 10.2 the value of α is given, together with the original radius

Table 10.2. Variation of the characteristics of the original Mercury with the mean density, ρ_s, of lost material. The lost material had α times the present volume of Mercury and a mass M_L. The original radius was R_I.

ρ_s (kg m^{-3})	α	R_I (km)	$M_L(10^{23}$ kg)
2800	1.175	3160	2.00
2900	1.288	3214	2.27
3000	1.426	3277	2.60
3100	1.595	3352	3.00
3200	1.811	3442	3.52

of Mercury and the total mass of lost material for various values of ρ_s. Since a great deal of material is lost, ρ_s is taken in a range appropriate to a mixture of mantle material and crust material.

Table 10.2 shows that the mass of lost material is similar to, but probably less than, that of Mercury now. Taking an average Mercury radius of 3000 km over the period of abrasion and an orbital distance of 1.5×10^5 km, greater than that of Miranda about Uranus, the loss of material, as deduced from (9.14) is 2.4×10^{23} kg, similar to the estimates in the final column of table 10.2. The implication of this result is that Mercury would have been in a close orbit and in a part of the orbit that would have exposed it fully to the debris from the collision. It also suggests that the original mass of Mercury was probably just less than that of Mars.

10.4 Neptune, Pluto and Triton

Neptune, Pluto and Triton are frequently regarded as being related by some unusual event in the evolution of the Solar System. Triton is a large body that might be expected to be a regular satellite but it has a retrograde orbit, which clearly makes it irregular. On the other hand, for a planet Pluto is anomalous in being so small; its mass is approximately one-sixth that of the Moon and is actually smaller and less massive than Triton. Despite its small size, with diameter 2302 km, Pluto has a satellite with diameter 1186 km, comparable in size to some of the larger satellites of Uranus. Pluto has the most eccentric orbit of any of the planets, $e = 0.249$, the largest semi-major axis, 39.46 AU, and the largest inclination, 17.3°. At perihelion it moves a distance 0.44 AU inside the orbit of Neptune but the commensurate orbits, described in section 1.2.3, ensure that the bodies never approach too closely. In fact, as seen in figure 1.3 the closest approach to Pluto is by Uranus, not Neptune.

10.4.1   Encounter scenarios for the Neptune–Triton–Pluto system

In 1977 Dormand and Woolfson proposed that Pluto had originally been a satellite of one of the colliding planets which had been released into a heliocentric orbit. They showed that if Triton had been a regular satellite of Neptune then a close interaction between Pluto and Neptune could have thrown Pluto into its present orbit while at the same time Triton would have been perturbed into a retrograde orbit. The interaction required a very close approach of Triton and Pluto. The final outcome was a Pluto orbit about the Sun with $a = 27.0$ AU, $e = 0.635$ and a retrograde Triton orbit with $a = 44R_N$ and $e = 0.957$, where R_N is the radius of Neptune.

It can be shown that, for this scenario, it would not have been possible for Pluto to have been transferred from an orbit spanning a distance from 2.8 AU, an assumed distance from the Sun of the planetary collision, to 30.1 AU, the orbital radius of Neptune. This conclusion comes from the use of Tisserand's criterion that is usually applied to the perturbation of comets by planets. It applies strictly to a planar system and states that the value of

$$T = \frac{1}{2a} + \left[\frac{a(1 - e^2)}{a_p^3} \right]^{1/2} \tag{10.12}$$

is unaffected by the interaction where (a, e) are the orbital elements of the comet and a_p is the radius of the planet's orbit, assumed circular. The value of T for Pluto's orbit in relation to Neptune is 0.0496 AU^{-1} while for an orbit with perihelion 2.8 AU and aphelion 30.1 AU it is 0.0422 AU^{-1}. This is a large discrepancy that is not removed by small modifications to the assumed original planetary orbit. Dormand and Woolfson showed that it would have been possible for Pluto to have started in an orbit that linked with a planet close in, then to have been transferred to an orbit linking with another planet further out and so on until it reached the region of Neptune. This staged transfer of the orbit outwards can be achieved with the satisfaction of Tisserand's criterion at each stage. The final orbit, when Pluto passed near Neptune, could have been with a perihelion near Uranus.

Harrington and Van Flandern (1979) produced an alternative model that began with both Pluto and Triton as regular satellites of Neptune. A planet of unknown provenance and mass $5M_\oplus$ passed through the Neptunian system both expelling Pluto into its heliocentric orbit and reversing the direction of Triton's orbit around Neptune. Such an event is almost certainly dynamically feasible. However, Harrington and Van Flandern do not place their planet within the pattern of the remainder of the Solar System or its origin and its only purpose is to disturb the Neptunian system.

In 1978 poor quality images from ground-based telescopes using CCD detectors clearly showed the presence of Pluto's satellite, Charon. This enabled a revised estimate of Pluto's mass to be made which, although not very precise, showed that the previously accepted value was much too high. The Dormand and Woolfson (1977) model had assumed what were then the estimated values

Table 10.3. Numerical data for the Pluto–Triton encounter (Dormand and Woolfson 1980).

Masses					
Neptune	$5.13 \times 10^{-5} M_\odot$				
Pluto	$5 \times 10^{-9} M_\odot$				
Triton	$10^{-7} M_\odot$				

Orbital elements					
Pre-encounter	*a* (AU)	*e*	*Post-encounter*	*a* (AU)	*e*
Triton:Sun	34.38	0.1633	Triton:Neptune	0.3212	0.9774
Pluto:Neptune	0.0125	0	Pluto:Sun	39.52	0.2643
Triton:Neptune	0.625	1.01			

of the masses of Triton and Pluto, 1.34×10^{23} kg and 7×10^{23} kg respectively. The mechanism for reversing the sense of Triton's orbit required a high ratio of Pluto's mass to that of Triton. The discovery of Charon and the better estimate of Pluto's mass changed the estimated value of $M_{\text{Pluto}}/M_{\text{Triton}}$ from 5.2 to 0.05 so that their original model was no longer tenable. They then described an alternative model (Dormand and Woolfson 1980) which reversed the roles of the two bodies. Pluto was now a regular satellite of Neptune and Triton played the role of the incoming body. They found an interaction which led to Pluto being expelled into a heliocentric orbit very similar to that of present-day Pluto while Triton was captured into an extended highly-elliptical orbit around Neptune. The data for, and results of, their calculation are shown in table 10.3

The reduction of the estimate of Pluto's mass required very stringent conditions to achieve the results given in table 10.3. Capture of Triton into a retrograde orbit from an original orbit that had a perihelion less than 3 AU, at the site of the proposed planetary collision, was not possible. It had to be assumed that Triton's orbit had been modified by repeated interactions with other planets, as previously described, so that when the interaction with Neptune took place the hyperbolic excess of the orbit was small. The retrograde Triton orbit about Neptune would have been quite stable even though it had high eccentricity and a large semi-major axis. The orbit would have been well within the sphere of influence of Neptune given by

$$S = r_{\text{N}} \left(\frac{M_{\text{N}}}{2M_\odot} \right)^{1/3} \tag{10.13}$$

where r_{N} and M_{N} are the orbital radius and mass of Neptune. This is about 0.9 AU so that the orbit of Triton about Neptune would be comfortably included. Another requirement of the interaction is the rather close approach of Triton and Pluto, 3020 km, which was less than the estimated sum of the radii of the two bodies at that time. This suggested to Dormand and Woolfson that a collision might have been involved, although this was not included in their model.

The idea that Pluto was once the Neptune satellite and Triton that of an inner planet better matches the expectations from the model proposed by Williams and Woolfson (1983) that, in general, satellite masses would be greater in the inner system. Again, a very close approach of Pluto and Triton during the interaction provides a very reasonable explanation of the formation of Charon, either through tidal forces, as suggested by Harrington and Van Flandern (1979) or through break-up by collision. A tidal origin for Charon was also suggested by Farinella *et al* (1979) who put forward a scenario for Neptune, Triton and Pluto similar in some ways to the Dormand and Woolfson (1980) scenario. They proposed that Pluto had been a satellite and Triton was captured by Neptune from a heliocentric orbit. Tidal friction then caused Triton's orbit to decay until it interacted with Pluto and ejected it into its present orbit around the Sun. This is quite a feasible mechanism and would not even require tidal decay of Triton's orbit since its initial perifocal distance could have been within the orbit of Pluto. The lower relative velocities of the two bodies would have enabled a significant interaction to take place at much larger distances than was found necessary by Dormand and Woolfson. Although feasible this mechanism has the difficulty, mentioned in relation to the Moon in section 9.1.3, that capture of Triton into a stable orbit is a difficult and extremely unlikely process.

When the spacecraft Voyager 2 visited Neptune in 1989 a much more precise estimate of Triton's mass became available. This was $1.11 \times 10^{-8} M_{\odot}$ (2.21×10^{22} kg) or one-ninth of the value used in table 10.3. This obviously affects the detailed calculation that led to table 10.3 but does not affect the general validity of the mechanism proposed. Since the sum of the revised radius of Triton, 1352 km, and that of Pluto, 1151 km, is lower than was previously thought an interaction without collision is more feasible. Nevertheless, a purely gravitational interaction, as envisaged in the Dormand and Woolfson (1980) model, is unable to remove sufficient energy from Triton to give capture unless Triton's orbit had only a small hyperbolic excess with respect to Neptune. This requires Triton to have gone through several stages of interaction with planets until its perihelion was close to the orbit of Uranus and its aphelion was close to that of Neptune.

Woolfson (1999) has taken into account the now-reliable estimates of the masses of Triton, Pluto and Charon to examine a modification of the Dormand and Woolfson (1980) model with a collision taking place instead of a purely gravitational interaction. The collision between Triton and Pluto was taken as that between two spherical bodies with an *elasticity parameter*, β, (Trulsen 1971). If Triton and Pluto are at positions r_T and r_P and prior to the collision have velocities v_T and v_P then the velocities after the collision are given by

$$v'_T = v_T + \frac{m_P}{m_P + m_T} \beta (dv \cdot k) k \qquad (10.14a)$$

and

$$v'_P = v_P - \frac{m_T}{m_P + m_T} \beta (dv \cdot k) k \qquad (10.14b)$$

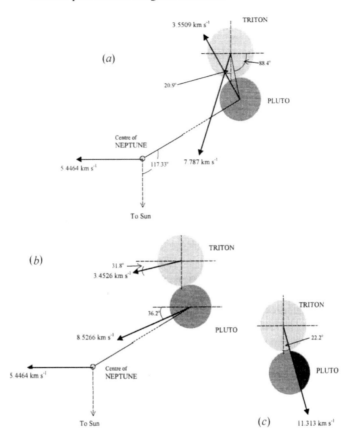

Figure 10.6. Positions and velocities of NTP before and after the collision. Angles and distances are not to scale for clarity of representation. The distance from the centre of Neptune to the centre of Pluto is 544 840 km. The radii of Triton and Pluto are 1350 km and 1200 km, respectively. Lines marked – – – are either in the direction of the Sun or perpendicular to that direction. (*a*) Velocities just before the collision. (*b*) Velocities just after the collision. (*c*) The velocity of Triton relative to Pluto just before the collision. There will be a tendency to shear off material on the darker-shaded side of Pluto.

In fact it is not absolutely necessary for a model to produce an orbit for Pluto that very closely resembles the present one—although it should be of similar extension and eccentricity. If the interaction takes place in the presence of a resisting medium then the orbit can evolve, although the calculations described in section 7.1.2 suggest that a body of low mass would not evolve quickly, as shown in table 7.1. However, there would have been some evolution of orbits and this could have given rise to the 3:2 resonance locking of the orbit of Pluto to that of Neptune and the near 3:1 commensurability with Uranus.

It could be argued that since the heliocentric orbit of Pluto began in the vicinity of Neptune it should return to that same place in space, close to Neptune's path. However, although round-off is slower for a small planet what is not slower is the precession of the orbit described in section 7.4.5 and illustrated in figure 7.11, as long as the resisting medium was in place so the orbit of Pluto would precess and it would end up wherever it happened to be when the medium disappeared.

The mechanism for producing Pluto in its present orbit, accompanied by Charon and also Triton as a retrograde satellite, cannot be uniquely defined by modelling. Nevertheless, an origin for Pluto as a regular satellite of Neptune and of Triton as a body originally in a heliocentric orbit is strongly indicated. Most models indicate that the orbit of Triton around Neptune originally had a large eccentricity. A tidal friction rounding-off mechanism would then act on Triton to give the circular orbit observed today (McCord 1966). While Triton was in its elongated orbit it would be able to interact with other Neptunian satellites. This provides a logical explanation for the unusual orbit of Nereid (diameter 340 km) which has an orbit with $(a, e) = (5.51 \times 10^6 \text{ km}, 0.749)$.

10.5 Irregular satellites

A regular satellite is one that is in an almost circular orbit in the equatorial plane of its parent planet. They tend to be the larger and more massive satellites but several smaller satellites, e.g. Amalthea, possess the orbital characteristics of regular satellites. There are some larger satellites, such as Triton and Iapetus, that are distinctly irregular and we have already seen that Triton's irregular relationship with Neptune can be explained in at least one plausible way.

From the calculation given by Woolfson (1999) it is clear that stable capture can come about by collisions in the vicinity of a planet. Colombo and Franklin (1971) considered the formation of the two outer satellite groups of Jupiter (table 1.4). They concluded that they could be the result of a collision between a pre-existing satellite of Jupiter and an asteroid. They also considered the possibility of a collision between two asteroids or between two satellites but concluded that these were less likely. Their argument against the collision of two asteroids rested on the observation that Jupiter's gravitational influence has cleared the Solar System of asteroids in its own vicinity. However, this seems to beg the question of when these satellite families formed and what the situation was around Jupiter soon after the Solar System formed. Dormand and Woolfson (1974) preferred a collision between asteroids. If one of the colliding bodies *had* been a satellite, and presumably a regular one, then it would have had an orbital radius of about 11 million kilometres, about six times the radius of Callisto's orbit.

The three outer satellites of Saturn—Hyperion, Iapetus and Phoebe—are all irregular in some way. Hyperion has a large eccentricity (0.104) but a small inclination and is clearly influenced by Titan with which it has a 4:3 resonance. Iapetus is in an inclined orbit ($i = 14.7°$) but it has a fairly small eccentricity. One could speculate that it was disturbed by a collision with a fairly large asteroid that tilted its orbit without greatly changing its shape. It does have hemispherical asymmetry, with the bright side ten times as reflective as the dark side. There are a number of theories to explain this and a collision event is one of these. Phoebe is a small satellite in a retrograde orbit of large radius and a collision origin is a distinct possibility.

Given the possibility of collisions, due to the large density of small objects in the early Solar System and the long period of time within which collisions could have taken place, plausible scenarios can doubtless be found for virtually all the irregular satellites. However, the two small satellites of Mars, Phobos and Deimos, may have a different origin. If Mars was a released satellite of a colliding planet, as suggested in section 10.2.2, then it could have captured small products of the collision into quite stable orbits since it was part of a many-body system with ample opportunity for energy exchange.

10.6 Summary

Looked at in a rather general way the Solar System has the following structure. At the centre is the Sun with a system of planets in co-planar and near-circular orbits with a trend of first increasing and then decreasing mass as one moves outwards. The main planets have satellites, the largest and most massive of which are mostly 'regular' in their characteristics. What this chapter has considered are the major departures from this simple structure. They have all been explained in terms of a single event, a collision between two early major planets and natural consequences of such a collision.

The three smallest planets, Mercury, Mars and Pluto, that are also those with the largest eccentricities, have all been suggested as having a satellite origin, the first two having been very large satellites of an early major planet orbiting well within 3 AU of the Sun. The pattern is then established of satellite sizes and masses being roughly correlated with distance from the Sun. An inner planet, or perhaps the two inner planets, would have had satellites Mars, Mercury, Moon, Triton and perhaps others that left the Solar System. This would have given satellite masses ranging up to more than eight times the mass of the Moon rather than just up to more than twice the mass as now represented by Ganymede.

The most massive of the irregular satellites is the Moon and it has been shown that it could have been acquired or retained by the Earth fragment of a shattered planet in a close and stable orbit. The many-body environment which enabled this to happen could also allow the retention of smaller collision fragments by the erstwhile satellite Mars to give its close attendants, Phobos and Deimos. Both the Moon and Mars show hemispherical asymmetry, the scars of bombardment by collision fragments towards which they presented one face. Features of both these bodies can be readily explained in terms of them having had material abraded off one hemisphere. Rather heavier damage to Mercury, presumed to be close in to the collision and directly in the path of the somewhat collimated stream of debris, led to a large loss of material, a complete reordering of the residual material and a final body of high density.

Various scenarios have been suggested from time to time concerning the formation of the Neptune–Triton–Pluto system. Here it is suggested, following Woolfson (1999), that Triton was originally a satellite of one of the colliding planets. Coming from an orbit with perihelion 2.55 AU, roughly where the collision was presumed to have taken place, Triton collided with Pluto, a satellite of Neptune, expelled it into its present orbit and itself was captured into a retrograde orbit. The geometry of the collision could have sheared off a portion of Pluto to give a satellite, Charon, in a retrograde orbit.

Although detailed scenarios for other prominent irregular satellites have not been suggested, what is fairly clear from the Neptune–Triton–Pluto calculation is that explanations in terms of energy-absorbing collisions in the vicinity of the appropriate planets can readily be found.

Chapter 11

Asteroids, meteorites and comets

11.1 Asteroid formation

The term *asteroid* is applied to a class of objects orbiting the Sun in a direct sense, sufficiently non-volatile not to be classed as comets and sufficiently small not to be classed as planets. Thus Ceres, which was sought and found in a position where a planet was expected to exist, is regarded as the largest known asteroid rather than as a planet. Its diameter is about 1000 km but Pluto, with a diameter about 2300 km, is readily accepted as a planet. However, the acceptance of Pluto as a planet may be due to the original lack of knowledge about it. First estimates of its mass were as high as $6M_\oplus$ but this gradually changed with time. Immediately before the discovery of Charon, and the consequent more reliable assessment of Pluto's mass, it was thought to be about $0.1M_\oplus$, which made it comparable to Mars.

The earliest idea about the origin of asteroids was that they were the products of the disruption of a planet. This idea has an obvious rationale since they are mainly observed in a region where it was thought that a planet ought to exist. Later knowledge gained about asteroids has, on the whole, tended to support the disruption idea. Asteroids with diameters less than about 300 km are known to be irregular in shape, which is what would be expected from the break-up of a large body. For asteroids larger than this, gravitational forces would be too great for the material strength to resist and the asteroid would go towards a sphere no matter what was its original shape (Hughes and Cole 1995). Another observation supporting the disrupted-planet origin is that asteroids seem to be of different compositions—some irons, some stones and some similar to carbonaceous chondrites (section 1.6.2). This would be consistent with asteroids being fragments from a differentiated body. A planet, with a high internal temperature capable of melting material and a sufficient gravitational field to separate material according to its density, would seem an obvious choice as a source.

Against the disrupted planet hypothesis there is the problem that no obvious source of energy is available spontaneously to disrupt an isolated planet (Napier and Dodd 1973). This gave support to the idea that asteroids are the products of a 'spoiled planet'—one that was in the process of forming but could not do so because of tidal effects due to Jupiter. It is clearly untenable to suppose that asteroids are formed by accumulation of even smaller bodies, as would be the pattern in the accretion of planetesimals in a solar nebula. This would imply that somehow the accreting bodies were dominantly iron or stone or even similar to the material of carbonaceous chondrites—a difficult scenario to justify. Instead it has been suggested that asteroids are the collision products of *parent bodies*, of sub-lunar mass which, had they survived, would have gone on to accumulate to form a planet (Anders 1971). In section 10.2.1 the delicate balance of conditions required for the accretion mechanism of planetary formation to operate was mentioned. The presence of Jupiter would have stirred up bodies in the asteroid region just too much to enable them to aggregate when they collided.

The parent-body origin of asteroids still has to cope with the problem of the differentiation of material since it is evident from equation (1.10) that the proposed parent bodies were too small to have melted and subsequently to have differentiated. The solution to this problem is suggested as the presence of radioactive ^{26}Al in the early Solar System. There is strong evidence that ^{26}Al occurred in some limited types of meteorite material (section 1.6.5) but no direct evidence that it was widespread throughout the early Solar System. However, if it *was* a small component of stable ^{27}Al at the time asteroids were formed, even at the one part in 10^8 level, it would have been sufficient to melt even small asteroids.

The planetary collision as a source of asteroids has no problems to solve. The gravitational energy of formation of the bodies will have given melting of material and the gravitational fields of the bodies themselves will have given differentiation on a short time-scale. There is no difficulty in finding the energy to disrupt the planets, which comes from the collision itself (but see also section 11.7.1.3 for a further source). Such an event would obviously have given collision products in direct orbits with a limited range of inclinations and a wide range of eccentricities but with perihelia at or closer in than the collision position. Those ejected bodies that went into the regions of the major planets would have been influenced in various ways. They could have collided directly with those planets—and evidence of damage on the surfaces of solid satellites suggests that a great deal of material must have been swept up in this way. Alternatively they could have undergone gravitational interactions which would either have sent them into different orbits or perhaps expelled them from the Solar System. The residue of asteroids we see today are those which either happened to start off in orbits which kept them out of the way of major planets or those which somehow evolved into stable orbits.

11.2   Meteorites

It has been suggested that meteorites, provided free by nature, are the most important single source of information concerning the Solar System and its origin—

exceeding in importance even the information provided by very expensive space research. This may well be true. The spectroscopic evidence described in section 1.5.3 indicates that they are related to asteroids and there can be little doubt that they are fragments of asteroids. No matter what cosmogonic theory is adopted there is a relationship between asteroids and planets so that the study of meteorites gives an opportunity of finding out about planetary material. They may be samples from throughout the solid or liquid parts of planets.

11.2.1 Stony meteorites

There are two types of stony meteorites *chondrites* and *achondrites* which differ chemically from each other. Most, but not all, chondrites contain *chondrules*, small glassy millimetre-size spheroids embedded in a fine-grain matrix that forms the main body of the meteorite. A section of a chondritic meteorite, illustrating clear chondrules, is shown in figure 1.20. Achondrites, as their name indicates, contain no chondrules and also virtually no metal or metal sulphides. In some ways they are similar to terrestrial surface rocks.

11.2.1.1 The systematics of chondritic meteorites

Olivine $(Mg, Fe)_2 SiO_4$ is the most common mineral accounting for some 45% of the commonest type of chondrite. The metal component can be anywhere from pure magnesium (fosterite) to pure iron (fayalite). Next in abundance are pyroxene (25%) another magnesium–iron silicate, $(Mg, Fe)SiO_3$, and plagioclase (10%), sometimes called plagioclase feldspar, a sodium–calcium–aluminium silicate with various compositions from $NaAlSi_3O_8$ (albite) to $CaAl_2Si_2O_8$ (anorthite). Native metal, in the form of iron with some nickel, occurs together with the mineral troilite, FeS. The native metal consists of two iron–nickel compounds— *taenite* that is nickel-rich with about 13% of nickel and *kamacite* which is nickel-poor with 5–6.5% of nickel. Where nickel has an overall level of between 6.5 and 13% then kamacite plates appear surrounded by a taenite matrix.

Within the chondrite classification there are three sub-types *ordinary, enstatite* and *carbonaceous*. The ordinary chondrites, so named because they are the most common type, are further subdivided according to their iron content as follows: H (high iron), L (low iron) and LL (low iron, low metal). The iron content and the form in which it occurs for these three types is shown in table 11.1. There are also differences in oxygen content where the amount of oxygen increases from H to LL. About 30% of the non-metallic iron in the H-type occurs as troilite whereas for the LL-type only 15% of the non-metallic iron is troilite.

Within each type of ordinary chondrite based on iron content a further *petrological classification* can be made on the basis of texture and mineral content. This is described in table 11.2, which also applies to the enstatite and carbonaceous chondrites. Petrographic type 3 ordinary chondrites show signs of rapid cooling from about 1700 K to 1100 K or less. Pyroxene crystals are very poorly

Table 11.1. Percentage of total iron and metallic iron in the three types of ordinary chondrites.

	H	L	LL
Total iron (%)	27	23	20
Metal iron (%)	12–20	5–10	2

formed, which shows that they formed quickly. Olivine crystals are also in a form indicative of rapid cooling. Their iron contents vary from 0 to 40% and the assemblages of minerals are non-equilibrated, which means that they were quenched before the minerals could rearrange themselves into a state closer to thermodynamic equilibrium. Another indication of rapid cooling is that the chondrules are clear and glassy. If sufficient time had been available then crystals would have formed and made the chondrules more opaque.. The total evidence points to petrographic type-3 ordinary chondrites as having been quenched at temperatures of 1000 K or less and were never subsequently heated to above the quenching temperature.

Petrographic type-4 ordinary chondrites have more opaque chondrules. Thus they either cooled more slowly or were subsequently heated thus enabling crystals to form within the chondrules. Olivine is more uniform in composition but pyroxene still cooled too quickly to give well-formed crystals of uniform composition. From the same considerations petrographic types 5 and 6 ordinary chondrites show signs either of very slow cooling or reheating to 1100 K and 1200 K respectively. Type 7 seems to have been heated over a long time to 1500 K so causing chondrules to melt and disappear and metal to melt and run away.

There are about 20 examples of enstatite chondrites available. They have such a low oxygen content that none of the iron present is combined with oxygen but either appears as sulphide or native metal. Another feature is the low Mg:Si ratio (<0.85) which means that there is no olivine present since in a pure magnesium olivine there would be twice as many magnesium as silicon atoms. Some 65% of these meteorites is the pure magnesium pyroxene, enstatite, from which they get their name. Plagioclase is also present and also many sulphides, including sulphides of sodium and potassium that are very rare components of meteorites. Enstatite chondrites seem to require for their formation a region that is metal rich but oxygen poor.

The final type of chondritic meteorite, the carbonaceous chondrites, are the richest source of information. They are distinguished by having a high Mg:Si ratio, approximately 1.05, and they are rich in carbon and water. With a single exception they contain very little free metal and they are dark in colour—almost black in most cases. They have the intriguing characteristic that they contain minerals that formed at very different temperatures; some materials are classed as high-temperature condensates while at the same time they are rich in volatile substances.

Table 11.2. Chondrite classification according to mineralogy and texture.

	Type 1	Type 2	Type 3	Type 4	Type 5	Type 6	Type 7
Chondrules	Absent	Sparse	Many and distinct		Visible	Indistinct	Absent
Chondrule glass	—	Clear and isotropic		Opaque	No glass present		
Matrix	Fine and opaque	Opaque		Transparent micro-crystals	Granular	Granular coarse-grained	
Silicate uniformity	—	>5% variation		Variation 0–5%	No variation		
Carbon (% mass)	3–5	0.8–2.6	0.2–1.0	<0.2%			
Water (% mass)	18–22	2–16	0.3–3.0	<1.5			
	Carbonaceous chondrites (Mg:Si = 1.05)						
				Ordinary chondrites (Mg:Si = 0.95)			
				Enstatite chondrites (Mg:Si < 0.85)			

Carbonaceous chondrites fall into four groups, each designated by the name of a representative member. One group, with five members, is CI (C ≡ carbonaceous; I ≡ Ivuna, the representative meteorite). They are all of petrographic group 1, and hence have no chondrules, but they are carbon rich and contain an abundance of hydrated minerals, e.g. serpentine, magnetite and even epsomite which actually dissolves in water. They contain about 20% of water in a bound form and if some very volatile materials are excluded then the remainder has a composition very close to that inferred for the Sun.

The largest group of carbonaceous chondrites, with 14 members, is CM2 (M ≡ Mighei with petrographic type 2). These also contain serpentine but have somewhat less magnetite, epsomite and water (10%) than the CI group. They contain chondrules, small olivine grains and some small regions containing high-temperature minerals.

There are four members of the CV2 group (V ≡ Vigarano) and another four designated CV3. Although they are quite dark they contain very little carbon. They contain chondrules and have a high Mg:Si ratio, which is what really characterizes carbonaceous chondrites. They also contain white inclusions of high-temperature minerals rich in calcium, aluminium and titanium.

The final group is CO3 (O ≡ Ornans) which are similar to the CV meteorites in being carbon poor and containing high-temperature inclusions. Their main characteristic is in containing an abundance of small (0.2 mm) chondrules. There is also one CO4 specimen.

There is a single carbonaceous chondrite meteorite, of petrographic type 5, that does not fit into these groups. It fell in Australia in 1972 and is characterized by very indistinct chondrules and olivine of uniform composition, both of which indicate either slow cooling or some reheating after formation.

11.2.1.2   Achondrites

Achondrites are mostly silicate-rich igneous rocks although some of them are mixtures of rocky fragments, sometimes referred to as 'soils'. They contain very little native metal or sulphides. Various ways of classifying them have been suggested but they mostly fall into five groups—*eucrutes, howardites, diogenites, ureilites* and *aubrites*. The chemical compositions of the five groups of achondrites are given, together with those of chondrites and the Sun for comparison, in table 11.3.

Diogenites are of material to that is similar to that usually assumed for a planetary mantle. The mantle of the Earth is mainly olivine with some pyroxene but diogenites are mainly pyroxene, about 25% of which is the Fe end member of that class of minerals. They are mostly crushed and then reassembled as *breccias*, i.e. rocks formed by assemblages of small rocky fragments. In the case of diogenites all the fragments are of the same kind of rock and such breccias are called *monomict*.

Eucrites are mostly more-or-less equal mixtures of plagioclase and a calcium-containing pyroxene and have been likened to lavas produced by volcanism. When mantle material melts a low-density calcium–aluminium–sodium-rich plagioclase will tend to accumulate at the top of the solidifying melt. It also has a low melting point which means that other denser minerals would solidify earlier and sink to the bottom of the cooling melt. Some eucrites are in an uncrushed state but most are monomict.

Howardites consist of aggregated fragments of different kinds of rock (*polymict*) with the component fragments similar to eucrites and diogenites. They can be best understood as a 'soil' formed on a body of mass sufficient to give an appreciable gravitational field. This gives enough compression to consolidate the material. They show radiation damage from the solar wind, similar to that found in lunar soils. The implication from the degree of radiation damage is that the material must have been somewhere in the inner part of the Solar System.

Aubrites consist mainly of enstatite and differ from enstatite chondrites in having a much smaller component of metal and sulphur. They are almost all brecciated.

The main interest in ureilites is their carbon content (up to 1%) much of which is in the form of micro-diamonds. It is thought that the high pressures and temperatures required to produce diamonds came from a collision in space between the ureilite parent body and some other object.

Some eucrites have a very coarse texture and a preferred orientation of crystallites that indicate that they formed by crystals dropping though a magma in a gravitational field. Such materials are called *cumulates*. There are some achondrites which are cumulates but are different from the five types previously described. They fall into three groups—the *shergottites, nakhlites and chassignites* and are collectively referred to as SNC meteorites. Whereas most stony meteorites have ages clustered around 4.5×10^9 years, the assumed age of the Solar

Table 11.3. The composition of stony meteorites. The sums do not total 100% as oxygen and some other elements are missing. The relative proportions of the elements for the Sun are normalized to Si = 20.

	Silicate component							Metal component			
	Si	Mg	Fe	Al	Ca	Na	H_2O	Fe	Ni	FeS	C
Ord. Chond.	19	14	9	1.4	1.3	0.70	0.3	11.7	1.3	5.9	—
CI	11	9	18	0.9	0.9	0.56	20.5	0.11	0.02	16.7	3.8
CM	13	11	21	1.1	1.2	0.40	13.2	—	0.16	8.6	2.4
CV	17	14	20	1.3	1.8	0.40	1.0	2.3	1.1	6.1	0.5
CO	17	15	22	1.4	1.4	0.41	0.7	1.9	1.1	5.7	0.3
Sun	20	21	16	1.7	0.7	1.21	—	—	1.0	—	235
Enst. Chond.	19	13	1	1.0	1.4	0.74	0.6	19.8	1.7	10.7	0.3
Eucrites	23	5	12	4.8	4.9	0.28	0.6	1.2	—	0.6	—
Howardites	24	7	12	4.7	4.8	0.25	0.3	0.4	0.1	0.6	—
Diogenites	26	16	12	0.8	1.0	0.03	0.1	0.8	0.03	1.1	—
Ureilites	19	21	10	0.2	0.6	0.13	1.1	8.1	0.15	—	0.7
Aubrites	27	22	9	0.3	0.7	0.09	1.1	2.3	0.2	1.3	—
Pallasite	8	12	5	0.2	0.2	0.05	0.2	49.0	4.7	0.5	—
Mesosiderite	10	4	4	2.2	2.1	0.13	0.7	46.0	4.4	2.8	—

System, the SNC meteorites are much younger. The crystallization ages, when they achieved closure, seem to be about 6.5×10^8, 1.4×10^9 and 1.3×10^9 years respectively. The source of these meteorites is suggested as the planet Mars since it could have been volcanically active recently enough to explain the ages. This idea is supported by the analysis of gas trapped in the meteorites which is rich in CO_2 and is similar in composition to the Martian atmosphere as measured by spacecraft. In 1996 it was announced that microfossils had been discovered within an SNC meteorite but that claim is not generally accepted.

11.3   Stony irons

Stony iron meteorites contain roughly equal proportions of stone and metallic iron, with associated nickel. There are two main groups—the *pallasites* and *mesosiderites*. The pallasites consist of olivine crystals set in a metal framework with a considerable amount of troilite also present (figure 11.1(a)). A probable formation mechanism for these meteorites is that the molten metal was forced into a region where olivine crystals had formed and were cooling, shrinking and cracking. This could take place in a cooling solid body in which a gravitational field had separated denser metal and less dense stone with pallasites deriving from the interface region. Troilite, with density 5000 kg m^{-3} that is between that of stone and iron, would naturally concentrate in such a region and so be more abundant than usual.

(a)

(b)

Figure 11.1. (*a*) A pallasite. Olivine is set in a framework of metal. (*b*) A mesosiderite. Blobs, fragments and veins of metal are distributed randomly throughout the meteorite. The mixing of metal and silicate probably occurred in a violent environment.

Mesosiderites have a completely different appearance (figure 11.1(*b*)). The rock is in fragments, mainly of plagioclase and calcium-bearing pyroxene, together with some olivine in the form of small spheroids. They contain minerals that are only stable at pressures below 3 kbar, which rules out an origin from deep within a massive body. The metal is present as globbules and also as veins running through the meteorite.

11.4   Iron meteorites

Most iron meteorites are iron–nickel mixtures which formed as cumulates from an initially liquid state although a few of them look as though they have never been completely molten. The metal exists in the form of two iron–nickel alloys, *taenite* and *kamecite*. Figure 11.2 shows the stability diagram for the formation of these two alloys in which α represents the nickel-poor body-centred cubic kamacite and γ the nickel-rich face-centred cubic taenite. For a meteorite with less than about 6.6% nickel the process of cooling can be followed along the line PT in figure 11.2. In the liquid phase between P and Q the metal will be a molten structureless mixture of iron and nickel. When it solidifies at Q it assembles itself into taenite until it has cooled to point R when plates of kamacite begin to appear in equilibrium with the taenite. The amount of kamacite increases with decreasing temperature until, at point S, it is all kamacite. According to the phase diagram at about 600 K some taenite should begin to reappear but in practice this will not happen. At such a low temperature the atoms cease to be mobile and whatever structure is present at about 650 K will be frozen in and will not subsequently change. There are about 50 iron meteorites with 5–6.5% nickel which are kamacite plus some troilite. The surfaces of such meteorites, examined with a microscope, show a characteristic pattern due to the cubic structure of kamacite and they are known as hexahedrites. A hexahedron is a regular solid with six faces, i.e. a cube. However, for a meteorite with, say, 20% nickel by the time a temperature was reached when some kamacite should appear the movement of the nickel atoms would be so sluggish that the material would end up as virtually pure taenite. Meteorites in this condition show no structure and so are called *ataxites*, meaning 'without form' in Greek.

For a meteorite with between 6.5% and 13% nickel the meteorite will contain a mixture of taenite and kamacite. When the kamacite plates begin to form in the cooling meteorite the nickel can diffuse faster out of the kamacite than it can into the neighbouring taenite and the taenite immediately outside the kamacite plates contains more nickel than taenite elsewhere. If the cooling is fast then the final solid meteorite shows this enrichment of nickel in taenite around the kamacite plates to a greater extent than if the cooling is slow. The rates of cooling can be assessed from the Widmanstätten figures (figure 1.21) and are usually in the range 1–10 K per million years.

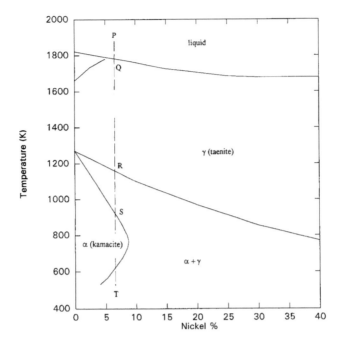

Figure 11.2. A stability diagram for the iron–nickel system.

11.5   Information from meteorites

The cooling rates indicated by iron meteorites enable a minimum size to be associated with the cooling body. If a body were less than about 200 km in radius then it would cool more rapidly, even in its central regions, than is compatible with deductions from most iron meteorites. However, it is not possible with certainty to define a maximum size of body from the cooling rate. Material cooling at a modest distance from the surface of a large body cannot be distinguished from material cooling at the centre of a small body.

Other indications of the sizes of the parent bodies of meteorites come from the minerals they contain. Some minerals can only be formed at high pressure and other minerals may be unstable at high pressure so an inventory of the minerals present in a meteorite can serve to place constraints on the pressure regime in which they formed. One way a high-pressure mineral may form is if the meteorite is subjected to a shock by a collision but this circumstance can usually be easily recognized. A high-velocity collision often produces a *shatter cone* a conical pattern of disturbance in the rock with its apex pointing towards the region of impact. When allowance is made for high-pressure minerals produced in this way it turns out that no other minerals found in stony meteorites could be produced over a period of time at a pressure over 12 kbar and for several minerals the limit would be as low as 3 kbar. Assuming that these pressures are characteristic at the centre of a body of stony composition then the sizes of these bodies would be between 500 and 1000 km in radius. Of course the meteorites could derive from the outer parts of larger bodies but most opinion is that the parent bodies from which meteorites were derived were much smaller than the Moon.

The number of possible source bodies, at least for iron meteorites, has been estimated by studying the concentrations of trace elements—e.g. Ga, Ge, Au—in relation to the amount of nickel present. From such studies the number of source bodies for iron meteorites has been estimated as somewhere between 10 and 16. It is less easy to estimate the possible number of parent bodies for chondritic meteorites or even stony meteorites in general. Analysis of chondritic meteorites show that they fall into eight or nine distinct chemical groups but this does not necessarily indicate the number of parent bodies. For example, minerals from different regions of the Moon may be chemically quite different and yet they clearly formed on the same body. Any body greater in extent than a few hundred kilometres is likely to show chemical heterogeneity. A better tool for estimating the number of possible parent bodies has been the measurement of oxygen isotopes, ^{16}O, ^{17}O and ^{18}O, in different chondritic meteorites. If two meteorites have different ratios of the three isotopes then it may be possible to derive one oxygen composition from the other by some process of physical or chemical fractionation (section 11.6.1). If one cannot be derived from the other in that way then it is reasonably certain that the source bodies of the two meteorites were different. From such analyses of oxygen isotopes it has been estimated that the number of parent bodies for chondritic meteorites is anywhere between 20 and 70.

11.6   Isotopic anomalies in meteorites

All naturally-occurring elements have at least one stable isotope and all elements have several unstable isotopes with half-lives varying from 10^{10} years down to microseconds or less. Where there is more than one isotope the ratios of one isotope to another can be an important diagnostic tool, either for determining that materials come from different source, or possibly the same source, or for indicating that the material has been processed or contaminated in some way. Here we shall be considering some of the more important isotopic anomalies that are found in meteorites. The term *anomalous* in this context implies that the isotope composition differs from that of terrestrial samples or some other cosmic standard.

Reynolds (1960) reported an excess of ^{129}Xe, which is a daughter product of ^{129}I in stony meteorites. From this observation it was inferred that the estimated time between some radiosynthetic event, that produced the ^{129}I, and the formation of cool rocks in the Solar System that could retain the released xenon was of order 1.7×10^8 years. Cameron (1978) suggested that the event was a supernova that triggered off the formation of the Solar System some 200 million years later.

Indeed this time-scale corresponds to the free-fall time of the interstellar medium, as given by (2.23), which adds plausibility to the idea.

Some isotopic anomalies in meteorites can be explained as the by-product of the decay of radioactive materials, but these are usually much more short-lived than ^{129}I so a second radiosynthetic event seems to be called for. Here we shall first describe the more important anomalies and then go through the ideas which have been put forward to explain them.

11.6.1   Oxygen isotopic anomalies

Oxygen has three stable isotopes, ^{16}O, ^{17}O and ^{18}O, and on Earth these generally occur in the proportions 0.9527:0.0071:0.0401, a composition referred to as SMOW (Standard Mean Ocean Water). The proportions measured in many samples differ from SMOW and a convenient way of defining the difference is through the δ notation. If the concentrations of ^{16}O and ^{17}O in the two samples are $n(^{16}O)$ and $n(^{17}O)$ then we write

$$\delta\,^{17}O(\text{‰}) = \frac{\{n(^{17}O)/n(^{16}O)\}_{\text{sample}} - \{n(^{17}O)/n(^{16}O)\}_{\text{SMOW}}}{\{n(^{17}O)/n(^{16}O)\}_{\text{SMOW}}} \times 1000$$

(11.1)

where the symbol ‰ indicates 'permille' or parts per thousand.

Terrestrial samples from various sources are found to vary from SMOW but in a very systematic way. A plot of $\delta\,^{17}O$ against $\delta\,^{18}O$ for different samples gives a straight line of slope 0.5 that may be explained by mass-dependent fractionation. Physical or chemical processes—e.g. diffusion in a thermal gradient or a rate of chemical reaction—may be linearly dependent on mass so that the change produced in the ratio of ^{17}O to ^{16}O may be just half that of ^{18}O to ^{16}O. This gives the line of slope 0.5, seen in figure 11.3 labelled Earth and Moon. This way of representing isotopic composition is known as a *three-isotope plot*. Three-isotope plots can also be produced for other elements with three stable isotopes.

Samples from the Earth and the Moon all fall on a single line of slope 0.5, seeming to indicate a common source. Samples of eucrites and other achondrites also give lines of slope 0.5 but displaced from the Earth–Moon (EM) line. This seems to indicate that mass-dependent fractionation of their material has occurred but that either they come from a different source than EM material or that they come from the same source which has been contaminated to give a different starting point for the fractionation. However, Clayton et al (1973) found that samples taken from carbonaceous chondrite anhydrous materials gave an oxygen three-isotope plot with slope almost unity. Later work refined the slope to 0.94 ± 0.01. This and other lines are shown in figure 11.3. Later it was found that some samples from ordinary chondrites gave a line of slope close to 1.0, also shown in figure 11.3. This line is well displaced from that given by carbonaceous chondrite material.

A slope of unity could be explained as the result of mixing some standard mixture of isotopes, which could be SMOW or something else, with various

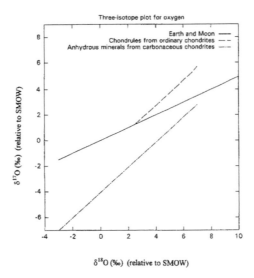

$\delta^{18}O$ (‰) (relative to SMOW)

Figure 11.3. Oxygen three-isotope plots for terrestrial and lunar materials, chondrules from ordinary chondrites and anhydrous materials from carbonaceous chondrites.

amounts of pure ^{16}O. If the amount of ^{16}O in the mixture was changed by a factor $1 + \alpha$ then, from (11.1),

$$\delta^{17}O = \frac{\frac{n(^{17}O)}{n(^{16}O)} \frac{1}{1+\alpha} - \frac{n(^{17}O)}{n(^{16}O)}}{\frac{n(^{17}O)}{n(^{16}O)}} \times 1000 = \frac{1000\alpha}{1+\alpha} \qquad (11.2)$$

and the same result would be found for $\delta^{18}O$. With different samples having different values of α the line of unit slope would follow automatically.

Various ideas have been advanced as to how the pure ^{16}O could have been produced and incorporated in the meteorites. The usual assumption is that the ^{16}O was produced by nuclear reactions in stars in which the common carbon isotope ^{12}C reacts with an alpha-particle, ^{4}He. This is then incorporated into grains that subsequently enter the Solar System. The normal oxygen in the Solar System then exchanges with the pure ^{16}O by diffusion processes in which most of the oxygen in the grain is replaced. Another model will be suggested in section 11.7.1.4.

11.6.2 Magnesium in meteorites

When the oxygen isotopic anomalies were discovered it became of interest to look for other anomalies, especially in those elements which are intimately linked to oxygen, for example, magnesium and silicon. As it turns out both of these elements also have three stable isotopes—for magnesium ^{24}Mg, ^{25}Mg and ^{26}Mg

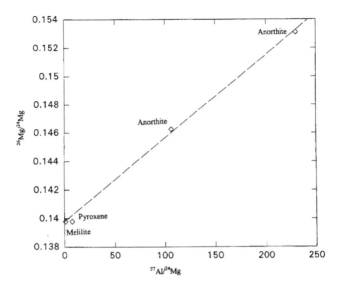

Figure 11.4. The excess of ^{26}Mg against total aluminium content for various minerals in a CAI inclusion.

which occur in the approximate proportions 0.790:0.100:0.110. For most samples the three-isotope plot showed the expected slope of 0.5. In 1976 Lee *et al* discovered that in some of the white calcium–aluminium-rich high-temperature inclusions (known as *CAI inclusions*) in carbonaceous chondrites there was an excess of ^{26}Mg which was proportional to the amount of aluminium in the sample. The effect was most easily measured in grains for which the minerals were rich in aluminium but contained comparatively little magnesium.

There is only one stable isotope of aluminium, ^{27}Al, but an unstable isotope ^{26}Al has the comparatively long half-life of 720 000 years. The interpretation of the meteorite measurements was that the aluminium present in the meteorite had originally contained a small proportion of ^{26}Al which decays according to

$$^{26}\text{Al} \rightarrow {}^{26}\text{Mg} + \beta^- + \nu. \tag{11.3}$$

In all parts of a particular meteorite the ratio of ^{26}Al to ^{27}Al was the same so that when different grains were taken with different amounts of aluminium and magnesium the ^{26}Mg excess was just dependent on the total aluminium content. This is shown by the linear relationship, illustrated in figure 11.4, between the ratios of ^{26}Mg/^{24}Mg and ^{27}Al/^{24}Mg.

The usual interpretation of this observation is that the rocks contained in the meteorite, and presumably in the whole Solar System, became cold closed systems within a few half-lives of a nucleosynthetic event, probably a supernova, which produced ^{26}Al. This is a time constraint, but not a very tight one, on the time-scale for forming the Solar System. Another implication is that if ^{26}Al was widespread in the early Solar System as a component of normal aluminium then it would have been an important source of heat. The inferred proportion of ^{26}Al in the original aluminium in the CAI inclusions varies from 2×10^{-5} down to 10^{-8} or less. An asteroid of radius 10 km containing 1.5% aluminium, a proportion of 10^{-6} of which was ^{26}Al, would become completely molten in its interior. This has implications for chondritic meteorites, assuming that they were all endowed with the ^{26}Al content of the CAI inclusions, since they show clear signs of melting and even re-melting. The melting of iron meteorites cannot be explained in this way since the iron does not contain any aluminium. It has been suggested that now-extinct 'super-heavy' elements which were soluble in iron could explain their early molten state although current evidence opposes rather than supports the suggestion.

There is also an anomaly associated with ^{25}Mg. Clayton et al (1988) measured δ^{25}Mg in various CAI specimens and found most of the values between -12.2 and $31.1\%o$. They also measured δ^{30}Si in the same specimens and found that the values were linearly related to δ^{25}Mg. Since magnesium and silicon have similar volatilities they concluded that what they were detecting was mass-dependent fractionation rather than an anomaly. However, one of their measurements gave δ^{25}Mg $= 350\%o$ that certainly cannot be explained as fractionation and must be a genuine anomaly.

11.6.3   Neon in meteorites

Many meteorites have gas trapped within them and these gases are an important source of information. The usual procedure is to step-heat the meteorite, whereby when all the gas has been released at a particular temperature the temperature is raised by a predetermined amount and the next gases released are collected and analysed. The temperature of the meteorite gives a measure of the energy with which the gas is bound to the rock and is taken to indicate entrapment at a particular kind of site within the crystal structure of the mineral.

Normal neon has three stable isotopes, ^{20}Ne, ^{21}Ne and ^{22}Ne, that are in the proportions 0.9051:0.0027:0.0922. Neon collected from different meteorites has a wide range of compositions, which are shown in figure 11.5. Since the compositions vary so greatly the ratios of isotopes relative to ^{22}Ne are given in figure 11.5 rather than the δ values used for oxygen in figure 11.3. Most samples fall in the triangle ABC; C corresponds to normal neon and it is assumed that points A and B correspond to two other sources. All possible mixtures of these three components then correspond to points within the triangle.

Of the greatest interest are the observations that fall outside the triangle. Some of these are very close to the origin at points corresponding to almost pure ^{22}Ne. An obvious interpretation of this neon highly enriched in ^{22}Ne, so-called neon-E, is that it was produced by the decay of radioactive sodium, ^{22}Na. The difficulty presented by this explanation is that the half-life of ^{22}Na is 2.6 years and

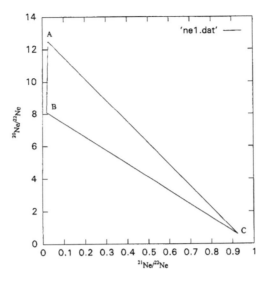

Figure 11.5. Most neon samples in the Solar System fall within the triangle ABC. Neon-E is close to the origin.

that this suggests the passage of, at most, tens of years between the production of the isotope and its incorporation into a cold rock, in particular a cold meteorite.

11.6.4 Anomalies in silicon carbide grains

Since the early 1980s there has been a systematic study of isotopic anomalies in silicon carbide, SiC, grains found in chondrites. This material can condense out of a vapour when the C/O ratio is sufficiently high, greater than 0.83 according to Larimer and Bartholomay (1979) but greater than unity according to Alexander (1993).

Silicon has three stable isotopes, ^{28}Si, ^{29}Si and ^{30}Si, that occur in the proportion 0.9223:0.0467:0.0310. For silicon from SiC a three-isotope plot, similar to that shown in figure 11.3 for oxygen, gives a slope of about 1.3 with a certain amount of scatter around the mean line. This cannot be explained by the addition of various proportions of pure ^{28}Si to normal silicon as this would give a slope of unity.

Carbon has two stable isotopes and the ratio $^{12}C/^{13}C$ in normal terrestrial carbon is 89.9. Some SiC grains contain 'heavy carbon' where the ratio is much smaller than the terrestrial value, down to about 20 or even less. However, there appears to be no correlation between the carbon and silicon anomalies; a plot of $^{12}C/^{13}C$ against $\delta\,^{29}Si$ shows no relationship whatsoever.

Nitrogen has two stable isotopes. The nitrogen present in SiC grains tends to be 'light nitrogen' with the $^{14}N/^{15}N$ ratio higher than the normal terrestrial value of 270. Most measurements by Zinner *et al* (1989) were of light nitrogen with the ratio up to about 2000 but a few samples are of 'heavy nitrogen' with ratios as low as 50.

The measurement of neon isotopes in SiC are interesting in that not only is there an enhancement of ^{22}Ne, the normal neon-E observation, but also an excess of ^{21}Ne (Zinner *et al* (1989). A plot of ^{20}Ne/^{22}Ne against ^{21}Ne/^{22}Ne gave a straight line relationship.

11.6.5   The deuterium anomaly

There are two stable hydrogen isotopes, ^{1}H (H) and deuterium, ^{2}H (D). The relative amounts of these isotopes vary widely in different bodies. In the atmosphere of Jupiter the ratio D/H is 2×10^{-5} and this is usually accepted as a Solar System standard. When Jupiter formed it would have contained a mixture of hydrogen isotopes characteristic of solar-system material at that time and, because of its large mass, it is expected that it retained everything with which it began its existence. On the Earth the ratio D/H is some eight times higher than on Jupiter, 1.6×10^{-4}, and in some meteorites, in particular carbonaceous chondrites, it is several times the Earth value.

The highest D/H ratio known in the Solar System occurs in the atmosphere of Venus which indicates a D/H ratio one hundred times greater than that of the Earth (Donahue *et al* 1982). The reason for this is well understood and is linked with the paucity of water on Venus. Early in its history Venus would have had much more water than now, much of it contained in its atmosphere because of its high ambient temperature. In the upper atmosphere water would have dissociated due to ultraviolet radiation from the Sun to give

$$H_2O + uv \rightarrow OH^- + H^+$$

or

$$HDO + uv \rightarrow OH^- + D^+.$$

The combination of temperature, that controlled the speeds of H or D atoms, and gravitational field, that controlled the escape speed from Venus, enabled hydrogen readily to escape while deuterium, with twice the atomic mass, was retained. Most of the oxygen released by this process is bound chemically in the crust.

Since the various ratios of D/H can be so easily related to differential atmospheric loss, either on present bodies or bodies that once existed, there is no need for exotic explanations of the observations. However, other anomalies cannot be so easily explained.

11.7   Explanations of isotopic anomalies in meteorites

Some, but not all, of the isotopic anomalies mentioned here can be associated with radioactive decay. In each such case a time-scale is indicated between the radiosynthetic event that produced the radionuclide and the formation of a closed system, a cold rocky material, within which the daughter product of the decay can reside and indicate an anomaly. The evidence from ^{129}Xe suggests that the interval between the production of ^{129}I and the formation of planets must have been less than 200 million years but probably similar to that interval of time (Hutchison 1982). The ^{26}Mg excesses in CAI samples suggest a maximum time-scale of a few million years from the production of ^{26}Al and the condensation of the CAI. Neon-E observations, interpreted as the result of the decay of ^{22}Na, suggest a maximum time-scale of a few tens of years between the production of the radioactive sodium and a rocky material cool enough to retain the daughter neon.

The obvious mechanism for forming new radioactive isotopes is a supernova and, as previously mentioned in section 11.6, a 200 million year interval between the production of ^{129}I in a supernova and the formation of the Solar System is quite plausible. What is not quite so plausible is the idea that there should have been two other supernovae a few million years and then a few tens of years before cold rocky materials were formed in the Solar System. The Mg–Al time-scale is supported by another isotopic anomaly that occurs in iron meteorites. This is an excess of the silver isotope ^{107}Ag due to the decay of radioactive palladium ^{107}Pd with a half-life of 6.5×10^6 years. This means that if one wishes to explain excess ^{26}Mg *without* a supernova then a second explanation is needed to explain an excess of ^{107}Ag, a completely different kind of element in a completely different kind of meteorite. Of course, the few million years estimate from ^{26}Mg and ^{107}Ag is an *upper* limit and the time between nucleosynthesis and cold-body formation could have been much less—perhaps even a few tens of years so that the neon-E observations could be explained by the same event. If this event had been a supernova, even as long as a few million years before the formation of the Solar System, then it would have produced ^{129}I and heavy elements such as ^{244}Pu (half-life 82 million years) that would have indicated in various ways that it had taken place at that time.

In view of the obvious difficulties in explaining the different time-scales by different supernovae, explanations have been offered in terms of grains from other parts of the galaxy coming into the Solar System carrying the isotopic anomalies with them. Black (1972) suggested for the neon-E anomaly that sodium-bearing grains could form in the cooling ejecta of a supernova in a few years and these could have been incorporated into solar-system material without any subsequent heating or any other modification. In view of the evidence for hot and turbulent histories of, at least, some chondrites, including carbonaceous chondrites, this scenario seems unlikely. Another suggestion is that irradiation of grains by energetic protons in the early Solar System could have led to the reaction ^{22}Ne(p, n) ^{22}Na, so that radioactive sodium was formed directly in the grains. If the grains grew quickly, and did not grow hot through irradiation and collisions, then the ^{22}Na could become buried in the interior of the grain thus enabling the ^{22}Ne to be retained.

11.7.1   A planetary collision origin for isotopic anomalies

In 1995 Holden and Woolfson (HW) put forward an alternative explanation for all
the lighter-atom isotopic anomalies we have considered here as a product of the
planetary collision, described in chapter 8. We shall now consider what happens
within the impact region when two planets collide.

Impacts of small projectile with more massive targets have been extensively
studied both theoretically and by modelling (Gault and Heitowit 1963). Such
modelling has also been extended to the fall of asteroid-size bodies on solid planets and the results compared with observations of craters on the Moon and other
solar-system bodies. The collision of two planetary-size bodies had never been
modelled in detail and, indeed, HW did not produce such a model. They first
made a rough estimate of the temperature that the material of the planets would
reach if all the kinetic energy involved in the collision were converted into heat.
If the relative speed of the planets at the moment of impact is V_c then relative to
the centre-of-mass the kinetic energy is

$$E = \frac{M_1 M_2}{2(M_1 + M_2)} V_c^2 \qquad (11.4)$$

where M_1 and M_2 are the masses of the two planets. Thus the temperature, T,
attained if all the energy of the collision goes into heating all the material of the
planets is given by

$$\frac{M_1 M_2}{2(M_1 + M_2)} V_c^2 = \frac{3}{2} kT \frac{M_1 + M_2}{\mu} \qquad (11.5)$$

where μ is the mean particle mass and it is assumed that the bodies are completely
vaporized. Writing

$$\frac{M_1}{M_2} = \varepsilon \qquad (11.6)$$

gives

$$T = \frac{V_c^2 \varepsilon \mu}{3(1 + \varepsilon)^2 k}. \qquad (11.7)$$

For values of ε about 0.5 and μ of order 2×10^{-27} kg, T is approximately 10^5 K.
There would be some extra energy available because of the merging of the two
planets that will release potential energy. In addition the material remote from
the collision region would barely rise in temperature at all and it is clear that a
small proportion of the material of the planets, that in the impact region, will take
up most of the released energy. In the other direction not all the kinetic energy
is transformed into thermal energy; some fraction of it remains as kinetic energy
throughout the collision process. In addition some of it will go into ionizing the
heated material which reduces the temperature in two ways. First, it decreases
the average value of μ that goes into (11.7) and second, by the production of
electrons, it increases the number of particles that share the released energy.

HW estimated that the highest temperature reached in the impact region would be somewhere in the range 2–4×10^6 K and in the calculations that follow the value 3×10^6 K was used. As we shall see in section 11.7.1.9 the estimate turns out to have been a reasonable one. Such a temperature is well below that at which significant nuclear reactions could take place with one notable exception—reactions involving deuterium. We have seen that there is a high D/H in the atmosphere of Venus, presumed to be due to the dissociation of water and the differential loss of hydrogen compared to deuterium. There are other ways that a high D/H ratio could occur. Early major planets of intermediate mass would have been hot soon after their formation and subsequently cooled. If they had extensive hydrogen atmospheres then there would be a period of time when the combination of temperature and escape speed would have given a heavy loss of hydrogen with little loss of deuterium. Michael (1990) investigated the conditions that would give a high D/H ratio and concluded that ratios similar to that on Venus could occur over a wide range of conditions for planets intermediate in mass between those of the Earth and Jupiter. This gives the possibility that in a collision between early proto-planets, one or both of which had high D/H ratios, a chain of nuclear reactions could be set off, triggered by those involving deuterium. This was the basis of the HW 1995 paper.

11.7.1.1 *The initial composition*

The material in the impact region was taken as a mixture of

(i)   minerals, mostly silicates, containing sodium, magnesium, aluminium and iron;
(ii) more iron either as free metal, oxide or sulphide;
(iii) water, ammonia and methane as icy components of surface material; and
(iv) an atmosphere containing hydrogen, helium, methane and neon.

The material would have been somewhat compressed by the collision although the Rankine–Hugoniot equations show that for adiabatic compression with very hot material, so that no molecules exist, the compression factor could be no more than a factor of four. Table 11.4 shows the composition in the impact region based on a density of 10^4 kg m^{-3}. This is probably an underestimate of the density so that the deduced nuclear reaction rates would tend to be too low. The ratios of isotopes are the usual terrestrial ones except for D/H which is taken at the Venus level and is within the range predicted by Michael's (1990) results. Small components of heavy atoms are excluded because attention is restricted to light-atom anomalies. However, iron is included because it is a large component; although it will not be involved in nuclear reactions significant for light-atom anomalies it will act as an important coolant. To have excluded iron would have overestimated the rates of the nuclear reactions of interest.

Table 11.4. The initial isotopic composition of material in the impact region.

^{1}H	1.4150×10^{30}	D	2.3008×10^{28}		
^{3}He	2.4332×10^{23}	^{4}He	1.7380×10^{29}		
^{12}C	7.2553×10^{26}	^{13}C	8.0696×10^{24}		
^{14}N	8.0182×10^{25}	^{15}N	2.9778×10^{23}		
^{16}O	8.5277×10^{26}	^{17}O	3.2482×10^{23}	^{18}O	1.7096×10^{24}
^{20}Ne	5.4559×10^{23}	^{21}Ne	1.6276×10^{21}	^{22}Ne	5.5578×10^{22}
^{23}Na	1.8046×10^{25}				
^{24}Mg	2.0142×10^{25}	^{25}Mg	2.5500×10^{24}	^{26}Mg	2.8076×10^{24}
^{27}Al	4.1680×10^{25}				
^{28}Si	1.6843×10^{26}	^{29}Si	8.5284×10^{24}	^{30}Si	5.6612×10^{24}
^{32}S	8.6353×10^{24}	^{33}S	6.8145×10^{22}	^{34}S	3.8252×10^{23}
^{56}Fe	2.4940×10^{25}				

11.7.1.2   Details of the model

An important feature to be put into any model involving high temperatures is ionization. This is an energy-absorbing process that reduces the increase of temperature in two ways; first, some energy is required to remove electrons from the atom and, second, those electrons add to the number of particles which are to be heated. HW incorporated ionization in their computational model by solving the standard Saha equations in a form suggested by Zel,dovich and Raizer (1966). For material with the composition given by table 11.4 the amount of energy involved in ionization is 35% of the total energy at a temperature of 2×10^5 K but only 3.5% at a temperature of 1.6×10^7 K. It can be seen that as the temperature approaches that where nuclear reactions take place ionization becomes less effective as a controller of temperature.

The assumption, mentioned in section 11.7.1.1, that the density remained constant at four times the uncompressed density in the reacting region, was based on the premise that the planetary collision resembled that of two similar uniform streams of gas, which is far from true. If the uncompressed density of the colliding streams of gas increases away from the collision region—and layering by density in the planet will ensure this—then much greater compression can take place. However, since this *increases* the reaction rates it reinforces rather than detracts from the results of the computation.

The expected behaviour of the model is that nuclear reactions would quickly build up to give an explosive event, at which stage the planets would be pushed apart and perhaps disrupted. During the period in which the planets approached each other the bulk of the impact region would be confined by the surrounding planets and although radiative cooling would occur to some extent it should not be an important factor. On the other hand, after the explosion the reaction region will expand rapidly and cooling will be rapid, so causing reactions to cease. HW simulated cooling by assuming that there was a maximum temperature, T_{max}, that was a parameter of their model. They then took the amount of energy available for ionization and heating material as a fraction z of that released by nuclear reactions where

$$z = \left(1 - \frac{T}{T_{max}}\right)^{\eta} \tag{11.8}$$

and η is another parameter, usually taken as 0.1. Considerable variation of the cooling parameters made no significant difference to the main results of the model.

There were 283 reactions, plus reverse reactions for most of them, plus 40 radioactive decay processes incorporated in the model. These are given in table 11.5. The analytical formulae used for the reaction rates were those given by Fowler *et al* (1967, 1975), Woosley *et al* (1978), Harris *et al* (1983) and Caughlan *et al* (1985). Although these reaction rates may have errors up to a factor of 2 they are certainly good enough to give a correct general pattern of behaviour in this application.

11.7.1.3 *Computational results—temperature and compositions*

HW solved the coupled differential equations describing the chains of nuclear reactions using a novel predictor–corrector approach in which analytical solutions over short time intervals were embedded in the numerical integration. This was found to be both stable and convergent. A variety of initial conditions were explored but the results presented here were based on $T_{max} = 10^9$ K, $\eta = 0.1$, D/H = 0.016 and an initial temperature of 3×10^6 K. The variation of temperature is shown in figure 11.6. There is a slow build up for 10.3 s, an explosive rise to about 5×10^8 K followed by a very slow rise thereafter. This can be understood in terms of the composition of hydrogen isotopes, shown in figure 11.7. The main early reactions involve D–D reactions forming tritium. When enough tritium has been formed then D–T and T–T reactions lead to an explosive generation of energy. Tritium and deuterium become exhausted when the temperature has reached more than 5×10^8 K and thereafter a slow increase in temperature is fuelled by reactions involving heavier elements.

New isotopes, especially lighter ones such as ^{7}Li, appear quite quickly. Nuclear reactions induced by neutrons (s-process reactions) are effective at lower temperatures and numbers of unstable isotopes soon appear although the most rapid changes in the concentration of isotopes occur during the explosive phase.

When the explosion takes place, material from the heart of the reaction region is violently expelled and mixes with other vaporized material that has either been less processed or perhaps not processed at all. HW simulated this mixing by taking a few per cent of the highly processed material (HPM—table 11.7) with the remainder being roughly equal but random mixtures of lightly processed material (LPM—table 11.6) and the original unprocessed material (UPM) as given in table 11.4.

Table 11.5. The reactions and radioactive decays incorporated in the collision model. For most reactions a reverse reaction can also occur. $^{26}\text{Al}_\text{m}$ and $^{26}\text{Al}_\text{g}$ are two differently-excited states of ^{26}Al.

$\text{H}(\text{e}^-,\nu)\text{n}$	$\text{H}(\text{p},\text{e}^+ + \nu)\text{D}$	$\text{H}(\text{p}+\text{e}^-,\nu)\text{D}$	$\text{D}(\text{D},n)^3\text{He}$
$\text{D}(\text{D},\text{p})\text{T}$	$\text{D}(\text{p},n)2\text{H}$	$\text{T}(\text{D},n)^4\text{He}$	$\text{T}(\text{T},2n)^4\text{He}$
$^3\text{He}(\text{e}^-,\nu)\text{T}$	$^3\text{He}(\text{p},\text{e}^+ + \nu)^4\text{He}$	$^3\text{He}(\text{D},\text{p})^4\text{He}$	$^3\text{He}(\text{T},\text{D})^4\text{He}$
$^3\text{He}(\text{T},n+\text{p})^4\text{He}$	$^3\text{He}(^3\text{He},2\text{p})^4\text{He}$	$^4\text{He}(2n,\gamma)^6\text{He}$	$^4\text{He}(n+\text{p},\gamma)^6\text{Li}$
$^4\text{He}(\text{D},\gamma)^6\text{Li}$	$^4\text{He}(\text{T},\gamma)^7\text{Li}$	$^4\text{He}(\text{T},n)^6\text{Li}$	$^4\text{He}(^3\text{He},\gamma)^7\text{B}$
$^4\text{He}(\alpha+n,\gamma)^9\text{Be}$	$^4\text{He}(2\alpha,\gamma)^{12}\text{C}$	$^6\text{Li}(\text{p},^3\text{He})^4\text{He}$	$^7\text{Li}(\text{D},n)2^4\text{He}$
$^7\text{Li}(\text{T},2n)2^4\text{He}$	$^7\text{Li}(^3\text{He},n+\text{p})2^4\text{He}$	$^7\text{Be}(\text{e}^-,\gamma+\nu)^7\text{Li}$	$^7\text{Be}(\text{D},\text{p})2^4\text{He}$
$^7\text{Be}(\text{T},n+\text{p})2^4\text{He}$	$^7\text{Be}(^3\text{He},2\text{p})2^4\text{He}$	$^9\text{Be}(\text{p},\text{D})2^4\text{He}$	$^{12}\text{C}+^{12}\text{C} \rightarrow {}^{24}\text{Mg}$
$^{12}\text{C}+^{16}\text{O} \rightarrow {}^{28}\text{Si}$	$^{16}\text{O}+^{16}\text{O} \rightarrow {}^{32}\text{S}$		

n–γ reactions $A(n,\gamma)B$ with $A = {}^{19}\text{Ne}, {}^{20}\text{Ne}, {}^{21}\text{Ne}, {}^{22}\text{Ne}, {}^{21}\text{Na}, {}^{22}\text{Na},$ $^{23}\text{Na}, {}^{24}\text{Na}, {}^{23}\text{Mg}, {}^{24}\text{Mg}, {}^{25}\text{Mg}, {}^{26}\text{Mg}, {}^{27}\text{Mg}, {}^{25}\text{Al}, {}^{26}\text{Al}_\text{m}, {}^{26}\text{Al}_\text{g}, {}^{27}\text{Al},$ $^{28}\text{Al}, {}^{29}\text{Al}, {}^{27}\text{Si}, {}^{28}\text{Si}, {}^{29}\text{Si}, {}^{30}\text{Si}, {}^{31}\text{Si}, {}^{28}\text{P}, {}^{29}\text{P}, {}^{30}\text{P}, {}^{31}\text{P}, {}^{32}\text{P}, {}^{33}\text{P},$ $^{31}\text{S}, {}^{32}\text{S}, {}^{33}\text{S}, {}^{34}\text{S}, {}^{35}\text{S}, {}^{36}\text{S}$

n–p reactions $A(n,\text{p})B$ with $A = {}^{21}\text{Na}, {}^{22}\text{Na}, {}^{25}\text{Al}, {}^{27}\text{Si}, {}^{28}\text{P}, {}^{29}\text{P}, {}^{30}\text{P}, {}^{32}\text{P},$ $^{31}\text{S}, {}^{33}\text{S}$

n–α reactions $A(n,\alpha)B$ with $A = {}^{19}\text{Ne}, {}^{21}\text{Ne}, {}^{20}\text{Na}, {}^{21}\text{Na}, {}^{22}\text{Na}, {}^{23}\text{Mg},$ $^{25}\text{Al}, {}^{27}\text{Si}, {}^{28}\text{P}, {}^{29}\text{P}, {}^{31}\text{S}, {}^{32}\text{S}, {}^{33}\text{S}, {}^{35}\text{S}$

p–γ reactions $A(\text{p},\gamma)B$ with $A = \text{D}, \text{T}, {}^6\text{Li}, {}^7\text{Be}, {}^9\text{Be}, {}^{10}\text{B}, {}^{11}\text{B}, {}^{11}\text{C}, {}^{12}\text{C},$ $^{13}\text{C}, {}^{14}\text{C}, {}^{13}\text{N}, {}^{14}\text{N}, {}^{15}\text{N}, {}^{16}\text{O}, {}^{17}\text{O}, {}^{18}\text{O}, {}^{19}\text{F}, {}^{20}\text{Ne}, {}^{21}\text{Ne}, {}^{22}\text{Ne}, {}^{23}\text{Ne}, {}^{24}\text{Ne},$ $^{23}\text{Na}, {}^{24}\text{Na}, {}^{25}\text{Na}, {}^{24}\text{Mg}, {}^{25}\text{Mg}^*, {}^{25}\text{Mg}^{**}, {}^{26}\text{Mg}, {}^{27}\text{Mg}, {}^{28}\text{Mg}, {}^{29}\text{Mg}, {}^{26}\text{Al},$ $^{27}\text{Al}, {}^{28}\text{Al}, {}^{29}\text{Al}, {}^{30}\text{Al}, {}^{28}\text{Si}, {}^{29}\text{Si}, {}^{30}\text{Si}, {}^{31}\text{Si}, {}^{32}\text{Si}, {}^{30}\text{P}, {}^{31}\text{P}, {}^{32}\text{P}, {}^{33}\text{P}, {}^{32}\text{S},$ $^{33}\text{S}, {}^{34}\text{S}, {}^{35}\text{S}, {}^{36}\text{S}$

$* \rightarrow {}^{26}\text{Al}_\text{m}$ $** \rightarrow {}^{26}\text{Al}_\text{g}$

p–n reactions $A(\text{p},n)B$ with $A = \text{T}, {}^7\text{Li}, {}^9\text{Be}, {}^{11}\text{B}, {}^{13}\text{C}, {}^{14}\text{C}, {}^{14}\text{N}, {}^{15}\text{N}, {}^{19}\text{F},$ $^{23}\text{Ne}, {}^{24}\text{Ne}, {}^{23}\text{Na}, {}^{24}\text{Na}, {}^{25}\text{Na}, {}^{26}\text{Mg}^*, {}^{26}\text{Mg}^{**}, {}^{27}\text{Mg}, {}^{28}\text{Mg}, {}^{29}\text{Mg}, {}^{28}\text{Al}, {}^{29}\text{Al},$ $^{30}\text{Al}, {}^{31}\text{Si}, {}^{32}\text{P}$

$* \rightarrow {}^{26}\text{Al}_\text{m}$ $** \rightarrow {}^{26}\text{Al}_\text{g}$

p–α reactions $A(\text{p},\alpha)B$ with $A = {}^7\text{Li}, {}^9\text{Be}, {}^{10}\text{B}, {}^{11}\text{B}, {}^{14}\text{N}, {}^{15}\text{N}, {}^{16}\text{O}, {}^{17}\text{O},$ $^{18}\text{O}, {}^{19}\text{F}, {}^{20}\text{Ne}, {}^{23}\text{Na}, {}^{24}\text{Na}, {}^{25}\text{Na}, {}^{24}\text{Mg}, {}^{27}\text{Al}, {}^{30}\text{Al}, {}^{31}\text{P}, {}^{32}\text{P}, {}^{33}\text{P}, {}^{35}\text{S}, {}^{36}\text{S}$

α–γ reactions $A(\alpha,\gamma)B$ with $A = {}^6\text{Li}, {}^7\text{Li}, {}^7\text{Be}, {}^{12}\text{C}, {}^{14}\text{N}, {}^{15}\text{N}, {}^{16}\text{O}, {}^{17}\text{O},$ $^{18}\text{O}, {}^{17}\text{F}, {}^{18}\text{F}, {}^{19}\text{F}, {}^{20}\text{Ne}, {}^{21}\text{Ne}, {}^{22}\text{Ne}, {}^{23}\text{Ne}, {}^{24}\text{Ne}, {}^{21}\text{Na}, {}^{22}\text{Na}, {}^{23}\text{Na}, {}^{24}\text{Na},$ $^{25}\text{Na}, {}^{24}\text{Mg}, {}^{25}\text{Mg}, {}^{26}\text{Mg}, {}^{27}\text{Mg}, {}^{28}\text{Mg}, {}^{25}\text{Al}, {}^{26}\text{Al}_\text{g}, {}^{27}\text{Al}, {}^{28}\text{Al}, {}^{29}\text{Al}, {}^{30}\text{Al},$ $^{28}\text{Si}, {}^{29}\text{Si}, {}^{30}\text{Si}, {}^{31}\text{Si}, {}^{32}\text{Si}, {}^{29}\text{P}, {}^{30}\text{P}, {}^{31}\text{P}, {}^{32}\text{P}, {}^{33}\text{P}, {}^{32}\text{S}, {}^{33}\text{S}, {}^{34}\text{S}, {}^{35}\text{S}, {}^{36}\text{S}$

α–n reactions $A(\alpha,n)B$ with $A = {}^7\text{Li}, {}^9\text{Be}, {}^{10}\text{B}, {}^{12}\text{C}, {}^{13}\text{C}, {}^{14}\text{N}, {}^{15}\text{N}, {}^{17}\text{O},$ $^{18}\text{O}, {}^{21}\text{Ne}, {}^{22}\text{Ne}, {}^{23}\text{Ne}, {}^{24}\text{Ne}, {}^{23}\text{Na}^*, {}^{23}\text{Na}^{**}, {}^{24}\text{Na}, {}^{25}\text{Na}, {}^{25}\text{Mg}, {}^{26}\text{Mg}, {}^{27}\text{Mg},$ $^{28}\text{Mg}, {}^{27}\text{Al}, {}^{28}\text{Al}, {}^{29}\text{Al}, {}^{30}\text{Al}, {}^{31}\text{Si}, {}^{35}\text{S}$

$* \rightarrow {}^{26}\text{Al}_\text{m}$ $** \rightarrow {}^{26}\text{Al}_\text{g}$

Table 11.5. (Continued)

α–p reactions $A(\alpha, \mathrm{p})B$ with $A = $ ^{18}F, ^{19}F, ^{22}Na, ^{23}Na, ^{24}Na, ^{25}Na, ^{25}Mg, ^{26}Mg, ^{25}Al, ^{26}Al$_g$, ^{27}Al, ^{28}Al, ^{29}Al, ^{29}P, ^{30}P, ^{31}P

Radioactive decays of T, ^{6}He, ^{7}Be, ^{8}Be, ^{8}B, ^{9}B, ^{11}C, ^{14}C, ^{12}N, ^{13}N, ^{14}O, ^{15}O, ^{17}F, ^{18}F, ^{19}Ne, ^{23}Ne, ^{24}Ne, ^{21}Na, ^{22}Na, ^{24}Na, ^{25}Na, ^{23}Mg, ^{27}Mg, ^{28}Mg, ^{29}Mg, ^{25}Al, ^{26}Al$_m$, ^{26}Al$_g$, ^{28}Al, ^{29}Al, ^{30}Al, ^{27}Si, ^{31}Si, ^{32}Si, ^{29}P, ^{30}P, ^{32}P, ^{33}P, ^{31}S, ^{35}S

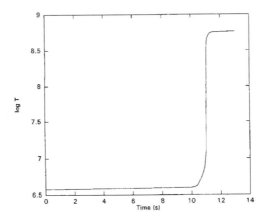

Figure 11.6. Temperature variation with time during the period of nuclear reactions.

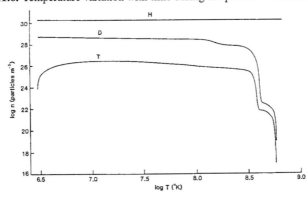

Figure 11.7. Concentration of hydrogen isotopes as a function of temperature during the period of nuclear reactions (Holden and Woolfson 1995).

11.7.1.4 *The oxygen anomaly*

The oxygen anomaly requires a source of pure ^{16}O that can be mixed with normal oxygen, taken as SMOW, in various proportions. A comparison of tables 11.4 and

Table 11.6. The isotopic composition of lightly processed material after 9.6 s with temperature 3.838×10^6 K.

^{1}H	1.4151×10^{30}	D	2.2876×10^{28}	T	5.9046×10^{25}
^{3}He	3.2610×10^{24}	^{4}He	1.7380×10^{29}		
^{6}Li	1.4593×10^{10}	^{7}Li	3.2891×10^{17}		
^{12}C	7.2553×10^{26}	^{13}C	8.0696×10^{24}	^{14}C	3.4091×10^{23}
^{14}N	7.9841×10^{25}	^{15}N	2.9778×10^{23}		
^{16}O	8.5277×10^{26}	^{17}O	3.2482×10^{23}	^{18}O	1.7096×10^{24}
^{17}F	6.1757×10^{15}	^{19}F	3.6244×10^{10}		
^{20}Ne	5.4423×10^{23}	^{21}Ne	2.9257×10^{21}	^{22}Ne	5.5597×10^{22}
^{23}Ne	4.3162×10^{19}				
^{22}Na	1.5162×10^{13}	^{23}Na	1.7725×10^{25}	^{24}Na	3.1845×10^{23}
^{25}Na	2.8174×10^{21}				
^{24}Mg	2.0047×10^{25}	^{25}Mg	2.5109×10^{24}	^{26}Mg	2.9319×10^{24}
^{27}Mg	1.0090×10^{22}	^{28}Mg	5.9855×10^{19}		
^{25}Al	1.1391×10^{16}	^{26}Al$_{\mathrm{g}}$	1.6346×10^{12}	^{27}Al	3.9867×10^{25}
^{28}Al	1.6871×10^{24}	^{29}Al	8.7540×10^{22}	^{30}Al	2.7286×10^{20}
^{28}Si	3.6370×10^{22}	^{29}Si	9.8799×10^{24}	^{30}Si	6.0157×10^{24}
^{31}Si	3.6370×10^{22}	^{32}Si	3.1443×10^{20}		
^{29}P	4.8825×10^{11}	^{31}P	1.0864×10^{19}	^{32}P	1.4459×10^{17}
^{33}P	1.8335×10^{20}				
^{32}S	8.4643021510^{18}	^{33}S	1.6246×10^{23}	^{34}S	3.8455×10^{23}
^{35}S	4.3461×10^{15}	^{36}S	8.4919×10^{19}		
^{56}Fe	1.7808×10^{25}				
n	3.1667×10^{15}	e^-	1.4716×10^{30}	e^+	3.3734×10^{15}

11.6 shows that LPM oxygen is the same as UPM oxygen. By contrast, for HPM from table 11.7 it appears that most of the ^{17}O and ^{18}O has been removed by nuclear reactions but that the ^{16}O has been little affected. However, in considering the oxygen isotope content of the processed material, either LPM or HPM, it is necessary to take into account the concentrations of ^{17}F and ^{18}F that decay with short half-lives (1.075 min and 1.83 hr respectively) into ^{17}O and ^{18}O. By the time that the vaporized material has cooled and begun to assemble itself into minerals these unstable fluorine isotopes will have been transformed into stable oxygen. Figure 11.8 shows the concentrations of ^{16}O, ^{17}O, ^{18}O, ^{17}F and ^{18}F as functions of temperature. It should be borne in mind that the majority of the temperature range corresponds to a short time interval. In the temperature range 4–5.8×10^8 K, even taking the decay of fluorine into account, the outcome is almost pure ^{16}O since ^{17}O and ^{18}O have been reduced by several orders of magnitude and there is also little ^{17}F and ^{18}F present.

The mixture of HPM with lower temperature material in different proportions will give the line of approximately unit slope seen in figure 11.3. A three-

Table 11.7. The isotopic composition of highly processed material after 11.2009 s at temperature 5.750×10^8 K.

^{1}H	1.4184×10^{30}	D	9.2436×10^{20}	T	1.6495×10^{20}		
^{3}He	2.8921×10^{19}	^{4}He	1.8366×10^{29}				
^{6}Li	7.9605×10^{11}	^{7}Li	1.8010×10^{16}				
^{7}Be	1.9096×10^{14}	^{9}Be	2.0410×10^{10}				
^{10}B	2.0911×10^{8}	^{11}B	2.0826×10^{10}				
^{11}C	9.6113×10^{10}	^{12}C	3.3958×10^{26}	^{13}C	1.5807×10^{26}		
^{14}C	2.3093×10^{26}						
^{13}N	8.1376×10^{24}	^{14}N	2.0321×10^{26}	^{15}N	3.1327×10^{22}		
^{14}O	2.3577×10^{22}	^{15}O	1.1263×10^{24}	^{16}O	8.2290×10^{26}		
^{17}O	2.5060×10^{16}	^{18}O	4.5648×10^{14}				
^{17}F	9.2942×10^{21}	^{18}F	4.6130×10^{18}	^{19}F	1.2791×10^{17}		
^{20}Ne	2.9907×10^{22}	^{21}Ne	1.0401×10^{21}	^{22}Ne	6.0983×10^{21}		
^{23}Ne	4.3410×10^{18}						
^{21}Na	2.5095×10^{17}	^{22}Na	4.2435×10^{19}	^{23}Na	6.4433×10^{21}		
^{24}Na	1.5782×10^{20}	^{25}Na	8.7919×10^{18}				
^{23}Mg	1.3019×10^{19}	^{24}Mg	5.5054×10^{22}	^{25}Mg	5.0443×10^{23}		
^{26}Mg	1.4613×10^{23}	^{27}Mg	2.2694×10^{22}	^{28}Mg	1.1276×10^{22}		
^{25}Al	1.3263×10^{21}	^{26}Al$_m$	1.9457×10^{21}	^{26}Al$_g$	2.2704×10^{22}		
^{27}Al	1.0426×10^{24}	^{28}Al	3.3025×10^{23}	^{29}Al	1.7306×10^{23}		
^{30}Al	6.4048×10^{23}						
^{27}Si	3.8188×10^{19}	^{28}Si	1.3822×10^{25}	^{29}Si	3.9766×10^{24}		
^{30}Si	1.5009×10^{25}	^{31}Si	5.0007×10^{24}	^{32}Si	1.0742×10^{26}		
^{29}P	1.0233×10^{21}	^{30}P	2.1932×10^{22}	^{31}P	1.8207×10^{24}		
^{32}P	6.2235×10^{24}	^{33}P	2.3360×10^{25}				
^{31}S	1.0185×10^{18}	^{32}S	4.0266×10^{23}	^{33}S	1.5016×10^{24}		
^{34}S	1.6737×10^{24}	^{35}S	1.5250×10^{23}	^{36}S	2.3223×10^{23}		
^{56}Fe	1.7808×10^{25}						
n	4.5398×10^{22}	e^-	1.7808×10^{30}	e^+	1.0438×10^{21}		

isotope plot of δ^{17}O and δ^{18}O for different grains from anhydrous C2 and C3 carbonaceous chondrites gives a straight line of slope 0.94 (Clayton 1981). Mixtures of UPM and different small proportions of HPM from the present model give a line with slope 0.97. If the HPM composition was taken at a slightly higher temperature a slope of 0.94 would be found.

The oxygen three-isotope plot from a mixture of HPM and UPM would go through the origin, which is inevitable since we have taken SMOW as the standard. However, the experimental plot, as seen in figure 11.3 always gives a value of δ^{18}O that is more positive (or less negative) than the value of δ^{17}O. This suggests that the UPM component should be 4–5% enriched in ^{18}O compared with SMOW. HW suggested that the Earth fragment of the less massive planet had been somewhat 'polluted' so that the isotopic composition of the UPM was not

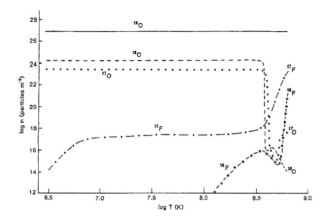

Figure 11.8. The concentrations of stable oxygen isotopes and radioactive fluorine isotopes as a function of temperature during the period of nuclear reactions (Holden and Woolfson 1995).

precisely that given by table 11.4, which, with the exception of hydrogen, gives the present Earth isotopic composition. Clearly there are possible explanations for the origin shift but the important characteristic of the anomaly, the near-unit slope, is well explained.

11.7.1.5   The magnesium anomaly and associated aluminium

In considering the final isotopic composition of the rocks which would condense from vaporized material produced by the collision it is necessary to take account of the relevant end products of radioactive isotopes. For magnesium anomalies there will be ^{26}Al that decays to ^{26}Mg but also ^{24}Ne and ^{24}Na that decay to ^{24}Mg and ^{25}Na and ^{25}Al that decay to ^{25}Mg. Also of interest is ^{27}Al, since the ratio of ^{26}Al/^{27}Al may be deduced from observations, and in this respect ^{27}Mg and ^{27}Si that decay to ^{27}Al must be taken into account.

Figure 11.9(a) shows the formation of isotopes giving ^{24}Mg and ^{25}Mg. The concentration of ^{24}Na briefly exceeds that of ^{24}Mg before it finally disappears and ^{24}Ne does not appear at all. The concentration of ^{25}Na is similar to that of ^{25}Mg for a short time before it disappears. However, ^{25}Al that occurs at a low level over the whole temperature range suddenly increases at the end and becomes the dominant source of ^{25}Mg. Figure 11.9(b) shows the isotopes leading to ^{26}Mg and ^{27}Al. Of particular significance are the large surges in the concentrations of ^{26}Al and ^{27}Si at the end of the temperature range.

In considering the oxygen anomaly, UPM and LPM were equivalent since low-temperature neutron-induced reactions had not affected the oxygen composition. This is not so for isotopes giving magnesium and aluminium so we consider

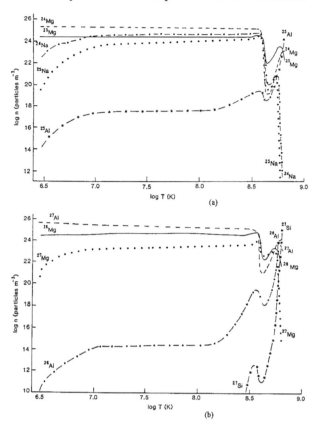

Figure 11.9. (*a*) Concentrations of ^{24}Mg and ^{25}Mg and short half-life isotopes decaying to ^{24}Mg and ^{25}Mg during the period of nuclear reactions (Holden and Woolfson 1995). (*b*) Concentrations of ^{26}Al, ^{26}Mg and ^{27}Al and short half-life isotopes decaying to ^{26}Mg and ^{27}Al during the period of nuclear reactions (Holden and Woolfson 1995).

a mixture of 50% UPM, 47% LPM and 3% HPM. With decay of all the relevant short half-life isotopes the mixture gives:

$$n(^{24}\text{Mg}) = 1.9644 \times 10^{25} \quad n(^{25}\text{Mg}) = 2.4716 \times 10^{24}$$

$$n(^{26}\text{Mg}) = 2.7862 \times 10^{24} \text{ m}^{-3}$$

$$n(^{26}\text{Al}) = 7.3949 \times 10^{20} \quad n(^{27}\text{Al}) = 2.0872 \times 10^{25} \text{ m}^{-3}.$$

The ratio $n(^{26}\text{Al})/n(^{27}\text{Al}) = 3.5 \times 10^{-5}$ which is comparable with the largest observed magnitude, about 5×10^{-5}. Smaller amounts of HPM or HPM material from a slightly lower temperature can give values down to the lowest observed, 5×10^{-8} (Clayton *et al* 1988).

Now consider a condensed mineral in which the ratio

$$c(^{27}\text{Al})/c(^{24}\text{Mg}) = R \tag{11.9}$$

where $c(x)$ is the concentration of component x in the mineral. From this and the relative isotopic concentrations in the mixture

$$c(^{24}\text{Mg}) = c(^{27}\text{Al})/R \quad c(^{25}\text{Mg}) = \frac{n(^{25}\text{Mg})c(^{27}\text{Al})}{n(^{24}\text{Mg})R}$$

$$c(^{26}\text{Mg}) = \frac{n(^{26}\text{Mg})c(^{27}\text{Al})}{n(^{24}\text{Mg})R} \quad c(^{26}\text{Al}) = \frac{c(^{27}\text{Al})n(^{26}\text{Al})}{n(^{27}\text{Al})}. \tag{11.10}$$

The ^{26}Mg anomaly is of the form that the excess of the isotope is proportional to the total amount of aluminium in the sample and manifests itself by a straight line relationship when $^{26}\text{Mg}/^{24}\text{Mg}$ $(=r)$ is plotted against $^{27}\text{Al}/^{24}\text{Mg}$. For the meteorite sample, once the ^{26}Al has decayed

$$r = \frac{c(^{26}\text{Mg}) + c(^{26}\text{Al})}{c(^{24}\text{Mg})} = \frac{n(^{26}\text{Mg})}{n(^{24}\text{Mg})} + R\frac{n(^{26}\text{Al})}{n(^{27}\text{Al})}. \tag{11.11}$$

The linear relationship between r and R was first observed by Lee *et al* (section 11.6.2) and subsequently confirmed by Steele *et al* (1978). Substituting the values for our mixture in (11.11)

$$r = 0.142 + 3.543 \times 10^{-5}R \tag{11.12}$$

that is similar to the observed relationship for this ratio of ^{26}Al to ^{27}Al.

Clayton *et al* (1988) measured ^{25}Mg anomalies in their specimens although they ascribed departures from the standard concentration as due to mass fractionation. They found for 20 CAI specimens that $\delta^{25}\text{Mg}$ was proportional to $\delta^{30}\text{Si}$ and since magnesium and silicon have similar volatility they assumed that fractionation in a vapour phase may have been involved. By taking various mixtures of UPM, LPM and HPM, such that the proportion of LPM was between 0.05 and 0.45 and that of HPM between 0.005 and 0.025, HW showed that a range of $\delta^{25}\text{Mg}$ values was found broadly compatible with those observed. One of the Clayton *et al* measurements gave $\delta^{25}\text{Mg} = 350\%_o$ that cannot be explained by fractionation. If the possibility is allowed of condensation from a HPM-rich vapour then, with HPM:UPM = 64:36, $\delta^{25}\text{Mg} = 350\%_o$.

The values of R, that express the ratios of the amount of aluminium to magnesium in the sample, were give as between 2.5 and 42 000 by Clayton *et al*. For one meteorite, Dhajala, for which $R = 15\,600$, the value of $\delta^{26}\text{Mg}$ was 920‰. For the UPM:LPM:HPM = 0.50:0.47:0.03 mixture it is found that

$$\delta^{26}\text{Mg} = (21.5 + 0.255R)\%_o. \tag{11.13}$$

This would indicate $\delta\,^{26}\text{Mg} = 4000\%o$, which is much too high. A lower proportion of UPM in the mixture, or HPM from a slightly lower temperature, would reduce the coefficient of R in (11.13) and allow the Dhajala value to be accommodated.

11.7.1.6 Neon-E

If ^{22}Na is present in the meteorite material when it is condensing then the neon released by staged heating will be pure ^{22}Ne or neon enriched in ^{22}Ne according to how much normal neon was trapped in the meteorite. This scenario requires not only that ^{22}Na should be present but also that the meteorite should cool and be able to retain neon on a time-scale of a few half-lives of ^{22}Na—say within 20 years.

A 1% component of HPM in the mixture would give a ^{22}Na concentration of 4.24×10^{17} m^{-3} which is about one part in 10^9 of the silicate-forming atoms in the mixture. This is an order of magnitude too low to explain the observations. However, the HMP material comes from a time when the concentration of ^{22}Na is increasing rapidly and 3 min later the concentration has increased by a factor of 12. If the condensing grains can cool fast enough then the neon-E observations can be explained.

11.7.1.7 Anomalies associated with SiC

The C/O ratios for UPM, LPM and HPM are 0.858, 0.859 and 0.895 respectively, each of which would satisfy the Larimer and Bartholomay criterion (section 11.6.4). Processed material at a slightly lower temperature than HPM would have C/O > 1 and some such material would undoubtedly be present. Given that the mixture of components would not be uniform, so that some regions would be richer in carbon than others, there is no problem in meeting the conditions for SiC formation.

The range of mixtures of UPM, LPM and HPM that were used by HW to find values of $\delta\,^{25}\text{Mg}$ were also used to find $\delta\,^{29}\text{Si}$ and $\delta\,^{30}\text{Si}$. The three-isotope plot is shown in figure 11.10(a) together with the line of slope 1.3 indicating the relationship given by Alexander (1993) using data from Zinner *et al* (1989), Alexander *et al* (1992), Stone *et al* (1990), Amari *et al* (1992) and Virag *et al* (1992). There is also given the line by Stone *et al* (1990) representing their data from the Orgueil and Murchison chondrites. The results from the HW model have a wider spread about the lines than the observations but the three-isotope plot for silicon is reasonably well explained.

For anomalies in SiC, HW suggested that the proportion of HPM might tend to be higher than for some other anomalies. This is because carbon compounds in the form of methane would form part of the atmosphere that would have been the seat of the explosive event. They took HPM carbon randomly between 1 and 12% of the total with LPM remaining between 5 and 45%. The ratios of

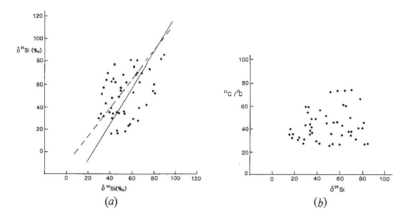

Figure 11.10. (*a*) A plot of $\delta\,^{29}$Si against $\delta\,^{30}$Si for 50 random mixtures of HPM, LPM and UPM. The dashed line, of slope 1.3, was given by Alexander (1993) and the full line by Stone (1991). (*b*) A plot of ^{12}C/^{13}C against ^{29}Si for various mixtures of HPM, LMP and UPM. There is no correlation, as is found from observations (Holden and Woolfson 1995).

^{12}C/^{13}C plotted against $\delta\,^{29}$Si are shown in figure 11.10(*b*). This clearly shows the existence of heavy carbon and the complete lack of correlation between the carbon and silicon anomalies, both consistent with observation.

For nitrogen anomalies it is necessary to consider the presence of ^{14}O and ^{15}O which decay into ^{14}N and ^{15}N on short time-scales and ^{14}C which decays into ^{14}N on a long time-scale. The development of the concentrations of these isotopes is shown in figure 11.11. The ^{14}C will be incorporated into SiC and the ^{14}N which comes from this would be tightly bound in the lattice. The normal nitrogen trapped in the meteorite as it condensed and cooled would be much more loosely bound. Excluding any contribution from ^{14}C the ratio of ^{14}N/^{15}N for the mixtures used for determining the ^{12}C/^{13}C ratios varied between 237 and 265, compared with the normal value of 270. In an SiC grain the contribution from ^{14}C is expected to be dominant and the majority of measurements show light nitrogen with ^{14}N/^{15}N as high as 2000 (Zinner *et al* 1989). Some measurements by Zinner *et al* were of heavy nitrogen with ^{14}N/^{15}N as low as 50. HW found that 5 min after their HPM the ratio ^{14}N/^{15}N had fallen to 6.9, even including ^{14}C in the ^{14}N inventory. A 6% mixture of this with UPM would give ^{14}N/^{15}N = 28.

The neon anomaly in SiC shows not only excess ^{22}Ne, which can be explained by the one-time presence of ^{22}Na, but also an excess of ^{21}Ne for which there is no radioactive parent. Zinner *et al* (1989) plotted the ratio ^{20}Ne/^{22}Ne against ^{21}Ne/^{22}Ne and found a linear relationship of the form

$$^{21}\text{Ne}/^{22}\text{Ne} = 0.002\,55(^{20}\text{Ne}/^{22}\text{Ne}) + 0.002\,49. \qquad (11.14)$$

Since neon can only be an atmospheric component HW considered the possibility

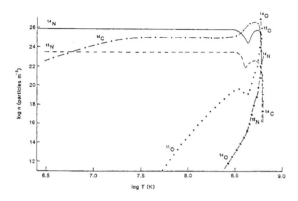

Figure 11.11. Concentrations of ^{14}N and ^{15}N and short half-life isotopes decaying to ^{14}N and ^{15}N during the period of nuclear reactions (Holden and Woolfson 1995).

that the neon component of SiC was nearly all of HPM origin with only small contributions of 5% UPM and 10% LPM. This gives relative contributions

$$^{20}\text{Ne} : {}^{21}\text{Ne} : {}^{22}\text{Ne} = 1.071 \times 10^{17} : 1.258 \times 10^{15} : 1.352 \times 10^{17}.$$

On a time-scale of a few tens of years ^{22}Na will add its contribution to give

$$^{20}\text{Ne} : {}^{21}\text{Ne} : {}^{22}\text{Ne} = 1.071 \times 10^{17} : 1.258 \times 10^{15} : 4.615 \times 10^{17}.$$

This will give the point A in figure 11.12. Over a longer time-scale there could be an exchange of neon; this means that the meteorite is slightly permeable to neon leading to a loss of the original neon and the gain of external normal neon. If a fraction of original neon to normal neon is $1 - x$:x then with $x = 0.875$ one has the point D. Points B and C correspond to $x = 0.5$ and 0.75, respectively. The points A, B, C and D do not fall exactly on a straight line but are quite close. The observational values also indicate a slightly steeper slope for lower points than for upper points.

11.7.1.8   *Cooling and condensation time-scales*

The most critical factor in the HW model is the time-scale for forming cool solid material that would be able to retain neon-E—less than 20 years or so. None of the other isotopic anomalies presents such a severe time constraint.

    The starting point of the HW analysis of cooling was a spherical volume of hot vapour of radius r_0 and temperature T_0. If this expands adiabatically to radius r then the temperature changes to T where

$$\frac{T}{T_0} = \left(\frac{r_0^3}{r^3}\right)^{\gamma-1} = \left(\frac{r_0}{r}\right)^2 \tag{11.15}$$

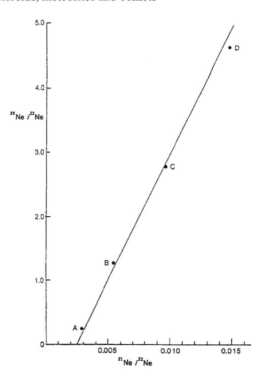

Figure 11.12. The points of the three-isotope plot for neon are from the Holden and Woolf-son model. The straight line is an average of two similar lines found by Zinner *et al* (1989) (Holden and Woolfson 1995).

and $\gamma = 5/3$ is the appropriate ratio of specific heats. The velocity of efflux of a gas into a vacuum is

$$v_e = \sqrt{\frac{2}{\gamma - 1}} c = \left[\frac{2\gamma kT}{(\gamma - 1)m} \right]^{1/2} \tag{11.16}$$

where c is the speed of sound in the gas and m is the mean particle mass. It was assumed that the boundary of the expanding sphere moved outwards with speed v_e so that

$$\frac{dr}{dt} = \left[\frac{2\gamma kT}{(\gamma - 1)m} \right]^{1/2}. \tag{11.17}$$

From (11.15) and (11.17)

$$\frac{dT}{dt} = -\frac{2}{r_0 T_0^{1/2}} \left[\frac{2\gamma kT}{(\gamma - 1)m} \right]^{1/2} T^2. \tag{11.18}$$

By integrating (11.18) the time can be found for the temperature to fall to T_c at which the first condensations should form. This gives, for $T_0 \gg T_c$,

$$t_c = \frac{r_0 T_0^{1/2}}{2T_c} \left[\frac{(\gamma - 1)m}{2\gamma k} \right]^{1/2}. \tag{11.19}$$

If the initial spherical region of hot vapour had a temperature of 10^7 K, corresponding to a few per cent HPM mixed with much cooler material, and radius 10^7 m then, with $T_c = 2000$ K and $m = 2 \times 10^{-27}$ kg, $t_c = 42\,600$ s or just under 12 hours. The actual time is not of much interest; despite the clear limitations in the simple model the cooling time is of the order of hours. Even if it were days, weeks or months it would fit in with the requirements of retaining neon-E, which is that cool grains could form reasonably quickly.

It was also shown by HW that the time for grains to form, based on the mobility of atoms and molecules and the target area of a growing grain, is also of the order of hours. It is evident from the HW analyses that the 2.6 year half-life of ^{22}Na presents no problems for their model.

11.7.1.9 Modelling a planetary collision

In the HW model it was assumed from general arguments that the temperature attained in the impact region of a planetary collision would be in the region 2–4×10^6 K. The calculations relating to the nuclear reactions were based on a triggering temperature of 3×10^6 K and this gave an explosion after an interval of about 11 s. If the triggering temperature had been 2.5×10^6 K then the interval between collision and explosion would have been over 1000 s which is too long relative to the duration of the collision event itself.

We now model a collision between two large planets using SPH (Appendix III). The calculation has much in common with the calculations by Benz *et al* (1986) and Benz *et al* (1987) on the single impact theory of Moon formation. The planets consisted of three layers—an inner iron core, a stony mantle and an extensive atmosphere consisting mainly of hydrogen with some helium. For the purposes of producing isotopic anomalies there should be a layer of ice-impregnated silicates on the solid surface but this could not be accommodated in the resolution of the simulation. It can be assumed that if sufficiently high temperatures are developed in the vicinity of the surface then the material would be available for the HW model.

The equations of state used for the planetary core and mantle were those used by Benz *et al* (1986, 1987) for iron and granite and described in section 9.1.4. For the atmosphere the equation used was

$$P = \tfrac{2}{3} u\rho (1 + 0.0358\rho) \tag{11.20}$$

that fits the low-density behaviour of hydrogen and has the right general behaviour at higher densities. The mean particle mass in the atmosphere was taken as 1.4

times the hydrogen atom mass. Other, more complex, equations of state have been considered by various workers and tabulated data for the equations of state of a number of materials are available from the Los Alamos National Laboratory which can be extracted by use of the SESAME program.

The planetary-collision model also includes the heating effect of thermonuclear reactions. The model atmosphere consisted of 80% hydrogen (with 1.6% D), 10% helium, 5% methane, 3.5% water and 1.5% ammonia. The rates of nuclear reactions and hence the rates of heat generation were found for a standard density and a variety of temperatures from 2×10^6–5×10^8 K. Reaction rates are proportional to density so the rate of heat generation could be found for all conditions of the atmosphere.

Each of the colliding planets was set up with SPH particles on a uniform close-packed hexagonal grid with particle masses giving a uniform density of 1.2×10^4 kg m^{-3} in the core and 4×10^3 kg m^{-3} in the mantle. The density of the atmosphere was taken of the form

$$\rho(x) = \rho_{max} \left(\frac{R_p - x}{R_p - R_s} \right)^2 \tag{11.21}$$

where $\rho(x)$ is the density at distance x from the centre of the planet, R_p is the total radius of the planet to the outside of the atmosphere and R_s is the radius of the solid part of the planet. The maximum atmospheric density, ρ_{max}, was selected on the basis of previous modelling of major planets (Stevenson 1978). The initial temperatures in the solid regions were taken as constant, 3000 K for the mantle and 6000 K for the core. The temperature in the atmosphere was modelled as

$$T = (200 + 2800\rho/\rho_{max})K \tag{11.22}$$

so that the temperature changes from 3000 K at the solid surface to 200 K at the upper boundary of the atmosphere.

The initial configuration gave an approximate model of a major planet but it was not an equilibrium model. Before starting the collision calculation each planet was allowed to relax under an SPH simulation until it settled into an equilibrium configuration. The two planets in equilibrium were then inserted into a collision program.

Several simulations were followed. One of those giving the required trigger temperature had the following initial planetary characteristics before relaxing to an equilibrium state:

Planet 1, mass 2.229×10^{26} kg ($\sim 37 M_\oplus$)
Radius, 6.000×10^7 m; radius of core, 6.000×10^6 m; radius of solid planet, 1.200×10^7 m.

Planet 2, mass 1.331×10^{26} kg ($\sim 22 M_\oplus$)
Radius, 5.241×10^7 m; radius of core, 5.241×10^6 m; radius of solid planet, 1.048×10^7 m.

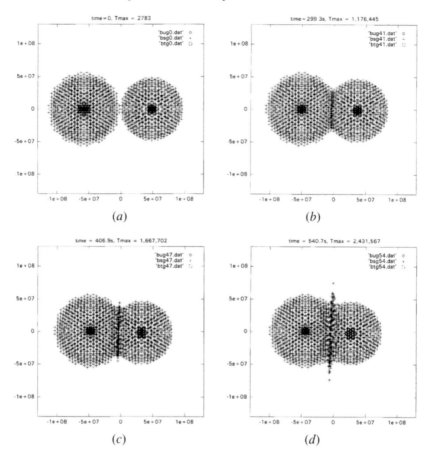

Figure 11.13. Stages in the computational model of the collision between planets.

The relative speed of the planets at the time of impact is 72 km s^{-1}. This comes about from a meeting close in to the Sun (at about 1.5–2 AU) in very eccentric orbits and includes the contribution of the potential energy released by the planets coming together. The results of the SPH collision calculation are shown in figure 11.13. The time is from the beginning of the simulation and the maximum temperature, T_{max}, is that in atmosphere material. Early in the simulation the highest temperature is in the core but it is the atmosphere that eventually gives temperature at which nuclear reactions can take place. The planets are each moving at an angle of 0.1 rad with respect to the x-axis, the larger planet upwards and the smaller one downwards. After 299.3 s (figure 11.13(b)) the collision interface is seen to be developing a high density and the temperature has risen to 1.176×10^{6} K. After 406.9 s (figure 11.13(c)) material is being thrown out in the plane of the collision interface although the main bulks of the two planets retain

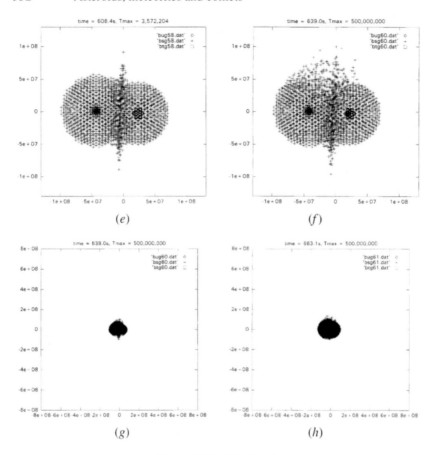

Figure 11.13. (Continued)

their integrity. The loss of material is clearly greater at 540.7 s (figure 11.13(*d*)) at which time the temperature has increased to 2.432×10^6 K and significant nuclear reactions will be occurring. Figure 11.13(*e*) shows the situation at 608.4 s where the temperature has reached 3.572×10^6 K and nuclear reactions are approaching the explosive stage. In figure 11.13(*f*) the explosive stage has been reached and the temperature is at the cut-off level of 5×10^8 K. Substantial amounts of the atmosphere are beginning to erupt outwards. Figure 11.13(*g*) is the same as figure 11.13(*f*) but re-scaled; at the highly reduced scale no detail can be seen. Figure 11.13(*h*) shows the situation 24 s later and some expansion is evident. After a further 34 s both planets have violently exploded into space (figure 11.13(*i*)). The details of what has happened to core material is shown at a larger scale in figure 11.13(*j*).

Although this model has supported the idea of a thermonuclear event being triggered by a planetary collision it has not shown that terrestrial planets could be

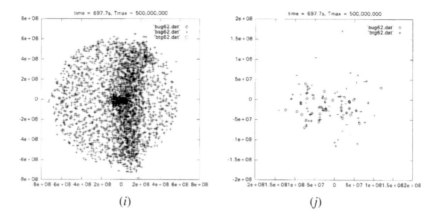

(i) *(j)*

Figure 11.13. (Continued)

an outcome. The explosion has been so violent that the combination of density and temperature of the previously solid material is unable to satisfy the Jeans critical mass criterion (2.22). This may be because inadequate cooling has been included in the model. The SPH equations give cooling when the material expands but no radiation cooling is included. Further modelling is required to explore the possibility of forming terrestrial planets from the solid residue of a collision between major planets.

11.7.1.10 *Theories of isotopic anomalies in meteorites; a discussion*

The study of the anomalous isotopic composition of meteorites has led to a great deal of speculation. There is general agreement that there was a major nucleosynthetic event in the region of formation of the Solar System about 200 million years before its birth. This is needed to explain the formation of heavier elements such as ^{244}Pu and ^{238}U. Cameron and Truran (1977) suggested that this might have been the event that triggered the formation of the Solar System by compressing a nebula cloud that was nearby. An alternative model has been presented by Golanski (1999). He showed that the injection of coolant material into the interstellar matter by a supernova would precipitate the formation of a cool dense cloud and this could lead to the formation of a stellar cluster.

The various major isotopic anomalies have given rise to a variety of explanations, some *ad hoc* but others that might seem to form a related consistent set. There is no doubt that neon-E presents a major challenge not only because of the short half-life of ^{22}Na, if that is the accepted source, but also because of the need to have a cool system to retain neon gas—again on a short time-scale. Clayton *et al* (1977) suggested that ^{22}Ne could be formed by particle irradiation of grains in the early Solar System. If, for example there was a strong flux of protons in the early Solar System then reactions such as ^{22}Ne(p, n) ^{22}Na and ^{26}Mg(p, n) ^{26}Al might be able to explain the neon-E and ^{26}Mg anomalies. On the other hand, Black (1978) has offered an explanation that the ^{22}Na and ^{26}Mg were formed in grains in cooler regions of a supernova and were only later transported into the Solar System. Clayton *et al* (1973) similarly suggested that pure ^{16}O could have come into the Solar System transported in grains—perhaps produced elsewhere by a reaction such as ^{12}C + ^{4}He → ^{16}O, requiring a temperature of about 10^8 K. However, the magnesium and oxygen anomalies occur in CAI material, which occurs in white inclusions some 2 mm in diameter. Lee *et al* (1976) have argued that objects of this size are unlikely to have come from outside the Solar System.

The heavy carbon content of SiC has given rise to other speculation about material coming into the Solar System from outside. Alexander (1993) has suggested that the wide range of ratios of ^{12}C/^{13}C can be explained by having material enter the Solar System from a number, perhaps six or more, of carbon stars.

The planetary collision event, which also explains so many other features of the Solar System, offers a coherent and consistent explanation for *all* the major anomalies that have been studied here. It provides a second nucleosynthetic event, which is clearly needed, in the framework of the way in which the planets were formed in highly eccentric orbits. This model is not necessarily dependent on the Capture Theory for the formation of the Solar System although it does require initially eccentric orbits so that planets can meet at a high relative speed.

11.8   Comets—a general survey

Comets are solid bodies with a high volatile content, the vaporization of which makes them conspicuous when they approach the Sun. Vaporized material escapes, sometimes from a limited region of the solid *nucleus* and forms a *coma* that can be up to 10^6 km in extent. In the out-gassing process some dust is also ejected. Due to the action of sunlight the coma becomes visible by fluorescence; outside the coma, with ten times its dimension, there is a cloud of hydrogen that emits ultraviolet radiation due to the action of sunlight. The comet often develops a tail, sometimes two tails, which stream out in the antisolar direction, pushed out by the action of the solar wind. The dust tail is usually much shorter than the plasma tail, 0.1 AU compared to 1 AU, and since gravitational forces are significant for dust particles the dust tail departs slightly from the anti-solar direction and so may be quite distinct.

Comets have a wide range of orbits with semi-major axes from a few AU to tens of thousands of AU and eccentricities mostly from about 0.5 to very close to unity. Those with periods greater than 20 years have more-or-less random inclinations so that their orbits are as likely to be retrograde as direct. There is a group of about 70 short-period comets, known as the *Jupiter family*, that all have direct orbits with inclinations less than about 30° and eccentricities in the range 0.5 to 0.7.

It is customary to divide comets into two categories—*short-period*, with pe-

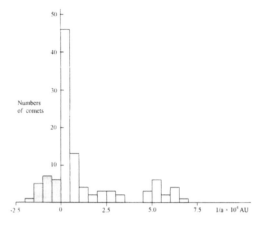

Figure 11.14. Numbers of comets with values of $1/a$ of small magnitude. Positive values correspond to bound (elliptical) orbits. The small number indicated as having hyperbolic orbits (negative values of $1/a$) are probably also in elliptical orbits (after Marsden *et al* 1978).

riods less than 200 years, and *long period* otherwise. This is really a distinction between comets with orbits mostly in the region of the planets and those with orbits mostly outside that region. Since the intrinsic energy of a body in orbit around the Sun is inversely proportional to the semi-major axis it is customary to give the intrinsic energies of comet orbits in units AU^{-1}. In figure 11.14 a histogram by Marsden *et al* (1978) is reproduced giving the numbers of long-period comets in intervals of $1/a$. The concentration of orbits with $0 < 1/a < 5 \times 10^{-5}$ AU^{-1} was previously noted by Oort (1948) who concluded that there was a spherically-symmetric cloud of comets, many in highly eccentric orbits, mostly at distances of several times 10^4 AU but with some maximum radius, r_0, about 2×10^5 AU. This is now known as the Oort cloud, estimated to contain 2×10^{11} comets. The basis of this estimate is that comets at a very great distance from the Sun would be subjected to significant stellar perturbations over a very long period of time so that their velocities become completely randomized in direction. Oort took as a model that at any very large distance, r, from the Sun the end point of the velocity vector of a comet would have a uniform distribution within a sphere of radius equal to

$$v(r)_{max} = \left\{ 2GM_\odot \left(\frac{1}{r} - \frac{1}{r_0} \right) \right\}^{1/2}. \tag{11.23}$$

It was also assumed that the swarm of comets would behave like a gas so that the gradient of some simulated pressure, dependent on the mean square speed of the comets and the mean density of the region they occupy, was in balance with gravitational forces. From this Oort showed that the number density of comets varied with distance from the Sun as

$$n(r) = A \left(\frac{r_0}{r} - 1\right)^{3/2}, \qquad (11.24)$$

where A is a normalizing constant proportional to the total number of comets. The model enables a prediction to be made of the number of near-parabolic comets that would be observed in one year in terms of the constant A and the radius r_0. The observed number of such comets penetrating to within 5 AU of the Sun is less than one per year but there are of the order of 10 per year those that can be detected with perihelia less than 15 AU. On this basis the estimated number of comets in the Oort cloud is 2×10^{11}. The estimate is based on a number of assumptions that would be difficult to justify convincingly but it is now so well accepted that it has acquired the status of a directly observed quantity.

The relationship of comets to other solar-system bodies is not known. Some believe that they are closely linked with asteroids so that they form a single family of small bodies with a common origin. They are then labelled according to a combination of the type of orbit they occupy and their volatile content. Others would argue that their physical properties are distinctly different from those of asteroids, as judged from meteorite samples, so that they are likely to come from a completely different source. Again there is uncertainty, as for asteroids, whether comets are the remains of material that did not become incorporated into planets when the Solar System formed or whether they are the result of the disruption of larger bodies such as the parent bodies of asteroids.

The Oort cloud presents two main problems to the theorist; the first is how it came into existence and the second is how it has endured over the lifetime of the Solar System. The Oort cloud extends about half way to the nearest star and, given the density and relative speeds of stars in the solar neighbourhood, it can be shown that many major perturbations by passing stars must have taken place. In addition the Solar System will have passed though a few Giant Molecular Clouds (GMCs) during its lifetime (Clube and Napier 1983) and these should have greatly disrupted or even completely removed the Oort cloud. However, Clube and Napier suggest that a GMC is not only a disrupter of the Oort cloud but also a source of new comets. At each passage of the Solar System through a GMC the old Oort cloud is removed and a new one is established. Akin to this idea is one that comets are formed at considerable distance from the Sun, either in the interstellar medium or in the outskirts of star-forming regions (Bailey 1987).

Many workers (e.g. Bailey 1983) have postulated the existence of a dense inner core of comets within the Oort cloud. Severe perturbations of the Oort cloud, that removed substantial numbers of comets from it, would also replenish the cloud from within. If such an inner core exists then it would be difficult to detect if it occurred at distances of hundreds of AU or more. The observation of *Kuiper-belt objects*, bodies with diameters from tens to hundreds of kilometres that are in orbit outside Neptune's orbit gives support to this idea. These objects may be the largest, and hence most visible, innermost harbingers of a large comet population stretching out hundreds or thousands of AU most of which may be of normal comet size—up to 10 km or so in diameter.

We shall now look at more detailed aspects of the possible origin of comets in the light of theories of Solar System formation.

11.8.1 New comets and the Oort cloud

Short-period comets and long-period comets with periods of a thousand years or so are of very limited interest in relation to the problem of the origin of the Solar System. For one thing every perihelion passage of a comet reduces its inventory of volatile material and by the time it has completed of the order of 1000 passages it will cease to be visible. It will then resemble a very small asteroid and probably be unobservable. Hence any visible comets of short period must have reached their present orbits in the recent past by the standards of the age of the Solar System. When comets enter the inner Solar System from the Oort cloud they are strongly perturbed by the planets, particularly the major planets, and some of them can become short-period comets. If the incoming comet is in an orbit close to the ecliptic and approaches a major planet closely then the orbital characteristics of the original and the new orbit are related by the Tisserand criterion (10.12). Thus the Jupiter family of planets would arise because of interactions of comets coming from the Oort cloud with Jupiter. For an incoming comet the orbit of which has a large value of a the first term in the Tisserand expression is negligible. Also with the initial eccentricity, e_i, very close to unity it is possible to write

$$T_{PC} = \left(\frac{2q_i}{a_P^3}\right)^{1/2} = \frac{1}{2a_f} + \left\{\frac{a_f(1 - e_f^2)}{a_P^3}\right\}^{1/2} \qquad (11.25)$$

where subscript P refers to the planet, i to the initial comet orbit and f to the final comet orbit. From (11.25) the initial perihelion distances for the various comets in Jupiter's family are found to be in the range 4.5–5.8 AU, compared with Jupiter's mean orbital radius of 5.2 AU. The strongest perturbations arise when the motion of the incoming comet is parallel to that of the planet and in a direct orbit so that they stay close together for a longer time.

There are three major influences perturbing comets in the Oort cloud and sending them into orbits which render them visible—stars, GMCs and the galactic tidal field. Once comets enter the inner Solar System they are then subject to perturbation by the planets. Thus a comet coming from the Oort cloud with $a = 20\,000$ AU would have an intrinsic negative energy with magnitude 5×10^{-5} AU^{-1}. When the comet enters the region of the major planets the perturbation it undergoes can either add to or subtract from its intrinsic energy. The actual change in intrinsic energy will obviously vary from one comet to another but the distribution of values can be approximated by a Gaussian function with standard deviation σ_E. The variation in σ_E as a function of the comet's perihelion distance and random inclinations is shown in figure 11.15. For those comets that

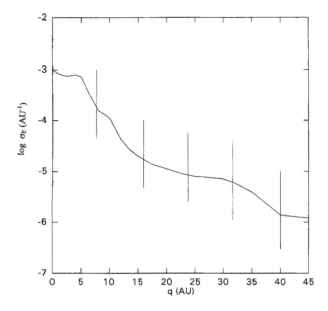

Figure 11.15. The value of σ_E for the perturbation of comets by planets as a function of the perihelion distance. The full line is the average for random inclinations. Vertical lines are the limits for inclinations from $0°$ (top) to $180°$ (bottom).

approach the Sun so closely that they are seen, the value of σ_E is 10^{-3} AU^{-1} or larger. Any comet coming from the Oort cloud that has an increase in energy due to planetary perturbation will almost certainly cease to be bound and escape from the Solar System. If, on the other hand, it loses energy then it will be much more tightly bound and go into an orbit with comparatively small a. For this reason any comet entering the inner Solar System which, after due allowance for planetary perturbation, is found to have come from the Oort cloud is referred to as a *new comet*. Because of the effect of planetary perturbation it is unlikely to have previously entered the inner Solar System on a similar orbit and it is unlikely ever to do so again.

We shall now consider in turn the perturbing influences that lead to new comets.

11.8.1.1 Stellar perturbations

For a comet in a very extended orbit such that $e \approx 1$ the intrinsic angular momentum

$$H = \{GM_\odot a(1 - e^2)\}^{1/2} \approx (2GM_\odot q)^{1/2} \qquad (11.26)$$

where q is the perihelion distance. The effect of a perturbation that changes a comet's orbit from being unobservable to observable must be that the perihelion

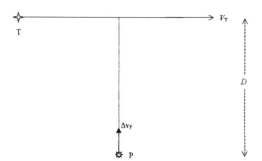

Figure 11.16. The geometry of the impulse approximation.

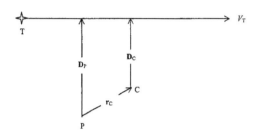

Figure 11.17. The geometry of the impulse approximation applied to the Sun at P and a comet at C. The system is not necessarily planar.

distance, and hence the magnitude of the intrinsic angular momentum, is reduced.

Most calculations on stellar perturbation have been made by the *impulse approximation*. In figure 11.16 there is shown the passage at speed V_T of a body, T, of mass M_T in a straight line past another body, P. This describes well the motion of a star passing by the Sun at a large distance, of order 1 pc. The relative speeds of field stars are usually in the range 20–30 km s^{-1} and at a distance of 1 pc this corresponds to a hyperbolic orbit of very large eccentricity. In the region of closest approach this is almost a straight line. It is easily shown that the net effect of the passage is to impart to the body P a velocity Δv_P perpendicular to the motion of the star in the plane defined by P and T's motion. This is given by

$$\Delta v_P = \frac{2GM_T}{DV_T} \tag{11.27}$$

where D is the nearest approach of the two bodies.

In figure 11.17 the passage of the star is considered relative to the Sun at S and the comet at C. The closest approaches of the star to the Sun and to the comet are given by the vectors D_S and D_C that are not necessarily co-planar. The comet is moving relative to the Sun during the interaction but to a first approximation the motion can be ignored. If the star passes at a distance 1 pc then appreciable inter-

Table 11.8. Minimum final perihelia for a stellar impact parameter of 0.5 pc with various values of r and initial q and stellar mass $M_\odot$.

	q_{init} (AU)			
r (AU)	40	30	20	10
80 000	23.47	15.68	8.31	1.73
60 000	33.25	24.15	15.23	6.62
40 000	37.59	27.92	18.30	8.80
20 000	39.49	29.56	19.64	9.75

actions will occur for less than 10 pc of the star's path. At a speed of 20 km s^{-1} relative to the Sun this will take 5×10^5 years. For a comet with $a = 40\,000$ AU the period is 6×10^6 years and the comet will not move very far in that time, especially near aphelion when the perturbation is strongest. Assuming that the comet is at rest during the perturbation is the basis of the impulse approximation.

Since Δv_S and Δv_C, the changes in velocity of the Sun and the comet, are different then the comet gains velocity $\Delta v_{CS} = \Delta v_C - \Delta v_S$ relative to the Sun. The consequent change in the intrinsic angular momentum of the comet's orbit is

$$\Delta H = \Delta v_{CS} \times r_C \tag{11.28}$$

where r_C is the position of the comet relative to the Sun. If ΔH is parallel to H then there is a maximum change of q found by differentiating (11.26)

$$\Delta q = \sqrt{\frac{2q}{GM_\odot}} \Delta H. \tag{11.29}$$

The direction of r_C relative to the star's motion is very important in determining Δq. Assuming a co-planar system if r_C is parallel to the star's motion then $D_C = D_S$ and there is no effect due to the impulse approximation. On the other hand if r_C is perpendicular to the star's motion then Δv_{CS} has the greatest possible value but since it is parallel to r_C there is no change of H and hence of q. For the maximum change of q, r_C should be at about 45° to V_T.

Table 11.8 shows the minimum possible perihelia for a number of initial values of q and r_C for the passage of a star of mass $M_\odot$ with impact parameter 0.5 pc.

If new comets have never penetrated to within, say, 20 AU from the Sun in a previous orbit then an encounter with a solar mass star at 0.5 pc can produce a detectable new comet within 15 AU but not a readily visible one within 5 AU. It is possible to estimate the number of stellar passages past the Sun within a distance D from the observed characteristics of the field stars in the solar neighbourhood. However, stars all have different masses and, as will be seen from (11.29), it is the ratio M_T/D that determines the extent of the perturbation, assuming that the relative velocities of field stars are not mass dependent.

We now determine the average interval between interactions of stars with the Sun corresponding to having

$$M_T/D = S. \tag{11.30}$$

From (2.10) we take the number density of stars in the solar vicinity as

$$N(M) = KM^{-2.5} \tag{11.31}$$

so that the number density with masses between M and $M + dM$ is $N(M)\,dM$. If the Sun is moving at a speed V relative to other stars in its neighbourhood then the number of interactions per unit time with stars of mass between M and $M + dM$ within a distance M/S is

$$dn = \frac{\pi M^2 V}{S^2} N(M)\,dM = \frac{K\pi M^{-1/2}V}{S^2}\,dM. \tag{11.32}$$

Thus the total number of interactions per unit time with a particular minimum value of S is

$$n = \frac{K\pi V}{S^2} \int_{M_{\min}}^{M_{\max}} M^{-1/2}\,dM = \frac{2K\pi V}{S^2}(M_{\max}^{1/2} - M_{\min}^{1/2}), \tag{11.33}$$

where $M_{\max}$ and $M_{\min}$ are the maximum and minimum masses of stars. The constant K can be found from the total number density of stars N_* so that

$$N_* = K \int_{M_{\min}}^{M_{\max}} M^{-2.5}\,dM = K(M_{\min}^{-1.5} - M_{\max}^{-1.5}). \tag{11.34}$$

Writing

$$S = M_\odot/D_\odot$$

so that $D_\odot$ is the distance at which a solar mass star gives a perturbation indicated by S we find for the mean time between such interactions

$$t_S = \frac{\mu_{\min} + \mu_{\max} + \sqrt{\mu_{\min}\mu_{\max}}}{2\pi V N_* D_\odot^2 \mu_{\min}^{1.5}\mu_{\max}^{1.5}} \tag{11.35}$$

where $\mu_{\min} = M_{\min}/M_\odot$ and $\mu_{\max} = M_{\max}/M_\odot$. Taking $\mu_{\min} = 0.1$, $\mu_{\max} = 10$, $N_* = 0.08$ pc^{-3} and $V = 20$ km s^{-1} (11.35) indicates a time interval of 4.3×10^6 years for the perturbation that gave the results in table 11.8. To bring a comet with a near-parabolic orbit from a perihelion outside the influence of the major planet to a perihelion distance of less than 15 AU requires values of S larger than $2M_\odot$ pc^{-1} that was used for table 11.8.

11.8.1.2   *Perturbations by Giant Molecular Clouds*

Giant molecular clouds (GMCs) were first detected through CO emission indicating the presence of large clouds of molecular gas. A typical mass of a GMC is $5 \times 10^5 M_\odot$ with a radius of 20 pc. Biermann (1978) was the first to point out the significance of GMCs as potential perturbers of comets. From their number density in the galaxy it might be expected that the Solar System has actually travelled through 10 ± 5 GMCs during its lifetime. By differentiating (11.27) the speed relative to the Sun imparted to a comet by an external perturber can be found. This is

$$\Delta v_{\mathrm{CS}} = \frac{2GM_{\mathrm{T}}}{D^2 V} \delta D \qquad (11.36)$$

where δD depends on the length and direction of the comet–Sun vector. For a grazing passage of a GMC and a comet–Sun vector of length 8×10^4 AU perpendicular to the relative motion of the GMC, $\Delta v_{\mathrm{CS}} = 211$ m s^{-1} where the escape speed of the comet from the Sun is 149 m s^{-1}. For a comet in a very extended orbit this comet–Sun distance corresponds to $a \approx 4 \times 10^4$ AU, within the probable outer limits of the Oort cloud.

Later observations revealed that GMCs have a hierarchical substructure containing clumps of masses typically $2 \times 10^4 M_\odot$ and radii 2 pc which in their turn contain smaller clumps of mass about $50 M_\odot$ and radius 16 000 AU. Clube and Napier (1984) have argued that these clumps would greatly enhance the disruptive power of a GMC, given that the Solar System was passing through it. Bailey (1983) found that a loosely bound Oort cloud would be disrupted by GMCs but then argues for an inner reservoir of comets at distances of $\sim 10^4$ AU that would be perturbed outwards to replenish the cloud.

11.8.1.3   *Perturbations by the galactic tidal field*

The Sun is situated in the disc of the Milky Way galaxy. At the position of the Sun the disc has a thickness of about 1 kpc with a more-or-less uniform distribution of stars. To a first approximation the Solar System can be thought of as existing within a slab of material of uniform density and infinite extent. The gravitational field external to an infinite plane of areal density σ is perpendicular to the plane and of magnitude

$$E = 2\pi G \sigma. \qquad (11.37)$$

Within a uniform slab there will be a field gradient perpendicular to the slab. At the point P in the slab shown in figure 11.18, a distance z from the mean plane, the field will be towards the mean plane and be due just to a thickness $2z$ since the gravitational fields of the two shaded regions will be equal and opposite. If the volume density of the slab is ρ then the field at P is

$$E_{\mathrm{P}} = 4\pi G \rho z$$

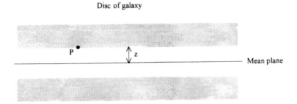

Figure 11.18. The gravitational field at P is due to the unshaded regions of the model galactic disc.

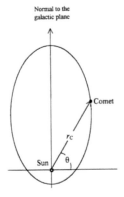

Figure 11.19. A comet orbit containing the normal to the galactic disc.

corresponding to a field gradient

$$\frac{\mathrm{d}E_P}{\mathrm{d}z} = 4\pi G\rho. \tag{11.38}$$

If the z-coordinate of a comet is different to that of the Sun then this gives an acceleration of the comet relative to the Sun that constitutes a perturbation.

In figure 11.19 we show a simple case where the plane of the comet's orbit contains the z direction. The magnitude of the acceleration, a, of the comet with respect to the Sun is

$$|a| = 4\pi G\rho r_C \sin\theta \tag{11.39}$$

giving a rate of change of intrinsic angular momentum

$$\frac{\mathrm{d}H}{\mathrm{d}t} = |a \times r_C| = 4\pi G\rho r_C^2 \sin\theta \cos\theta = 2\pi G\rho r_C^2 \sin 2\theta. \tag{11.40}$$

The rate of change of angular momentum depends on the comet's position in its orbit but, for an extended orbit, it will spend most of the orbital period near aphelion. Assuming that it spends the whole period *at* aphelion will give an upper bound for the change of angular momentum per period, δH, but one that will probably be within a factor of two of the true value. Assuming that the aphelion distance $r_C = 2a$ for an extended orbit this assumption gives

$$\delta H = 8\pi G\rho P a^2 \sin 2\theta_a \qquad (11.41)$$

where P is the period and θ_a the value of θ at aphelion. This clearly has a maximum value when $\theta_a = \pi/4$.

To estimate ρ we take the mass density of stars in the solar region of the galaxy as $0.08 M_\odot$ pc^{-3}. Taking a factor of 10 to allow for the so-called *missing mass* gives a mean density of disc material 5.5×10^{-20} kg m^{-3}. For a comet with $a = 20\,000$ AU the period is $(20\,000)^{3/2}$ years. Substituting these values in (11.41) gives a maximum change of intrinsic angular momentum 7.4×10^{16} m^2 s^{-1} that corresponds to a perihelion distance of 136 AU for an extended comet orbit. This result indicates that a comet may easily be transferred by this effect from one with a perihelion well outside the planetary region to one with a perihelion close to the Sun in a single orbit.

All the quantities in this calculation have been taken at their extreme values to enhance the effectiveness of the mechanism. The missing-mass factor of 10 is required to explain motions in clusters of galaxies and somewhat lower factors are required to explain the rotation periods of stars within a galaxy. This would be consistent with most, but not all, the missing mass being in the form of massive black holes within the galactic nucleus. Another extreme condition taken was that the plane of the comet's orbit contained the z-direction. If the plane was perpendicular to the z-direction then there would be no torque and hence no effect.

If the galactic field was the dominant one in producing new comets then it might be expected that this would show itself statistically in that the directions of the perihelia should concentrate round an angle $\pi/4$ and semi-major axes should be shorter on average in the $\pi/4$ direction. No such correlation has been found; on the contrary there is a slight tendency for perihelion directions to be aligned with the galactic plane. If in figure 11.19 the direction of the comet's motion was reversed then the galactic tidal field would increase rather than diminish the intrinsic angular momentum. Since new comets have almost equal numbers of direct and retrograde orbits this again suggests that the galactic tidal field may not be a dominant effect in producing new comets.

11.9   The inner-cloud scenario

It is clear that the three types of perturbation dealt with here can all be very effective in changing the orbits of comets especially under extreme conditions—e.g. a very close stellar passage or a trip through a GMC that passed close to one of the sub-condensations within it. Under less extreme conditions, that would have almost certainly occurred from time to time during the lifetime of the Solar System, the outermost comets of the Oort cloud would have gained enough energy to have escaped. The model of having an inner cloud which gradually, or perhaps spasmodically, is perturbed outwards to repopulate the Oort cloud is an attractive one to maintain the Oort cloud as a permanent, rather than transient, feature of the Solar System.

We now examine the Oort picture of outer comets being constantly perturbed by stellar passages that randomized their motions. An impulsive change of velocity of a comet due to a passing star will be added vectorially to the original velocity. From some arbitrary starting point when the comet velocity was v_{CS} with respect to the Sun after an impulsive addition the velocity will have become

$$v'_{CS} = v_{CS} + \Delta v_{CS}. \tag{11.42}$$

The intrinsic kinetic energy will have changed from $|v_{CS}|^2$ to $|v_{CS} + \Delta v_{CS}|^2$. The intrinsic energy of the orbit will change from E to E'; since Δv_{CS} may be inclined at any angle to v_{CS}, taking expectation values we may write

$$\langle E'_{CS} \rangle = E_{CS} + \Delta E_{CS} \tag{11.43}$$

where $\Delta E_{CS} = \frac{1}{2}|\Delta v_{CS}|^2$. If m independent perturbations are applied to a single comet then

$$\langle E'_{CS} \rangle = E_{CS} + \sum_{i=1}^{m} (\Delta E_{CS})_i. \tag{11.44}$$

The expression (11.44) has only statistical significance since any individual perturbation may reduce rather than increase the intrinsic energy of the orbit. However, after many interactions the energy may become positive and then the comet becomes detached from the Solar System.

A detailed analysis of how the Oort cloud is affected by stellar perturbations is not possible using the impulse approximation. The greatest effects are for very close passages of stars when the impulse approximation is invalid although Eggers and Woolfson (1996) have devised a modification of the impulse approximation that can be used in computational applications. The number of stellar passages at distances between $D + \frac{1}{2}\delta D$ and $D - \frac{1}{2}\delta D$ during the lifetime of the Solar System, T_{SS}, is given by

$$N(D)\delta D = 2\pi DV T_{SS} N_* \delta D \tag{11.45}$$

and the numbers in intervals of 1000 AU around various values of D are given in table 11.9.

It can be seen that the Oort cloud has had large numbers of stars ploughing through it, greatly perturbing comets in their vicinity. For example, if a comet with $a = 2500$ AU is at distance 4000 AU from the Sun then a solar-mass star passing with speed 20 km s^{-1} at distance 242 AU could give it enough energy to leave the Solar System completely. That would require the impulse to be exactly in the direction of motion of the comet; otherwise the comet could be thrown inwards or outwards although, statistically, its energy would increase and the value of a become greater.

Table 11.9. Expected numbers of passing stars in distance interval of 1000 AU around various distances, D.

D (AU)	2000	3000	4000	5000	7000	9000	15 000	20 000	30 000	50 000
$N(D)\delta D$	2.2	3.3	4.4	5.5	7.7	9.9	16.4	21.9	32.9	54.8

From the information in table 11.9 it can be deduced that, over the lifetime of the Solar System, more than 1000 stars have passed through the volume occupied by the Oort cloud comets. Outer comets would not only have been subjected to large numbers of small perturbations by distant star passages but to the ravages of the occasional close passage. Inner comets, including those at only a few thousand AU from the Sun, will also be heavily perturbed but at much longer intervals. It seems possible that feeding from an inner reservoir of comets can maintain the population of the outer parts of the Oort cloud from which the new comets are derived.

In this discussion no mention has been made of perturbations by GMCs or the galactic tidal field. This is not because they are considered unimportant but rather because stellar perturbations alone give credibility to the inner-cloud hypothesis and the addition of the other sources of perturbation merely reinforce the message.

11.10   Kuiper-belt objects

In 1951 Gerard Kuiper suggested that beyond the orbit of Neptune there was a region occupied by comet-type bodies orbiting close to the mean plane of the Solar System. His main reasons were, first, that it seemed to him to be unlikely that the Solar System should abruptly end beyond the orbits of Neptune and Pluto and, second, that material in the outer regions of the Solar System would have been too dispersed to form planets. In the 1970s Kuiper's idea received reinforcement from consideration of the Jupiter family of comets and other short-period comets. There are about 100 short-period comets with an average lifetime of order 10 000 years. To maintain the population requires the formation of one short-period comet per 100 years by planetary perturbation of an incoming long-period comet. With the availability of suitable computers it could be shown that the mechanism for perturbing long-period comets was very inefficient and also the orbits of the resulting short-period comets would have almost random inclinations. Since short-period comets tend to be close to the ecliptic it was suggested that the source of the short-period comets consisted of bodies in orbit close to the ecliptic. Oort cloud comets have their inclinations randomized by stellar perturbations so that a source closer in was postulated. A search began for bodies beyond Neptune in what became known as the *Kuiper belt*.

In 1992 the first such body, 1992 QB_1, was discovered. It has an orbit with $a = 44.35$ AU, $e = 0.078$ and $i = 2.2°$. By mid-1998 some 68 Kuiper-belt objects had been discovered with diameters up to 400 km. Their semi-major axes are mostly in the range 39–46 AU with eccentricities less than 0.34 and inclinations up to about 30°. An exception to this pattern is object TL66 that has $a = 85.22$ AU and $e = 0.59$. Almost one-half of the bodies are trapped in a 3:2 resonance with Neptune and one appears to be in a 4:3 resonance. These bodies will be 'protected' but the others will be unstable with respect to perturbation by Neptune and may become sources of short-period comets in due course.

It is believed that all the Kuiper-belt objects have a large volatile content and are more related to comets than asteroids. As suggested in section 11.8 these bodies may just be the larger outriders of an inner system of comets or even a single system stretching from the Kuiper belt right out to the outer boundary of the Oort cloud.

11.11   Comets from the planetary collision

In considering the planetary collision as a source of isotopic anomalies in meteorites the solid surface material of the colliding planets was taken to be ice-impregnated silicates, at least in part. Such material is an ideal candidate as a source of comets.

The region of highest temperature in which the nuclear processes were happening was restricted to a dimension of a few thousand kilometres and the rear portions of the planets, furthest from the impact region, would have been shielded from the high-temperature source and stayed quite cool. Shock waves passing through the planets would have material ablated off the rear surface and much of this would have attained escape velocity from the planet. The planet itself could have been thrown into either a highly eccentric orbit or even out of the Solar System. Outermost material would have left the planet with the greatest speed and so ended up further out in the Solar System. Material from further below the surface would have left the planet at lower speed and ended up closer to the Sun. We saw in section 1.4.3 that the distribution of C-type asteroids peaks further out from the Sun than that of the S-type asteroids. This is again suggestive that sub-surface material with a volatile content, similar to carbonaceous chondrites, ended further out than material from lower regions of the planet that would have given rise to the S-type asteroids. Much material of all types would have been left in the inner Solar System after the collision. Most of this would have had a short lifetime before colliding and merging with larger bodies. That which happened to move on safe orbits has survived to give the smaller bodies we see today.

Although this pattern of behaviour is purely speculative and not supported by modelling at present it does seem to be intuitively reasonable.

11.12   Ideas about the origin and features of small bodies

The prevalent view places the origin of small bodies in the Solar System in the context of the Solar Nebula Theory. The presence of the Sun, perhaps more luminous than now, prevented volatile materials from condensing towards the centre but further out in the system icy materials were able to condense together with silicates and iron. The composition of the planetesimals that formed thus reflected the location of their formation and explained the overall pattern of rocky-metal bodies in the terrestrial region and icy bodies with a high volatile content elsewhere. The general distributions of C- and S-type asteroids and also the icy nature of Kuiper-belt objects are explained by this model.

The inferred compositions of asteroids and the observed compositions of meteorites suggests that they may be the product of the collisions of parent bodies of sub-lunar size. The molten material produced by collisions would give rise to chondrules that would cool quickly so that they contained non-equilibrated minerals as observations suggest. The parent bodies would also give the gravitational fields to explain the characteristics of cumulates in eucrites.

The parent bodies would have been too small to have become molten by the release of gravitational energy. The separation of material by density to give stones and irons and the presence of Wittmanstätten figures in iron meteorites demands that molten material should have been present. The ^{26}Mg isotopic anomaly in CAI material, that certainly indicated the former presence of ^{26}Al in the Solar System is taken to provide an answer to this problem. If 1% of a body was aluminium and the proportion of the aluminium that was ^{26}Al was the same as in CAI material then bodies more than a few kilometres in size would have melted completely. While there is no evidence for the former presence of ^{26}Al in other than CAI material it is possible that in large bodies, where the temperature stayed higher for longer, migration of the ^{26}Mg daughter product, mixing with magnesium in high concentration elsewhere in the specimen would mask the excess.

The Solar Nebula Theory offers no scheme for explaining isotopic anomalies. Some are explained in terms of a second nearby supernova just as the Solar System was forming. Others are explained as products from another part of the galaxy or from a number of carbon stars that were transported into the Solar System in grains and yet others as due to particle bombardment of material within the Solar System. It seems doubtful that these various explanations really can explain the wide range of anomalies that occur and a series of *ad hoc* explanations is far from satisfactory.

The planetary-collision hypothesis has the advantage that a large number of diverse features of the Solar System are explained in terms of a single event. For the event to be plausible it requires that the colliding planets should be massive, several tens of Earth-masses each, and moving on very eccentric orbits; otherwise the triggering temperature for nuclear reactions involving deuterium will not be reached. This has been confirmed by modelling over a wide range of conditions. It is not absolutely necessary for it to be linked to the Capture Theory although that theory is the only one at present that unequivocally produces the right conditions—although the Proto-planet Theory might do so.

Restricting attention to asteroids, meteorites and comets, the topics of the present chapter, the gross features of the three types of body may be readily understood as having come from different regions of a planetary body, internally molten from the release of gravitational energy on formation. The radial distributions of C- and S-type asteroids and the presence of volatile Kuiper-belt objects and comets even further out agree with what would be expected for the final destinations of near-surface and interior material after the collision.

The planetary collision together with the presence of some deuterium-rich hydrogen, already known to exist in the Solar System, gives very satisfactory explanations for a number of important isotopic anomalies. These come in terms of mixing three products of the nuclear explosion—highly processed material, lightly processed material and unprocessed material. If the freedom had been taken to include other intermediate states of processing even better agreement could have come about—although the present agreement is convincing enough. Since the laws of physics are universal it could be argued that the processes that were shown to give the observed anomalies had occurred elsewhere than in a planetary collision. This presumably goes back to postulating a second supernova. The small scale of the planetary collision scenario, both in space and time, gives cooling and collection of material into fairly large cool bodies on a short time-scale, which is required for a ^{22}Na-based explanation of neon-E.

PART 4

THE CURRENT STATE OF THEORIES

Chapter 12

Comparisons of the main theories

12.1 The basis of making comparisons

In making comparisons between different theories there are a number of factors to be taken into consideration. The four theories to be considered here are the Proto-planet Theory, the Modern Laplacian theory, the Solar Nebula Theory and the Capture Theory. They have all been under development for more than 20 years—more than 30 in the case of the Capture Theory. Even if development has not been continuous, as is true for the Proto-planet Theory, it can still be argued that the theories are all mature and that they have all been investigated in some depth. In developing a theory the pros and cons of this or that mechanism soon become evident, as are the major problems than need to be solved. Thus one should be tolerant of new ideas even if they seem to be groping around for answers for that is the way that many good theories develop. However, eventually there comes a time when persistent lack of progress may be telling the theorist that he/she cannot find the answers because he/she is asking the wrong questions. That must be the position for the theories here being considered—if there are basic problems that have resisted decades and hundreds of person-years of well-directed work then the plausibility of the theory must be in doubt. Good theories give answers to questions and then suggest new questions and give new answers!

By and large the extent of development of the theories reflect the difficulties they have been having. As problems are solved in a satisfactory way so attention moves towards new aspects of the Solar System. A theory that presents explanations of the major features of the Solar System to the theorist enables him/her inevitably to move to consider the finer details of the system. Conversely a theory that is forever failing to solve the major problems will not move forward or, if it attempts to do so, it is tantamount to building a structure on a foundation of quicksand.

Another aspect of the theories that needs to be considered is that which is usually included under the Occam's-razor principle. If every feature of the Solar System requires a different *ad hoc* explanation then the theory will lack cohesion.

The Solar System obviously came about through some major event, and all the theories agree about that. It might be expected that a major event would leave behind not a single sign or clue that it had happened but several, all pointing in the same direction. Greater credibility will be attached to any theory that can link together many features of the system either through a single mechanism or event or, perhaps, through different mechanisms that are causally related.

With these considerations we now consider the four theories in turn elaborating on points previously made in section 6.6.

12.2   The Proto-planet Theory reviewed

Through the formation of proto-planet blobs in a turbulent collapsing cloud the Proto-planet Theory links together the formation of a slowly spinning Sun and the existence of the planets themselves. It is certainly a very positive feature of the theory that two major features, even *the* two major features, of the Solar System are intimately linked in this way.

The theory has, as its basic scenario, the formation of a galactic cluster of stars and the model for star formation in a galactic cluster (Woolfson 1979) is clearly related to it. In determining the conditions in the star-forming cloud Mc-Crea (1988) took as his starting point a particular temperature, 40 K, and the requirement that the density of the cloud should be that for which the Jeans critical mass would be that of a proto-planet blob. By contrast the Woolfson star-forming model took as its starting point a cool dense interstellar cloud with characteristics based on observations and then followed the progress of the cloud by an analytical approach. This indicated that the first condensations formed by the collision of turbulent elements were of greater than solar mass and that they should be spinning slowly. This throws into doubt the whole basis of the Proto-planet Theory—first, the need to explain the slow rotation of late-type stars is removed and, second, it throws into question whether the conditions in the cloud would ever be suitable for producing planetary-mass condensations.

In the original 1960 model for the Proto-planet Theory the original condensations—floccules—had three times the mass of the Earth and several of them had to accumulate to form a major planet. In doing so they acquired spin angular momentum far greater than that observed at present and McCrea put forward an attractive model for both disposing of the surplus angular momentum and also producing satellites (section 5.2). The same mechanism is feasible if Jupiter is produced by a combination of three or four of the larger proto-planet blobs of the 1988 model. The collapsing spinning proto-planet would break up into two unequal parts. The smaller part would escape from the Solar System, taking with it most of the spin angular momentum, and droplets in the neck between the separating parts would leave natural satellites attached to the major portion. However, for the masses of the proto-planet blobs in the revised 1988 theory the less massive major planets require at most one blob to form them. If the mechanism to produce satellites is similar to that described for Jupiter then the excess spin angular momentum has to reside in a single blob; that this is so has not been demonstrated.

The orbital angular momentum of the final planetary system has been equated to the difference between the original angular momentum associated with the proto-planet blobs when separated in a star-forming region and that when they combine to form a star. It seems unlikely that, when a blob enters the forming star, angular momentum is somehow transferred to outside small bodies. Indeed the question could be asked that if there were no other small bodies around at that time then where would the angular momentum end up? The obvious answer is that the motions of blobs are not restricted to small star-forming regions that are isolated in such a way that each can be regarded as a depository of a fixed amount of angular momentum. Blobs actually move throughout the cloud and if the cloud is considered as a set of star-forming regions then an important part of the total angular momentum of the cloud is in the form of the relative motions of the regions themselves. The Proto-planet Theory assumes that the final angular momentum associated with the relative motions of stars is the same as that for the relative motions of regions. Then, inevitably, the angular momentum associated with material *within* each region can be equated to the angular momentum of sub-condensations around the central star. In fact the motion of blobs between regions would ensure that they were strongly coupled so that some, at least, of the original angular momentum within regions would be transformed into the relative motions of the stars. If small collections of blobs formed within the cluster then these would be sharing the general motion and some may be captured by the stars. A planetary system *could* form but not in the rather systematic way suggested by the Proto-planet Theory.

The criticism that the system formed by the Proto-planet Theory will not be even approximately planar and may include planets both in direct and retrograde orbits may not be a very damaging one. If there is a preponderance of direct orbits, and an exact balance is very unlikely, then collisions may remove the retrograde objects. This would also provide the material to explain the high level of bombardment in the early Solar System. The debris from such collisions would also add to the resisting medium that would inevitably surround the early Sun. Since planets would almost certainly form on eccentric and inclined orbits in the presence of the resisting medium, leading to orbital precession as orbits rounded off, the collision scenario and its consequences, as described in part IV, could be an outcome of this theory. In addition the mechanism described by Melita and Woolfson (1996) giving commensurabilities between the orbits of major planets could also be applied.

To summarize, the Proto-planet Theory is implausible but not impossible. It has some good features but there are a number of important assumptions that cannot be justified in the present state of knowledge. Some of these assumptions, for example relating to the way that angular momentum appears in the final star cluster, are amenable to investigation by numerical modelling. On the other hand, the assumption relating to the formation of proto-planet blobs has implicitly been investigated by Woolfson (1979) and this indicates that stars are formed earlier and otherwise in the development of the star-forming cloud. Some answers to this and other criticisms would be necessary to promote the Proto-planet Theory to the status of plausibility.

12.3   The Modern Laplacian Theory reviewed

The Modern Laplacian Theory is by far the most complicated of those under discussion here. It begins on a very shaky premise that few would find acceptable, that the Sun formed initially from grains of solid molecular hydrogen. Maser emission from star-forming regions show widths of emission lines corresponding to temperatures between 50–100 K (Cook 1977), a commonly accepted range for cool dense clouds and one much higher than could possibly support solid hydrogen. The next stage in the development of the star involves creating a very small moment-of-inertia factor as it collapses. This is done through the formation of needle-like elements that extend the main body of the star while at the same time creating a higher density shell in the outer parts. Material in one of these shells migrates towards the equatorial plane and then, spasmodically, the ring so formed separates from the main body of the collapsing star. Many rings are formed, one-half of them within the orbit of Mercury, and they all have similar masses, about $M_\odot/300$. The theory developed by Prentice (1978) gives a set of 16 or 17 rings, the largest having the radius of Neptune's orbit and the smallest just outside the radius of the present Sun. The ratio of the radii of successive rings is a constant, equal to about 1.73, which gives a rough match to the Titius-Bode law. The total mass contained in the rings is somewhat over 5% of the total mass of the original nebula but the ratio of the angular momentum left in the central condensation, H_S, to that contained in the rings, H_R, is

$$\frac{H_S}{H_R} = \left(\frac{R_n}{R_0}\right)^{1/2} \tag{12.1}$$

where R_0 is the radius of the outermost ring and R_n is the radius of the nth ring, the one just outside the Sun's surface. Putting R_0 equal to the radius of the orbit of Neptune and R_n equal to the radius of the Sun we find $H_S \approx 0.012H_R$. This means that nearly 99% of the angular momentum of the condensation can be removed with just over 5% of the mass. It is quite difficult to envisage that all the material inside Mercury's orbit, accounting for 20 to 30 Jupiter masses, and much of the material outside Mercury's orbit can be removed completely from the system. If any of it is reabsorbed by the Sun then it carries its quota of angular momentum with it. It is assumed that the early Sun is very active, goes through a T-Tauri stage and removes much of the material by heating it or blowing it away.

The formation of planets, or planetary cores, by the accumulation of material at the centre of an annular ring first depends on producing a concentration of material at the axis of the ring cross-section and then for there to be a sufficiently long lifetime for the ring. In view of the results given in table 6.4 it seems impossible that the rings could survive for a sufficiently long time.

The Modern Laplacian theory depends on a number of assumptions or theoretical results with probabilities varying from small to vanishingly small. Its total scope is to explain the existence of the slowly rotating Sun, the planets and of satellites. Applying the general model of satellite formation Prentice (1989) successfully predicted the existence of four previously undiscovered satellites of Neptune before the Voyager visitation in 1989. In addition he has been able to explain and predict the chemical compositions of material in various parts of the Solar System. Despite this, and the fact that it is an intricate, closely argued theory, the considerable doubts about many important aspects of it must make it very implausible. The predictions may just relate to patterns of temperature and density in the early Solar System that some other and more plausible theory could also suggest. Certainly by any application of Occam's razor, in view of its convoluted and complicated nature, the Modern Laplacian Theory would be ruled out entirely.

12.4   The Solar Nebula Theory reviewed

The revival of interest in nebula ideas, now enshrined in the Solar Nebula Theory, came about because meteorite studies suggest that a hot vapour must have existed in the early Solar System. The task of attempting to solve the problems of nebula-based theories, encountered by Laplace and others subsequently, has been in progress since the 1970s. It was felt that with much greater knowledge about the Solar System and with all the new analytical and computational tools that are now available, solutions would be found for the basic problems. The two basic problems of the Solar Nebula Theory, neither of which has yet been convincingly solved, are, first, the formation of a slowly-rotating Sun and, second, the formation of planets from diffuse nebula material.

If material spirals inwards to form the Sun then the Sun could not form at all as a condensed body since material would all be in Keplerian orbit about interior mass. The loss of *some* angular momentum, while the body is accumulating, say by the action of a magnetic field or a gravitational torque (section 6.3.1.2), could lead to a coherent rapidly-spinning body. The problem is that the mechanisms suggested, except that involving the solar magnetic field, involve a spiralling in process. The magnetic field mechanism involves material flowing *outwards* and it is difficult to conceptualize a process whereby material moves inwards to join the growing Sun while losing angular momentum by ionized material flowing outwards. If a rapidly rotating body could somehow form then a mechanism would then be required to remove over time about 99.9% of the angular momentum it contained. The only possible mechanism for doing this is described in section 6.2.1 but to remove so much angular momentum requires a solar magnetic field some 1000 times as strong as the present one with a loss of fully ionized material 10^6 times the present rate of loss for 10^6 years. The inferred rates of loss from T-Tauri stars would be suitable but observations suggest that the material lost from T-Tauri stars is only partially ionized. There is also no evidence for the existence of such strong magnetic fields associated with T-Tauri stars. The situation with respect to the slow rotation of the Sun is that, if the Sun could form in the first place, a mechanism exists to remove the excess angular momentum but that observational evidence that the correct conditions could occur is lacking.

Observational evidence taken to support the Solar Nebula Theory has come from observations either showing or inferring the existence of dusty discs around new stars. Stars tend to spin somewhat more rapidly when they are newly formed, perhaps a factor of 10 or so more rapidly, but this is not a great problem; removing 90% of angular momentum is far less of a problem than moving 99.99%. Since a new star spinning at an acceptable rate is seen with a surrounding disc, that is taken to imply that the angular momentum problem can be solved somehow. What it may alternatively suggest is that discs can form around stars formed by a mechanism other than an evolving solar nebula. Indeed, it may be rather difficult to think of any process of star formation so efficient that it is left with a clean boundary surrounded by material little more dense than the interstellar matter. If a star forms by any process, say the collision of turbulent elements in a cloud, it will certainly start its existence surrounded by a dusty cloud. Energy-removing interactions in the cloud will then produce a dusty disc in the equatorial plane.

The Solar Nebula Theory would lead to almost all stars having planets around them—or at least stars resembling the Sun. Observations of planets around a number of other stars has lent weight to the idea that planet formation is a common occurrence and that a plausible model should give rise to many planetary systems. So far what has been observed are mostly single planetary companions rather than *systems* of planets but one system with three planets has been detected. The most readily detected bodies around stars are those with large mass and short-period orbits and those seen so far have been selected by their observability. Indeed some are Jupiter-type planets with orbital radii less than that of Mercury around the Sun so whatever else exists around these particular stars it will not closely resemble the Solar System.

The second basic problem not convincingly solved by the Solar Nebula Theory is that of planet formation from diffuse material. Actually with a very massive nebula, equal in mass to the Sun itself, planets can form very easily by spontaneous collapse of Jeans-mass units to form a large number of Jupiter-type objects. The problem of disposing of the surplus planets is so intractable that solar-nebula theorists do not seriously consider this possibility. With a low-mass nebula the stage of getting solid material close to the mean plane on a short time-scale is reasonably straightforward if the sticking mechanism suggested by Weidenschilling *et al* (1989) is valid. The gravitational instability of such a dense disc of solid material also presents no theoretical difficulty to give the formation of planetesimals—although there is some dispute about their sizes. It is the stage in going from planetesimals to terrestrial planets or planetary cores that presents the difficulty. There are two contradictory requirements. The first is that the planetesimals should approach each other as slowly as possible, which means that their relative speed at a large distance should be small. In this way they will come together with a small hyperbolic excess so that a small loss of energy due to the collision itself can lead to mutual capture. The difficulty here is that the rate at which the bodies get together is low so that the time-scales for forming planets or cores is large. Forming a core for Jupiter takes 150 million years and a core for Neptune cannot be formed in the lifetime of the Solar System. Although the formation time for a Jupiter core is a tiny fraction of the age of the Solar System it is far too long. The observations of discs that were taken to give credibility to the Solar Nebula Theory also gave it a time constraint since the discs seem not to be present after a few million years. This is taken as a limit for the time of formation of planets or cores. The visibility of a dust disc depends very much on the form in which the dust exists. If it has *all* collected into large objects then it will become virtually invisible as a source of infrared radiation. On the other hand if it consisted of planetesimals that were constantly colliding then this would be a source of new dust which would render the region visible again

Attempts have been made to overcome the time-scale problem by increasing the density of the disc with perhaps local density enhancements and by assuming an equipartition of energy for objects of different mass. This gives rise to what Stewart and Wetherill (1988) refer to as runaway growth. The observational evidence and theoretical considerations do not support the conditions postulated. In addition Wetherill (1989) showed that when Earth-mass bodies were produced in the Jupiter region they would be scattered throughout the Solar System rather than collect together into a single body.

Given the amount of effort that has been put into the Solar Nebula Theory the lack of success in giving clear answers to the most basic questions indicates that it is almost certainly wrong. Although it is the theory that has had the greatest amount of effort expended on it and has the greatest number of adherents it is unable to answer convincingly the most basic questions concerning the origin of the Solar System.

12.5   The Capture Theory reviewed

The Capture Theory is the most highly developed of those presented here and describes not only the formation of a slowly-spinning Sun, the planets and regular satellites but is linked causally to descriptions of the origins of many other features of the Solar System. Basically it depends on mechanisms that are believable and well understood—tidal mechanisms, gravitational instability and collisions—all part of the standard armoury of the theoretical cosmogonist.

The scenario begins with the interstellar matter (ISM) and a nearby supernova. Coolant material is injected into a particular region that may also be compressed somewhat by a shock wave passing though it. With some compression and extra coolant the region of the ISM becomes unstable and collapses. The density increases and the temperature decreases to form a dense cool cloud in approximate pressure equilibrium with the ISM. At this stage, if the cool cloud is more massive than the Jeans critical mass, it will continue to collapse, slowly at first but then more quickly as is the pattern for free-fall collapse. Turbulence within the cloud is fed by released gravitational energy, slowing down the collapse to a fairly constant rate (figure 2.22(a)) although the turbulent energy steadily increases. Eventually the collision of turbulent elements with subsequent cooling can produce a high-density low-temperature region capable of collapsing to form a star on a time-scale shorter than the coherence time associated with reordering of the cloud material. Work by Whitworth *et al* (1995) and others has shown that multiple star systems can also arise in the collision of turbulent streams of material. The first stars produced by this primary process have masses about $1.4M_\odot$ and as the density in the cloud increases so stars of lesser mass are formed. Such stars have comparatively little angular momentum—typically a factor of ten or so more than is observed and modest removal of excess angular momentum presents no theoretical difficulties. Stars formed by the primary process which happen to move into dense regions of the evolving cloud accrete matter, become more massive but also acquire large amounts of angular momentum in so doing. The relationship between angular momentum and mass derived from this model agrees well with what may be deduced for actual stars (figure 2.25)

Stars formed in a collapsing cloud will tend to follow the general direction of their constituent material—i.e. inwards. The stellar system will go through a period of high density before it re-expands (Gaidos 1995, Lada and Lada 1991, Kroupa 1995). For a few million years the *embedded cluster* is held together by the gaseous matter of the cloud but once this gas is dispersed the cloud will re-expand and will eventually evaporate leaving behind a stable small multiple stellar system. During the embedded stage the stellar density may be as high as 10^5 pc^{-3}, compared with less than 0.1 pc^{-3} in the present solar environment. In the embedded stage close interactions between stars will occur and the outcome of an interaction will depend on the relative orbit at closest approach and the masses and states of development of the two stars. If one of the stars is diffuse then material from it may be captured by the other star in the form of an enveloping cloud or as one or more condensed bodies to form a planetary companion or a planetary system.

The Capture Theory envisages an interaction between a condensed Sun and a rapidly collapsing, but still diffuse, proto-star of lesser mass. A filament of material is drawn from the proto-star, condensations form in it due to gravitational instability and these condensations, with greater than the appropriate Jeans critical mass, collapse to form a planetary system. The form of the interaction is such that the proto-planets initially move on highly eccentric orbits away from the Sun, that gives them several tens of years to collapse away from the most severe effects of the solar tidal field. They are affected by the tidal field of the Sun to the extent that they collapse in a non-spherically symmetric form. Tidal bulges at early stages of collapse develop into tongues or filaments in the later more rapid stage of collapse and condensations within filaments give the regular satellites of the major planets. The relationships of the intrinsic orbital angular motion of the secondary bodies to the intrinsic spin angular momentum of the primary bodies, as given in table 6.6, are well explained by the proposed mechanisms. The slow spin of the Sun is due to its mode of formation through the collision of turbulent streams of gas (section 2.6.2.6). The angular momentum of the planetary orbits derives from the original Sun–proto-star hyperbolic orbit. The solar tidal field imparted more angular momentum in the outer parts of the proto-planets than the inner parts so that outer material could be left behind in satellite orbits. However, with basically the same original source for angular momentum the ratios of the intrinsic angular momentum in satellite orbits to that of planetary spins is much less extreme than the corresponding ratios for the planets and the Sun.

The Capture Theory envisages an initial system of six major planets in eccentric approximately co-planar orbits rounding off in a resisting medium surrounding the Sun. Early modelling suggested that, notwithstanding the eccentric orbits, the proto-planets would not survive if on their first perihelion passage they were closer to the Sun than about the present orbit of Mars. These original major planets would all have regular satellite companions with the trend that the closer in the planet was the larger and more numerous were its satellites. This was a natural consequence of the strengths of the solar tidal fields to which they were subjected.

Because of non-central gravitational forces due to the resisting medium the eccentric orbits of the planets precessed so that the non-co-planar orbits intersected from time to time. This gave the possibility of interactions of various kinds between the condensing proto-planets. A fairly common form of interaction would be that of a fairly close passage so that each proto-planet exerted a considerable tidal force on the other. This would lead to tilts of the spin axes of the planets. In section 7.4.5 an interaction between proto-Jupiter and proto-Uranus that would have given Uranus its present relationship of spin axis to orbital plane was illustrated. Other less extreme axial tilts can be explained by similar interactions. Closer interactions are also possible and the results of calculations given in table 8.2 show that there is a high probability of major interactions between planets before the round-off of orbits is complete. The consequences of a collision between two major planets that would have rounded off approximately in the regions of the asteroid belt and Mars has been investigated and described.

It was shown in section 8.2 that one possible outcome is that one of the planets could have gained sufficient energy for it to have been expelled from the Solar System while the other lost energy and was sheared into two parts that rounded off in the regions of Venus and the Earth. Another possibility that has not been explored in detail is that Venus and the Earth are the residues of planetary cores, one from each of the colliding planets. Debris from the collision would have been the source of asteroids and comet material. More volatile material would have come from the outer regions of the solid parts of the colliding planets and so would have been thrown out further. In the central region of the Solar System this would have given the different statistical spreads of C and S asteroids, the C-type, resembling volatile-containing carbonaceous chondrites in composition being the further out. The most volatile material could have been thrown out to distances from hundreds to thousands of AU to create an inner comet cloud. Interactions with planets still on highly eccentric orbits could have perturbed many of these comets into orbits with perihelia well outside the present planetary region, so ensuring their survival. Passages of the Solar System near or through Giant Molecular Clouds or occasional close passages of stars (table 11.9) would at first move an inner comet cloud outwards to create the Oort cloud and later remove Oort cloud comets but restore the numbers from the residual inner cloud.

As a result of the collision, satellites of the colliding planets could end up in a number of destinations. In chapter 9 it was shown that the Moon could have been one such satellite that was either retained by the planet giving the Earth fragment or captured from the other planet. The hemispherical asymmetry and other features of the Moon are consistent with a loss of 20–40 km thickness of crust on one side that could be caused by abrasion due to debris from the collision. Mars was also suggested as an ex-satellite that had gone into an independent heliocentric orbit. Its hemispherical asymmetry is suggested to be of similar origin to that of the Moon and the relationship of the spin axis to the plane of asymmetry is well explained by polar wander (section 10.2.5) associated with its major surface features. Mercury, with its very high intrinsic density, is explained as a highly abraded satellite originally similar to Mars both in mass and general structure. The comparatively high eccentricities of Mars and Mercury would be due to their late arrival in heliocentric orbits. First, the gaseous component of the resisting medium would move outwards with the passage of time and, second, round-off is less efficient for small bodies (table 7.1).

The Neptune–Pluto–Triton system has also been explained in terms of one of the satellites of a colliding planet. It has been shown that an escaped satellite, identified as Triton, with an orbit passing through the region of the supposed collision, could collide with a Neptune regular satellite (identified as Pluto). A consequence of such a collision could be that Triton was left as a retrograde satellite of Neptune while Pluto went into an orbit similar to its present one. The form of the collision was such that a part of Pluto could be sheared off to form the satellite Charon. Although the orbit of Pluto did not round-off, as it is a small body and may have gone into orbit rather late, its orbit would have precessed so that its orbit became well separated from that of Neptune.

Meteorites are mostly fragments of asteroids, with the exception of SNC meteorites that may have come from Mars. A very distinguishing feature possessed by some of them is the presence of isotopic anomalies that make them different from terrestrial and lunar material and most other meteorites. The ratio of D/H varies greatly for bodies in the Solar System with Venus having the highest known ratio of 0.016. It has been shown that high D/H ratios could come about from a variety of regimes of temperature and gravitational field in the early Solar System. If a colliding planet had a D/H ratio similar to that of Venus then the collision would have given a triggering temperature sufficient to ignite nuclear reactions involving deuterium (section 11.7.1.9). This will then set off a chain of reactions thereafter involving many other heavier elements. It was demonstrated in section 11.7 that all the important isotopic anomalies considered could be explained in terms of mixtures of three components—heavily processed material, lightly processed material and unprocessed material.

There are aspects of the Solar System that do not need to be explained in terms of any particular model for its origin. Given that by some means or other a population of asteroids is established then energy-absorbing collisions in the vicinity of major planets can give rise to irregular satellites. The Mars companions, Phobos and Deimos, may be an exception because of the small mass of Mars and capture of the asteroid-size satellites may perhaps be understood in terms of their capture in the many-body environment accompanying the planetary collision.

12.6   General conclusion

There can be little doubt that the Capture Theory is the only one to offer explanations for a wide variety of features of the Solar System in terms of a coherent and self-consistent model. It has become associated with ideas about the evolution of the Solar System in which many of the minor features of the system are almost all linked to a single event—a collision between early proto-planets. The evolutionary story may be attached to any other theory that can give rise to planets on initially highly elliptical orbits in the presence of a resisting medium. At present no other theory seems able to do this convincingly, although the Proto-planet Theory comes the closest to doing so.

The plausibility of the Capture Theory and the evolutionary pattern that follows is based not on separate disjointed explanations of individual features of the Solar System but more on the way that each component of the overall theory leads naturally to the next. This is illustrated in figure 12.1. There is a logical path from the interstellar medium through to the formation of planets in elliptical orbits around the Sun in the presence of a resisting medium. There are three distinct outcomes from the presence of a resisting medium, one of which is a planetary collision which, in its turn, gives rise to six distinct outcomes. Thus features of the Solar System that may not have any obvious connection are actually related by a common source event. The trigger which sets the whole scenario in motion is a supernova, an event of a kind that has been observed and is reasonably well understood.

One aspect of the Capture Theory that has not been thoroughly explored is the frequency it predicts for planetary systems. The requirement is that some part of the disrupted proto-star, that is to give the filament, is capable of condensing

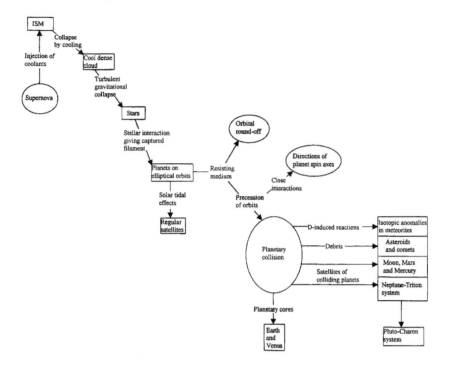

Figure 12.1. A schematic representation of the Capture Theory and related events.

into planetary-mass objects. If the mass of the proto-star was not much more than the Jeans critical mass when it began to collapse then this implies that it should have collapsed to substantially higher density when the capture-theory interaction took place. The final stages of the gravitational collapse of objects, while not free-fall because of pressure-induced forces, are still fairly rapid so that the time window for producing a system with many planets is restricted. In relation to his star-forming model Woolfson (1979) predicted that about one star in 10^5 would have a planetary system resembling our own.

Other considerations might greatly affect the estimate of the frequency of planetary systems—for example the realization that clusters go through an embedded stage. In this respect it is of interest that Whitworth *et al* (1998), being apparently unaware of the Capture Theory, have independently produced a smoothed-particle-hydrodynamics model of a capture event. These workers considered interactions between stars and proto-stars and between pairs of proto-stars. A feature of their work, not present in the Capture Theory, is that the proto-stars have very massive and extensive discs; the proto-star model they used consisted of a $0.5M_\odot$ central condensation with a rotationally supported disc, also of mass $0.5M_\odot$, of extent 1000 AU and surface density profile varying as $R^{-3/2}$. One of their models consisted of an interaction between a condensed star of mass

$M_\odot$ and a proto-star in which the periastron of the orbit is 1000 AU. A dense tidal bridge formed between the two bodies and a tidal filament was produced on the side of the proto-star away from the star. Three planetary condensations, each about four times the mass of Jupiter, formed in this filament. At the end of the simulation all the bodies, including the star which had acquired a disc of mass $0.1M_\odot$, were bound to the proto-star. This capture event differs from the standard Capture Theory model in that the proto-star has a disc that is the source of the tidal bridge. Since the disc is rotationally supported it has a long lifetime and the window of opportunity for the proto-planet to take part in a capture event is therefore much longer. Again, since the effective extent of the proto-stars is so great, the implication drawn by Whitworth *et al* is that interactions of this kind should be very common.

Even with a greater predicted frequency it seems unlikely that the Capture Theory process would lead to planetary *systems* being so common that a high proportion of stars should possess one. What should be more common, however, are planetary *companions* of stars where the outcome is to be a single fairly massive planet or perhaps a pair of planets. This requires much less stringent conditions for the tidal interaction between stars. So far observations of extra-solar planets have been of up to three large companions. This could well be the result of selection since the planets observed are those easiest to observe. Future observations might enable more complex systems to be detected or, alternatively, enable it to be said with some authority that complex systems do not exist.

Despite its plausibility in the light of present knowledge, the Capture Theory is no more than just a theory. It may be modified, for example by having the proto-star with a disc as suggested by Whitworth *et al*. On the other hand it may always be refuted by new observations or some theoretical argument previously overlooked but, until that happens, it is the best we have.

Appendix I

The Chandrasekhar limit, neutron stars and black holes

For a normal main-sequence star the material behaves for the most part like a perfect gas although at very high temperatures radiation pressure may have an important role in determining the state of equilibrium. However, a body consisting of electrons and positive ions (neutral on the whole) will, at high densities, have its properties constrained by quantum mechanical considerations. In a one-dimensional model, if the uncertainty in the position of a particle is Δx, then the corresponding uncertainty in momentum, Δp, has a lower bound given by the Heisenberg uncertainty principle

$$\Delta x \Delta p \geq h \tag{I.1}$$

where h is Planck's constant. If the material is at high density then Δx will be small and Δp correspondingly large. The kinetic energy of the particle of mass m, in the non-relativistic case, is thus

$$E_{\mathrm{K}} \approx \frac{(\Delta p)^2}{2m} \geq \frac{h^2}{2(\Delta x)^2 m}. \tag{I.2}$$

This kinetic energy creates a pressure (energy per unit volume); because of the way that m occurs in (I.2) it is clear that the dominant contributors to the pressure will be electrons. The pressure caused in this way is referred to as *electron degeneracy pressure*. This pressure occurs even at absolute zero temperature and, indeed, it can dominate over kinetic pressure so that the pressure is virtually independent of temperature. Material in such a state is said to be *degenerate*.

There is another property of electrons that we have so far not mentioned—that they are spin-$\frac{1}{2}$ particles, i.e. fermions, and are governed by the Pauli exclusion principle. This means that two electrons in the same quantum-mechanical state cannot occupy the same space and 'space' here is defined in terms of the uncertainty principle. Thus if we define the state of an electron by its position and momentum (r, p) then the smallest region of position–momentum space that the particle can occupy is h^3. This means that the maximum number of fermions that can be contained within a sphere of radius r having momentum magnitude in the range p to $p + dp$ is

$$dN = \frac{\frac{4}{3}\pi r^3 \times 4\pi p^2\, dp}{h^3} \times 2. \tag{I.3}$$

The factor 2 in (I.3) allows for electrons with opposite spins, and hence different states, to be in the same position-momentum cell. Hence the total number within a sphere of radius r and with momentum magnitude from 0 to P is

$$N = \frac{32\pi^2 r^3}{3h^3} \int_0^P p^2\, dp = \frac{32\pi^2 r^3 P^3}{9h^3}. \tag{I.4}$$

In the non-relativistic case the total kinetic energy for all particles with momentum magnitude between p and $p + dp$ is given by

$$dE_K = \frac{p^2}{2m}\, dN = \frac{16\pi^2 r^3 p^4}{3h^3 m}\, dp \tag{I.5}$$

or, for all the particles,

$$E_K = \int_0^P \frac{16\pi^2 r^3 p^4}{3h^3 m}\, dp = \frac{16\pi^2 r^3 P^5}{15h^3 m}. \tag{I.6}$$

Substituting for P in terms of N from (I.4)

$$E_K = \frac{16}{15}\left(\frac{9}{32}\right)^{5/3}\frac{h^2 N^{5/3}}{r^2 m\pi^{4/3}}. \tag{I.7}$$

The gravitational potential energy of a uniform sphere of mass M and radius r is

$$E_V = -\frac{3GM^2}{5r}$$

so that the total energy associated with the spherical mass is

$$E_T = E_K + E_V = \frac{16}{15}\left(\frac{9}{32}\right)^{5/3}\frac{h^2 N^{5/3}}{r^2 m\pi^{4/3}} - \frac{3GM^2}{5r}. \tag{I.8}$$

The star is stable, i.e. will neither collapse nor expand, when $dE_T/dr = 0$ or

$$-\frac{32}{15}\left(\frac{9}{32}\right)^{5/3}\frac{h^2 N^{5/3}}{m\pi^{4/3}r^3} + \frac{3GM^2}{5r^2} = 0$$

which gives

$$r = \left(\frac{9}{32}\right)^{2/3}\frac{h^2 N^{5/3}}{\pi^{4/3}GM^2 m}. \tag{I.9}$$

We now apply this result to a white dwarf, a small star which is supported by electron degeneracy pressure. If the average mass of a nucleon is m_N then the total number of nucleons in the star is M/m_N. A white dwarf is the final stage of a well-processed star in which much of the material has been converted into atoms such as carbon and oxygen for which the number of electrons is equal to one-half of the number of nucleons. Taking this fraction

$$r = \left(\frac{9}{32}\right)^{2/3} \frac{h^2}{\pi^{4/3}Gm(2m_N)^{5/3}} \frac{1}{M^{1/3}}. \tag{I.10}$$

Substituting for the various physical constants

$$r = 9.013 \times 10^{16}/M^{1/3}$$

where r is in metres if M is in kilograms. Thus the radius of a white dwarf with the mass of the Sun is 7.15×10^6 m or about 10% more than the radius of the Earth. It will be seen that the relationship gives the interesting result that the radius *decreases* with increasing mass of the white dwarf. This raises the question of whether there is some limiting mass for a white dwarf at which its radius becomes vanishingly small.

Actually, under the conditions we are specifying the kinetic energies will be so high that we have to abandon classical mechanics and move to relativistic mechanics. The classical kinetic energy expression $p^2/2m$ in (I.5) must be replaced by the relativistic expression $(p^2c^2 + m^2c^4)^{1/2} - mc^2$ so that the expression for the kinetic energy of a star of mass M and radius r is

$$_M E_K(r) = \frac{32\pi^2 r^3}{3h^3} \int_0^P \{(p^2c^2 + m^2c^4)^{1/2} - mc^2\}p^2 \, dp \tag{I.11}$$

with P given by (I.4). By a change of variable this becomes

$$_M E_K(r) = Ar^3 \int_0^B \{(1+q^2) - 1\}q^2 \, dq \tag{I.12}$$

where

$$A = \frac{32\pi^2 m^4 c^5}{3h^3} \quad \text{and} \quad B = \frac{h}{mc} \left(\frac{9}{64\pi^2 m_N}\right)^{1/3} \frac{M^{1/3}}{r}.$$

The total energy of the star is thus

$$E_T = {}_M E_K(r) - GM^2/r \tag{I.13}$$

and the condition for stability is $dE_T/dr = 0$, corresponding to a minimum energy. If dE_T/dr is positive for all values of r then the star will shrink without limit. The simplest way to explore the conditions for stability, or its lack, is numerically. The finite-difference approximation

$$\frac{d({}_M E_K(r))}{dr} = \frac{{}_M E_K(r+\delta) - {}_M E_K(r-\delta)}{2\delta}$$

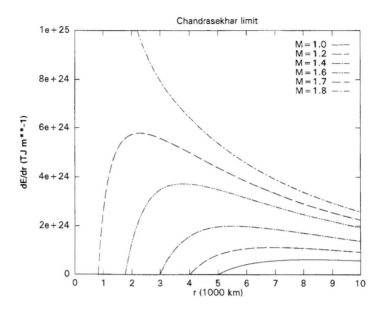

Figure AI.1. Evaluation of (I.12) for various combinations of M and r. Above about $M = 1.7M_\odot$ no radius gives stability. This corresponds to the Chandrasekhar limit.

with integrals evaluated by numerical quadrature enables the derivative of the first term on the right-hand side of (I.13) to be easily found for any combination of M and r.

The results of such a calculation are seen in figure AI.1. For a mass of $1.0M_\odot$ the radius for stability is 5000 km—somewhat smaller than the value found using the classical expression for kinetic energy. For a mass of $1.7M_\odot$ the stability radius is about 800 km but for a mass of $1.8M_\odot$ it can be seen that dE_T/dr is always positive. For the treatment given here the *Chandrasekhar limit* is somewhat over $1.7M_\odot$. The value deduced by Chandrasekhar (1935) was $1.44M_\odot$; his analysis was much more complicated than that given here and included consideration of the density variation within the white dwarf so that some difference in the deduced values is inevitable. However, the analysis given here brings out the essential physics involved and gives a quite sensible estimate for the limiting mass.

The question is now what happens to a collapsing star consisting of degenerate material if its mass exceeds the Chandrasekhar limit. The first thing that happens is that the electrons in the star combine with the protons in the ions to form neutrons so that the star becomes a *neutron star*. Neutrons are also spin-$\frac{1}{2}$ particles, fermions, so they are able to generate a *neutron degeneracy pressure*. Since they are much more massive than electrons they must be much more highly compressed before they exert the pressure necessary to resist gravity and, conse-

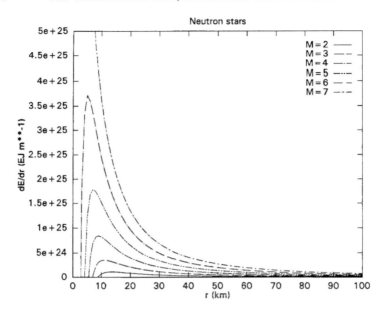

Figure AI.2. Evaluation of (I.12) with parameters corresponding to a neutron star. The limiting mass for a neutron star is indicated as about $6M_\odot$. (Note: 1 EJ $= 10^{18}$ J).

quently, neutron stars are much smaller even than white dwarfs. Investigating the size of a neutron star can be done by the same sort of analysis as has been given above with minor modifications. The number of degenerate particles is now equal to the number of nucleons (not one-half as many as in the neutron case) since all the particles are now neutrons. The other change is that the neutron mass must be used where previously the electron mass appeared. Equation (I.11) is evaluated to find the stability conditions but now with

$$A = \frac{32\pi^2 m_n^4 c^5}{3h^3} \quad \text{and} \quad B = \frac{h}{mc}\left(\frac{9}{32\pi^2 m_n}\right)^{1/3}\frac{M^{1/3}}{r}$$

where m_n is the mass of the neutron. Calculations for the neutron star are shown in figure AI.2. A neutron star of mass $2M_\odot$ has a radius of about 9 km and this falls to less than 3 km for a mass of $6M_\odot$. For a mass of $7M_\odot$ there is no stable configuration and the star will collapse without limit to a *black hole*. Most published estimates of the maximum mass of a neutron star are $3M_\odot$ or so but some are as high as $6M_\odot$, which is suggested by the present analysis.

Appendix II

The Virial Theorem

The Virial Theorem applies to any system of particles with pair interactions for which the total volume of occupied space is constant and where the distribution of particles, in a statistical sense, does not vary with time. The theorem states that

$$2T + \Omega = 0 \tag{II.1}$$

where T is the total translational kinetic energy and Ω is the potential energy. Here we show the validity of the theorem for a system of gravitationally interacting bodies.

We take a system of N bodies for which the ith has mass m_i, coordinates (x_i, y_i, z_i) and velocity components (u_i, v_i, w_i). We define the *geometrical moment of inertia* as

$$I = \sum_{i=1}^{N} m_i(x_i^2 + y_i^2 + z_i^2). \tag{II.2}$$

Differentiating I twice with respect to time and dividing by two

$$\frac{1}{2}\ddot{I} = \sum_{i=1}^{N} m_i(\dot{x}_i^2 + \dot{y}_i^2 + \dot{z}_i^2) + \sum_{i=1}^{N} m_i(x_i\ddot{x}_i + y_i\ddot{y}_i + z_i\ddot{z}_i). \tag{II.3}$$

The first term is $2T$; the second can be transformed by noting that $m_i\ddot{x}_i$ is the x component of the total force on the body i due to all the other particles or

$$m_i x_i \ddot{x}_i = \sum_{i=1}^{N} Gm_i m_j \frac{x_i(x_j - x_i)}{r_{ij}^3}, \tag{II.4}$$

where r_{ij} is the distance between particle i and particle j.

Combining the force on i due to j with the force on j due to i the second term on the right-hand side of (II.3) becomes

$$\sum_{i=1}^{N} m_i(x_i\ddot{x}_i + y_i\ddot{y}_i + z_i\ddot{z}_i)$$

$$= -\sum_{\text{pairs}} Gm_im_j \frac{(x_i - x_j)^2 + (y_i - y_j)^2 + (z_i - z_j)^2}{r_{ij}^3}$$

$$= -\sum_{\text{pairs}} \frac{Gm_im_j}{r_{ij}} = \Omega. \tag{II.5}$$

Equation (II.3) now appears as

$$\tfrac{1}{2}\ddot{I} = 2T + \Omega. \tag{II.6}$$

If the system stays within the same volume with the same general distribution of matter, at least in a time-averaged sense, then $\langle \ddot{I} \rangle = 0$ and the Virial Theorem is verified. The Virial Theorem has a wide range of applicability and can be applied to the motions of stars within a cluster of stars or to an individual star where the translational kinetic energy is the thermal motion of the material.

Appendix III

Smoothed particle hydrodynamics

Smoothed particle hydrodynamics (SPH) is a method developed by Lucy (1977) and Gingold and Monoghan (1977) which has been widely used for astrophysical problems. It is a scheme involving the integration of the motion of n-particles, without the need for a grid and a finite-difference approach, but which brings in the equation of state of the fluid being modelled, the forces due to pressure gradients within the fluid and viscosity. In some sense it is a Lagrangian code where each particle represents a cell of the Lagrangian mesh but where the need to define the shapes of highly distorted cells does not arise. The ith particle has associated with it a mass m_i, velocity v_i and a quantity of internal energy u_i. At each point in the fluid being modelled the properties are a weighted average of properties contributed by neighbouring particles, with the weight function, called the *kernel* or *smoothing function*, monotonically decreasing with distance. The kernel is usually terminated at some distance and is a function of h, the *smoothing length*, which is a scale factor defining the range and standard deviation of the kernel. In an ideal situation it is desirable to have at least 20 to 30 particles contributing to the averaging process. Since the density of the fluid, and hence the number density of the particles, will be a function both of position and time the smoothing length is adjusted at each time-step for each particle. There have been many modifications of SPH since it was first introduced giving considerable improvements in the simulations of astrophysical situations. These improvements have also been at the expense of increasing complexity; here we shall restrict ourselves to the very basic ideas that underlie the method.

The kernel is a normalized function so that

$$\int W(\boldsymbol{r}, h)\, \mathrm{d}V_{\mathrm{r}} = 1 \tag{III.1}$$

where the integral is over the volume occupied by the kernel. As an example of its use the density at point j in the fluid due to all the surrounding particles is given by

$$_{\mathrm{s}}\rho_j = \sum_i m_i W(r_{ij}, h_i) \tag{III.2}$$

where the pre-subscript s on the left-hand side indicates that it refers to a general point in space (which *could* be the position of a particle) and where r_{ij} is the distance from particle i to the point j. For the value of a general quantity, q, at point j we may write

$$_sq_j = \sum_i \frac{m_i}{\rho_i} q_i W(r_{ij}, h_i). \qquad \text{(III.3)}$$

On the right-hand side ρ_i refers to the density at the point i and q_i the amount of quantity q associated with that point. For example the velocity of the material at point j is estimated as

$$_s\boldsymbol{v}_j = \sum_i \frac{m_i}{\rho_i} \boldsymbol{v}_i W(r_{ij}, h_i). \qquad \text{(III.4)}$$

Several analytical forms of kernel have been suggested, the necessary condition being that as $h \to 0$ the kernel should become a delta function and that it should also be differentiable. In the original work of Gingold and Monoghan (1977) they suggested a Gaussian form

$$W(r, h) = \left(\frac{1}{\pi h^2}\right)^{3/2} \exp\left(-\frac{r^2}{h^2}\right) \qquad \text{(III.5)}$$

that was usually truncated when $r = 2h$, which slightly disturbed the normalization. A cubic spline form, which is smooth and continuous through the second derivative, was suggested later by Monoghan and Lattanzio (1985) and is more commonly used.

This is

$$W(r, h) = W_0 \begin{cases} 4 - 6(r/h)^3 + 3(r/h)^2 & 0 \leq r < h \\ [2 - (r/h)]^3 & h \leq r < 2h \\ 0 & r \geq 2h \end{cases} \qquad \text{(III.6)}$$

where $W_0 = 1/(4\pi h^3)$.

The appearance of the two forms of kernel is shown in figure AIII.1. The property of differentiability of the kernel is important when it comes to calculating quantities such as the divergence or gradient of some property at a point. As an example if we want to find ∇q then we evaluate

$$_s\nabla q_j = \sum_i \frac{m_i}{\rho_i} q_i \nabla W(r_{ij}, h_i). \qquad \text{(III.7)}$$

On the other hand if we wish to find $\nabla \cdot \boldsymbol{s}$, where $\boldsymbol{s}$ is a vector quantity then we find

$$_s\nabla \cdot \boldsymbol{s}_j = \sum_i \frac{m_i}{\rho_i} \boldsymbol{s}_i \cdot \nabla W(r_{ij}, h_i). \qquad \text{(III.8)}$$

It should be noted that $\nabla W(r_{ij}, h_i)$ is a vector quantity and is given by

$$\nabla W(r_{ij}, h_i) = \frac{\partial W(r_{ij}, h_i)}{\partial r_{ij}} \hat{r}_{ij} \qquad \text{(III.9)}$$

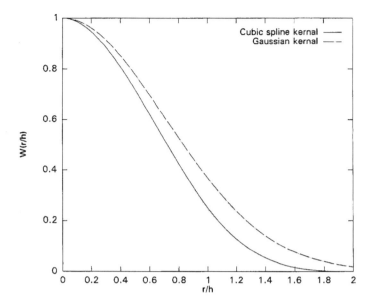

Figure AIII.1. Two types of SPH kernal.

where

$$\hat{r}_{ij} = \frac{r_i - r_j}{|r_i - r_j|}$$

is the unit vector in the direction point j to point i.

A feature that often arises in dealing with compressible fluids is the existence of shocks where densities undergo abrupt changes. In mesh systems these cannot be represented because of the finite size of the mesh and in the numerical solutions wild oscillations occur at grid points adjacent to the shock. In SPH, which is a Lagrangian-based method, the distance between particles is equivalent to a mesh size and the introduction of large density, and hence pressure, gradients causes non-physical behaviour of the system. To deal with this, as previously explained, it is customary to introduce *artificial viscosity* which broadens the shocks to a width which is compatible with the mesh, or effective mesh, and retains physically-plausible behaviour of the system. This is only necessary if the natural viscosity is too small to give the required effect.

Including contributions from gravitational forces, pressure gradients and artificial viscosity, the equations of motion that have to be solved in the SPH process are

$$\frac{\mathrm{d}r_j}{\mathrm{d}t} = v_j \qquad (\mathrm{III}.10)$$

and

$$\frac{\mathrm{d}v_j}{\mathrm{d}t} = -\nabla\phi_j - \frac{1}{\rho_j}\nabla P_j + \nu_j^{\mathrm{a}} \qquad (\mathrm{III}.11)$$

where ϕ_j is the gravitational potential at the position of particle j and ν_j^a is the artificial viscosity at that position.

From (III.2) we find

$$\frac{1}{\rho_j}\nabla P_j = \sum_i m_i \frac{P_i}{\rho_i^2} W(r_{ij}, h_i) \qquad \text{(III.12)}$$

but this form has some disadvantages. The contribution to the force on particle j due to particle i will not be equal and opposite to the contribution of the force on particle i on particle j because h_i and h_j will, in general, be different. To avoid the consequent non-conservation of linear and angular momentum that this would give, a symmetric form of the pressure-gradient term is used. There are various ways of doing this. Monoghan (1992) suggests writing

$$\frac{1}{\rho_j}\nabla P_j = \nabla\left(\frac{P_j}{\rho_j}\right) + \frac{P_j}{\rho_j^2}\nabla\rho_j.$$

Using (III.7) for the gradient terms and taking an average smoothing length in the kernel we find

$$\frac{1}{\rho_j}\nabla P_j = \sum_i m_i \left(\frac{P_j}{\rho_j^2} + \frac{P_i}{\rho_i^2}\right)\nabla W\left(r_{ij}, \frac{h_i + h_j}{2}\right). \qquad \text{(III.13)}$$

Artificial viscosity is usually incorporated in the pressure term and a common form of pressure plus artificial viscosity, which preserves symmetry is

$$\frac{1}{\rho_j}\nabla P_j = \sum_i m_i \left(\frac{P_j}{\rho_j^2} + \frac{P_i}{\rho_i^2} + \Pi_{ij}\right)\nabla W\left(r_{ij}, \frac{h_i + h_j}{2}\right). \qquad \text{(III.14)}$$

In (III.14)

$$\Pi_{ij} = \begin{cases} (-\alpha\bar{c}_{ij}\mu_{ij} + \beta\mu_{ij}^2)/\bar{\rho}_{ij} & v_{ij}\cdot r_{ij} \leq 0 \\ 0 & v_{ij}\cdot r_{ij} > 0 \end{cases} \qquad \text{(III.15)}$$

where

$$\mu_{ij} = \frac{\bar{h}_{ij}v_{ij}\cdot r_{ij}}{|r_{ij}|^2 + \eta^2}, \qquad \text{(III.16)}$$

$\bar{c}_{ij}$, $\bar{\rho}_{ij}$ and $\bar{h}_{ij}$ are the means of the sound speeds, densities and smoothing lengths at the positions of the particles i and j, the numerical factors α and β are usually taken as 1 and 2, respectively, and $\eta^2 \sim 0.01\bar{h}_{ij}^2$ is included to prevent numerical divergences.

The gravitational potential at the position of particle j due to the surrounding points cannot best be simulated by point-mass gravitational effects because each particle actually represents a distribution of matter. A common way of handling this has been by using a form such as

$$\phi_j = -\sum_i \frac{Gm_i}{(r_{ij}^2 + \gamma^2)^{1/2}} \qquad \text{(III.17)}$$

where γ is a softening parameter. By making $\gamma = 5h/7$ the correct potential at $r = 0$ is obtained for the distribution of density corresponding to kernel (III.6). Ideally the potential should be a continuous function of r and vary as r^{-1} for $r > 2h$. Various ways of doing this have been suggested and a fairly simple formula suggested here is

$$\phi = \begin{cases} -\dfrac{Gm}{\left\{r^2 + \gamma^2\left(1 - \frac{r}{2h}\right)^2\right\}^{1/2}} & r \leq 2h \\ -\dfrac{Gm}{r} & r > 2h. \end{cases} \tag{III.18}$$

This has the characteristic that it is both continuous and smooth through the transition at $r = 2h$ and so gives a continuously varying field.

All the components are now in place to be inserted in (A25) and it only remains to consider the change in the internal energy associated with each particle. For a polytropic gas the equation of state is given by

$$P = (\gamma - 1)\varepsilon, \tag{III.19}$$

where ε is the specific internal energy. The internal energy is changed by compression of the gas and by viscous dissipation. There could also be loss or gain by radiation to or from the region outside the system being modelled but we shall not include that factor here. The rate of change in specific internal energy associated with particle j is given by

$$\frac{d\varepsilon_j}{dt} = \sum_i m_i \left(\frac{P_j}{\rho_j^2} + \frac{P_i}{\rho_i^2} + \Pi_{ij}\right) v_{ij} \cdot \nabla W \left(r_{ij}, \frac{h_i + h_j}{2}\right). \tag{III.20}$$

From the density and internal energy the pressure can be found from (III.19) and then the temperature can be found from

$$\theta = \frac{P\mu}{k\rho}, \tag{III.21}$$

where μ is the mean molecular mass of the material being modelled.

One final feature of SPH we mention here is to have variable and individual smoothing lengths for the particles. Some workers in this field prefer to choose a smoothing length adjusted to give a fixed number, say 30, of particles within a distance of $2h$. Another, and convenient, way of adjusting the smoothing length is to relate it to the local density so that

$$h_j = \chi \rho_j^{-1/3}. \tag{III.22}$$

The constant χ can be initially determined on the basis of having a desired number of other particles within a distance $2h$.

The SPH method does have some features which are more-or-less arbitrary. Nevertheless it is a process which works very well in practice and does give very good simulations of the behaviour of astrophysical systems.

Appendix IV

The Bondi and Hoyle accretion mechanism

In figure AIV.1 a star of mass M is shown embedded in a gaseous medium which is locally moving at a uniform speed V relative to the star. With the star taken at rest, material moving with an impact parameter R_V with respect to the star is deflected by the gravitational field of the star and cuts the axis, defined by the direction of V passing through the star, at the point G, distance D from the star. At G the speed of the material is U and it moves in a direction making an angle γ with the axis.

From conservation of angular momentum

$$V R_V = U D \sin \gamma \qquad \text{(IV.1)}$$

and from conservation of energy

$$\frac{1}{2} V^2 = \frac{1}{2} U^2 - \frac{GM}{D}. \qquad \text{(IV.2)}$$

Material cutting the axis at G interacts with other material so destroying the component of motion perpendicular to the axis. For the limiting case where the residual horizontal component is just equal to the escape speed from the star

$$(U \cos \gamma)^2 = \frac{2GM}{D}. \qquad \text{(IV.3)}$$

Substituting the expression for V obtained from (IV.1) and the expression for GM/D obtained from (IV.3) into (IV.2)

$$U^2 \frac{D^2}{R_V^2} \sin^2 \gamma = U^2 - U^2 \cos^2 \gamma$$

from which

$$D = R_V. \qquad \text{(IV.4)}$$

At this stage we must refer to the geometry of the hyperbola which describes the motion of the gas relative to the star before it interacts on the axis. The path

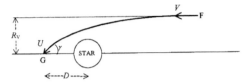

Figure AIV.1. A stream of material with impact parameter R_V arriving at the axis at point G.

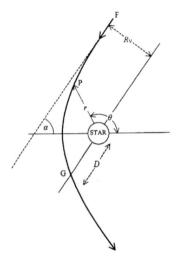

Figure AIV.2. The hyperbolic motion of the material. Points F and G correspond to those in figure AIV.1.

of the gas is represented in figure AIV.2 with points F and G and distances D and R_V shown corresponding to those in figure AIV.1. Any conic section is described in polar coordinates by

$$r = \frac{q(1+e)}{1+e\cos\theta} \qquad (\text{IV.5})$$

where q is the closest distance from the origin, the position of the star in the case we are taking. The asymptote to the curve at infinity, which is the direction of the axis in figure AIV.1, makes an angle $\alpha = \cos^{-1}(-1/e)$ with the x-axis in figure AIV.2. The point P on the curve with coordinates (r,θ) is distance T from the line LG where

$$T = r\sin(\alpha - \theta) = \frac{q(1+e)\sin(\alpha - \theta)}{1+e\cos\theta} \qquad (\text{IV.6})$$

and

$$R_V = (T)_{\theta \to \alpha}.$$

Substituting $\theta = \alpha$ in (IV.6) gives zero in both numerator and divisor but Hôpital's rule can be used to give

$$R_V = \left\{ \frac{-q(1+e)\cos(\alpha - \theta)}{-e\sin\theta} \right\}_{\theta \to \alpha} = \frac{-q(1+e)}{-e\sqrt{1 - \frac{1}{e^2}}} = \frac{q(1+e)}{\sqrt{e^2 - 1}}. \qquad \text{(IV.7)}$$

From figure AIV.2, D is the value of r when $\theta = \pi - \alpha$ and, from (IV.4), $D = R_V$. Thus from (IV.5) and (IV.7)

$$\frac{q(1+e)}{\sqrt{e^2 - 1}} = \frac{q(1+e)}{1 + e\cos(\pi - \alpha)} = \frac{q(1+e)}{1 - e\cos\alpha} = \frac{q(1+e)}{2}.$$

This shows that $e^2 = 5$ for the limiting gas stream.

From the basic theory of orbital motion

$$e^2 = 1 + \frac{2H^2 E}{G^2 M^2} \qquad \text{(IV.8)}$$

where H and E are the intrinsic angular momentum and intrinsic energy of the orbiting material. For the critical gas stream

$$H = R_V V \quad \text{and} \quad E = \tfrac{1}{2}V^2$$

which inserted in (IV.8) gives the required result

$$R_V = \frac{2GM}{V^2}.$$

Appenbix V

The Poynting–Robertson effect

We consider a perfectly absorbing spherical particle of radius a and density ρ at distance R from the Sun. The total energy absorbed per unit time is

$$P = L_\odot \frac{a^2}{4R^2}. \tag{V.1}$$

The energy comes to the particle radially but then is re-emitted by the particle moving in its orbit at speed v. The radiated energy thus possesses momentum which must be taken from the particle. Considering the mass equivalent of the radiated energy the rate of change of momentum of the particle, or the tangential force on it, is

$$F = L_\odot \frac{a^2 v}{4R^2 c^2}. \tag{V.2}$$

This force exerts a torque that changes the angular momentum of the particle at a rate

$$\frac{dh}{dt} = -FR = L_\odot \frac{a^2 v}{4Rc^2}. \tag{V.3}$$

If the mass of the particle is m then its angular momentum is

$$h = m\sqrt{GM_\odot R}$$

so that

$$\frac{dh}{dt} = \frac{1}{2}m\sqrt{\frac{GM_\odot}{R}}\frac{dR}{dt} = \frac{1}{2}mv\frac{dR}{dt}. \tag{V.4}$$

From (V.3) and (V.4) and expressing m in terms of a and ρ we find

$$\frac{dR}{dt} = -\frac{3L_\odot}{8\pi Rac^2 \rho}. \tag{V.5}$$

This can be integrated to give the total time for the particle to be absorbed by the Sun

$$t = \frac{4\pi R^2 ac^2 \rho}{3L_\odot}. \tag{V.6}$$

References

Aarseth S J 1968 *Bull. Astron.* **3** 105–25

Alexander C M O'D 1993 *Geochim. Cosmochim. Acta* **57** 2869–88

Alexander C M O'D, Swann P and Walker R M 1992 *Lunar Planet. Sci.* **XXIII** 9–10

Alfvén H 1978 *The origin of the Solar System* ed S F Dermott (Chichester: Wiley) pp 41–8

Allan R R 1969 *Astron. J.* **74** 497–507

Allen C C 1979 *Icarus* **39** 111–23

Allinson D J 1986 The early evolution of a protoplanet *PhD Thesis* CNAA (University of Middlesbrough)

Amari S, Hoppe P, Zinner E and Lewis R S 1992 *Lunar Planet. Sci.* **XXIII** 27–8

Anders E 1971 *Ann. Rev. Astron. Astrophys.* **9** 1–34

Anders E and Owen T 1977 *Science* **198** 453–65

Arrhenius S 1901 *Archives Néelandaises* **6** 862–73

Arvidson R E, Goettel K A and Hohenberg C M 1980 *Rev. Geophys. Space Sci.* **18** 565–603

Aust C and Woolfson M M 1973 *Mon. Not. R. Astron. Soc.* **161** 7–13

Bailey M E 1983 *Mon. Not. R. Astron. Soc.* **204** 603–33

——1987 *Icarus* **69** 70–82

Beaugé C and Ferraz-Mello S 1993 *Icarus* **103** 301–18

Beck S C and Beckwith S V 1984 *Mon. Not. R. Astron. Soc.* **207** 671–7

Benz W, Slattery W L and Cameron A G W 1986 *Icarus* **66** 515–35

——1987 *Icarus* **71** 30–45

Biermann L 1978 *Moon and Planets* **18** 447–64

Bills B G and Ferrari A J 1977 *J. Geophys. Res.* **82** 1306–14

Binder A B and Langer M A 1980 *J. Geophys. Res.* **85** 3194–208

Black D C 1972 *Geochim. Cosmochim. Acta* **36** 377–94

——1978 *The origin of the Solar System* ed S F Dermott (Chichester: Wiley) p 583

Black D C and Bodenheimer P 1976 *Astrophys. J.* **206** 138–49

Bondi H 1952 *Mon. Not. R. Astron. Soc.* **112** 195–204

Bondi H and Hoyle F 1944 *Mon. Not. R. Astron. Soc.* **104** 273–82

Borra E F, Landstreet J D and Mestel L 1982 *Ann. Rev. Astron. Ap.* **20** 191–220

Boss A P 1988 *Astrophys. J.* **331** 370–6

Brush S G 1996 *Fruitful Encounters: The Origin of the Solar System and of the Moon from Chamberlin to Apollo* (Cambridge: Cambridge University Press)

Buffon G L L 1745 De la formation des planètes *Oevres Complets (1852)* vol 1 (Brussels: Adolf Deros) pp 120–40

Butler R P and Marcy G W 1996 *Astrophys. J.* **464** L153–6

Cameron A G W 1969 *Meteorite Research* ed P M Millman (New York: Springer) pp 7–15

Cameron A G W 1973 *Icarus* **18** 407–50

——1978 *The Origin of the Solar System* ed S F Dermott (Chichester: Wiley) pp 49–74

Cameron A G W and Pine M R 1973 *Icarus* **18** 377–406

Cameron A G W and Truran J W 1977 *Icarus* **30** 447–61

Cameron A G W and Ward W R 1976 *Lunar Planet. Sci. VII* 120–2

Caughlan G R, Fowler W A, Harris M J and Zimmerman B A 1985 *At. Data Nucl. Data Tables* **32** 197–233

Chamberlin T C 1901 *Astrophys. J.* **14** 17–40

Chandrasekhar S 1939 *An Introduction to the Study of Stellar Structure* (Chicago, IL: University of Chicago Press)

Chapman S, Pongracic H, Disney M, Nelson A, Turner J and Whitworth A 1992 *Nature* **359** 207–10

Clayton D D, Dwek E and Woosley S E 1977 *Astrophys. J.* **203** 300–15

Clayton R N 1981 *Phil. Trans. R. Soc.* A **303** 339–49

Clayton R N, Grossman L and Mayeda T K 1973 *Science* **187** 485–8

Clayton R N, Hinton R W and Davis A M 1988 *Phil. Trans. R. Soc.* A **325** 483–501

Clube S V M and Napier W M 1984 *Mon. Not. R. Astron. Soc.* **208** 575–88

Colombo G and Franklin F A 1971 *Icarus* **15** 186–9

Connell A J and Woolfson M M 1983 *Mon. Not. R. Astron. Soc.* **204** 1221–30

Cook A H 1977 *Celestial Masers* (Cambridge: Cambridge University Press)

Cox A N and Stewart J N 1970 *Astrophys. J. Suppl. No 174* **19** 243–59

Darwin G H *Nature* **18** 580–2

De Campli W M and Cameron A G W 1979 *Icarus* **38** 367–91

De Hon R A 1979 *Proc. Lunar Sci. Conf.* **10** 2935–55

De Hon R A and Waskom J D 1976 *Proc. Lunar Sci. Conf.* **7** 2729–46

Descartes R 1664 *Le Monde, ou Traité de la Lumière* (Paris: Jacques le Gras)

Disney M J, McNally D and Wright A E 1969 *Mon. Not. R. Astron. Soc.* **146** 123–60

Dodd K N and McCrea W H 1952 *Mon. Not. R. Astron. Soc.* **112** 307–31

Donahue T M, Hoffman J H, Hodges R R and Watson A J 1982 *Science* **216** 630–3

Dormand J R and Woolfson M M 1971 *Mon. Not. R. Astron. Soc.* **151** 307–31

——1974 *Proc. R. Soc.* A **340** 349–65

——1977 *Mon. Not. R. Astron. Soc.* **180** 243–79

——1980 *Mon. Not. R. Astron. Soc.* **193** 171–4

——1988 *The Physics of the Planets* ed S K Runcorn (Chichester: Wiley) pp 371–83

——1989 *The Origin of the Solar System: The Capture Theory* (Chichester: Ellis Horwood)

Duncan M and Quinn T 1993 *Ann. Rev. Astron. Astrophys.* **31** 265–95

Eddington A S 1926 *The Internal Constitution of the Stars* (Cambridge: Cambridge University Press)

Eggers S and Woolfson M M 1996 *Mon. Not. R. Astron. Soc.* **282** 13–18

Ezer D and Cameron A G W 1965 *Can. J. Phys.* **43** 1497–517

Farinella P, Milani A, Nobili A M and Valsacchi G B 1979 *Moon Planets* **20** 415–21

Ferrari A J, Sinclair W S, Sjogren W L, Williams J G and Yoder C F 1980 *J. Geophys. Res.* **85** 3939–51

Field G B, Rather J D G and Aannestad P A 1968 *Astrophys. J.* **151** 953–75

Fowler W A, Caughlan G R and Zimmerman B A 1967 *Ann. Rev. Astron. Ap.* **5** 525–70

——1975 *Ann. Rev. Astron. Ap.* **13** 69–112

Freeman J W 1978 *The Origin of the Solar System* ed S F Dermott (Chichester: Wiley)
    pp 635–40
Gaidos E J 1995 *Icarus* **114** 258–68
Gault D E and Heitowit E D 1963 *Proc. Sixth Hypervelocity Impact Symp. (Cleveland, OH,
    April 30-May 2)*
Gausted J E 1963 *Astrophys. J.* **138** 1050–73
Gingold R A and Monaghan J J 1977 *Mon. Not. R. Astron. Soc.* **181** 375–89
Goettel K A 1980 *Reports of the Planetary Geology Program 1979–1980* NASA Tech.
    Memo TM-81776, p 404
Golanski Y 1999 Modelling the formation of dense cold clouds *DPhil Thesis* University of
    York
Goldreich P 1963 *Mon. Not. R. Astron. Soc.* **126** 257–68
——1965 *Mon. Not. R. Astron. Soc.* **130** 159–81
——1966 *Rev. Geophys.* **4** 411–39
Goldreich P and Ward W R 1973 *Astrophys. J.* **183** 1051–61
Gomes R S 1995 *Icarus* **115** 47–59
Grasdalen G L, Strom S E, Strom K M, Capps R W, Thompson D and Castelaz M 1984
    Astrophys. J. **283** L57 *et seq*
Greenberg R 1973 *Astron. J.* **78** 338–46
Grossman L 1972 *Geochim. Cosmochim. Acta* **36** 597–619
Grzedzielski S 1966 *Mon. Not. R. Astron. Soc.* **134** 109–34
Guest J, Butterworth P, Murray J and O'Donnell W 1979 *Planetary Geology* (London:
    David and Charles)
Harrington R S and Van Flandern T C 1979 *Icarus* **39** 131–6
Harris M J, Fowler W A, Caughlan G R and Zimmerman B A 1983 *Ann. Rev. Astron. Ap.*
    21 165–76
Hartmann W K and Davis D R 1975 *Icarus* **24** 504–15
Hayashi C 1961 *Publ. Astron. Soc. Japan* **13** 450–2
——1966 *Ann. Rev. Astron. Astrophys.* **4** 171–92
Hayashi C and Nakano T 1965 *Prog. Theor. Phys.* **33** 554–6
Herbig G H 1962 *Adv. Astron. Astrophys.* **1** 47–103
Holden P and Woolfson M M 1995 *Earth, Moon and Planets* **69** 201–36
Hood L L, Coleman P J and Russell C T 1978 *J. Geophys. Res. Lett.* **5** 305–8
Hoyle F 1960 *Q. J. R. Astron. Soc.* **1** 28–55
Hughes D W and Cole G H A 1995 *Mon. Not. R. Astron. Soc.* **277** 99–105
Hunter C 1962 *Astrophys. J.* **136** 594–608
——1964 *Astrophys. J.* **139** 570–86
Hutchison R 1982 *Physics of the Earth and Planetary Interiors* **29** 199–208
Iben I 1965 *Astrophys. J.* **141** 993–1018
Iben I and Talbot R J 1966 *Astrophys. J.* **144** 968–77
Ingersoll A P 1990 *Science* **248** 308–15
Jeans J H 1902 *Phil. Trans. R. Soc.* A **199** 1–53
——1919 *Problems of Cosmogony and Stellar Dynamics* (Cambridge: Cambridge Univer-
    sity Press)
——1929 *Mon. Not. R. Astron. Soc.* **89** 636–41
Jeffreys H 1929 *Mon. Not. R. Astron. Soc.* **89** 636–41
——1930 *Mon. Not. R. Astron. Soc.* **91** 169–73
——1952 *Proc. R. Soc.* A **214** 281–91

Jones J H 1984 *Geochim. Geocosmica Acta* **48** 641–8

Kant I 1755 *Allgemeine Naturgeschichte und Theorie des Himmels* (Konigsberg: Petersen)

Kiang T 1962 *Mon. Not. R. Astron. Soc.* **123** 359–82

Kroupa P 1995 *Mon. Not. R. Astron. Soc.* **277** 1491–506

Kuhn T S 1970 *The Structure of Scientific Revolutions* 2nd edn (Chicago, IL: University of Chicago Press)

Kuiper G P 1951 *Astrophysics: A Topical Symposium* ed J A Hynek (New York: McGraw-Hill) p 357

Lada C J and Lada E A 1991 *The Formation and Evolution of Star Clusters (ASP Conf. Series 13)* ed K Janes (San Fransisco, CA: Astronomical Society of the Pacific) pp 3–22

Lamb H 1932 *Hydrodynamics* 6th edn (Cambridge: Cambridge University Press)

Lamy P L and Burns J A 1972 *Am. J. Phys.* **40** 441–5

Laplace P S de 1796 *Exposition du Système du Monde* (Paris: Imprimerie Cercle-Social)

Larimer J W 1967 *Geochim. Cosmochim. Acta* **31** 1215–38

Larimer J W and Bartholomay M 1979 *Geochim. Cosmochim. Acta* **43** 1455–66

Larson R B 1969 *Mon. Not. R. Astron. Soc.* **145** 271–95

——1984 *Mon. Not. R. Astron. Soc.* **206** 197–207

——1989 *The Formation and Evolution of Planetary Systems* ed H A Weaver and L Danley (Cambridge: Cambridge University Press) pp 31–54

Lee T, Papanastassiou D A and Wasserberg G T 1976 *Geophys. Res. Lett.* **3** 109–12

Lewis J S 1972 *Earth Planet. Sci. Lett.* **15** 286–90

Lin D N C, Bodenheimer P and Richardson D C 1996 *Nature* **380** 606–7

Lin D N C and Papaloizu J 1980 *Mon. Not. R. Astron. Soc.* **191** 37–48

——1985 *Protostars and Planets* ed D C Black and M S Matthews (Tucson, AR: University of Arizona Press) pp 980–1072

Lucy L B 1977 *Astron. J.* **82** 1013–24

Lynden-Bell D and Pringle J E 1974 *Mon. Not. R. Astron. Soc.* **168** 603–37

Lyttleton R A 1960 *Mon. Not. R. Astron. Soc.* **121** 551–69

——1961 *Mon. Not. R. Astron. Soc.* **122** 399–407

Malhotra R 1993 *Icarus* **106** 264–73

Marsden B G, Sekanina Z and Everhart E 1978 *Astron. J.* **83** 64–71

Mayor M and Queloz D 1995 *Nature* **378** 355–9

McCord T B 1966 *Astron. J.* **71** 585–90

McCrea W H 1960 *Proc. R. Soc.* **256** 245–66

——1988 *The Physics of the Planets* ed S K Runcorn (Chichester: Wiley) pp 421–39

McCue J, Dormand J R and Gadian A M 1992 *Earth, Moon and Planets* **57** 1–11

McElroy M B, Ten Ying Kong and Yuk Ling Yung 1977 *J. Geophys. Rev.* **82** 4379–488

McNally D 1971 *Rep. Prog. Phys.* **34** 71–108

Melita M D and Woolfson M M 1996 *Mon. Not. R. Astron. Soc.* **280** 854–62

Michael D M 1990 Evidence of a planetary collision in the early solar system and its implications for the origin of the solar system *DPhil Thesis* University of York

Mizutani H, Mutsui T and Takeuchi H 1972 *Moon* **4** 476–89

Miyama S M 1989 *The Formation and Evolution of Planetary Systems* ed H A Weaver and L Danley (Cambridge: Cambridge University Press) pp 284–90

Monaghan J J 1992 *Ann. Rev. Astron. Astrophys.* **30** 543–74

Monaghan J J and Lattanzio J C 1985 *Astron. Astrophys.* **149** 135–43

Moulton F R 1905 *Astrophys. J.* **22** 165–81

Mullis A M 1991 *Geophys. J. Int.* **105** 778–81
——1992 *Geophys. J. Int.* **109** 233–39
——1993 *Geophys. J. Int.* **114** 196–208
Murray B C and Mallin M C 1973 *Science* **179** 997–1000
Nakamura Y, Latham G and Dorman H J 1982 *J. Geophys. Res.* **87** (suppl.) A117–23
Napier W McD and Dodd R J 1973 *Nature* **224** 250
Nolan J 1885 *Darwin's Theory of the Genesis of the Moon* (Melbourne: Robertson)
Nölke F 1908 *Das Problem der Entwicklung unseres Planetensystems. Aufstellung einer neuen Theorie nach vorgehender Kritik der Theorien von Kant, Laplace, Poincaré, Moulton, Arrhenius u.a.* (Berlin: Springer)
Oort J H 1948 *Bull. Astron. Inst. Neth.* **11** 91–110
Owen T and Bieman K 1976 *Science* **193** 801–3
Peale S J, Cassen P and Reynolds R T 1979 *Science* **203** 892–4
Pike R J 1967 *J. Geophys. Res.* **72** 2099–106
Poincaré H 1911 *Leçons sur les Hypothèse Cosmogonique* (Paris: Hermann)
Pollack J B, Roush T, Witterborn F, Bregman J, Wooden D, Stoker C, Toon O B, Rank D, Dalton B and Freedman R 1990 *J. Geophys. Res.—Solid Earth and Planets* **95** 14 595–627
Pongracic H, Chapman S, Davies R, Nelson A, Disney M and Whitworth A 1991 *Mem. S. A. It.* **62** 851–8
Prentice A J R 1974 *In the Beginning. . .* ed J P Wild (Canberra: Australian Academy of Science)
——1978 *The Origin of the Solar System* ed S F Dermott (Chichester: Wiley) pp 111–61
——1989 *Phys. Lett.* A **140** 265–70
Reddish V C and Wickramasinghe N C 1969 *Mon. Not. R. Astron. Soc.* **143** 189–208
Reynolds J 1960 *Phys. Rev. Lett.* **4** 8–10
Roche E 1854 *Mem. Acad. Montpellier* **2** 399–439
——1873 *Mem. Acad. Montpellier* **8** 235–324
Roy A E 1977 *Orbital Motion* (Bristol: Adam Hilger)
Runcorn S K 1975 *Nature* **253** 701–3
——1980 *Geochim. Cosmochim.* **44** 1867–77
——1988 *The Physics of the Planets* ed S K Runcorn (Chichester: Wiley)
Ruskol E L 1960 *Sov. Astron. AJ* **4** 657–68
Russell H N 1935 *The Solar System and its Origin* (New York: MacMillan)
Safronov V S 1972 *Evolution of the Protoplanetary Cloud and Formation of the Earth and Planets* (Jerusalem: Israel Program for Scientific Translations)
Schmidt O Y 1944 *Dokl. Acad. Nauk. USSR* **45** 229–33
Schofield N and Woolfson M M 1982a *Mon. Not. R. Astron. Soc.* **198** 947–61
——1982b *Mon. Not. R. Astron. Soc.* **198** 963–73
Seaton M J 1955 *Ann. Astrophys.* **18** 188–205
Sidlichovský M and Nesvorný D 1994 *Astron. Astrophys.* **289** 972–82
Singer S F 1968 *Geophys. J. R. Astonr. Soc.* **15** 205–26
——1970 *Eos* **51** 637–41
Smith F J 1966 *Planet. Space Sci.* **14** 929–37
Spitzer L 1939 *Astrophys. J.* **90** 675–88
Steele I M, Smith J V, Hutcheon I D and Clayton R N 1978 *Lunar Planet. Sci.* **9** 1104–6
Stevenson D J 1978 *The Origin of the Solar System* ed S F Dermott (Chichester: Wiley) pp 395–431

Stewart G R and Wetherill G W 1988 *Icarus* **74** 542–53

Stock J D R and Woolfson M M 1983a *Mon. Not. R. Astron. Soc.* **202** 287–91

——1983b *Mon. Not. R. Astron. Soc.* **202** 511–30

Stone J, Hutcheon I D, Epstein S and Wasserberg G J 1990 *Lunar Planet. Sci.* **XXI** 1212–13

Strom S E, Strom K M, Grasdalen G L, Capps R W and Thompson D 1985 *Astron. J.* **90** 2575 *et seq*

Sullivan W T 1971 *Astrophys. J.* **166** 321–2

Toksöz M N and Solomon S C 1973 *Moon* **7** 251–78

Tremaine S 1991 *Icarus* **89** 85–92

Trulsen J 1971 *Plasma to Planets* ed E A Elvius (London: Wiley) p 179

Turner J A, Chapman S J, Bhattal A S, Disney M J, Pongracic H and Whitworth A P 1995 *Mon. Not. R. Astron. Soc.* **277** 705–26

Vargaftik N B 1975 *Tables on Thermodynamic Properties of Liquids and Gases* (London: Wiley)

Virag A, Wopenka B, Amari S, Zinner E, Anders E and Lewis R S 1992 *Geochim. Cosmochim. Acta* **56** 1715–33

von Sengbusch K and Temesvary S 1966 *Stellar Evolution* ed R F Stein and A G W Cameron (New York: Plenum) p 209

von Weizsäcker C F 1944 *Z. Astrophys.* **22** 319–55

Weidenschilling S J and Davis D R 1985 *Icarus* **62** 16–29

Weidenschilling S J, Donn B and Meakin P 1989 *The Formation and Evolution of Planetary Systems* ed H A Weaver and L Danley (Cambridge: Cambridge University Press) pp 131–50

Wetherill G W 1986 *Origin of the Moon* ed W K Hartmann, R J Phillips and G J Taylor (Huston, TX: Lunar and Planetary Institute) pp 519–50

——1989 *The Formation and Evolution of Planetary Systems* ed H A Weaver and L Danley (Cambridge: Cambridge University Press) pp 1–30

Whitworth A P, Bhattal A S, Chapman S J, Disney M J and Turner J A 1994a *Astron. Astrophys.* **290** 421–7

——1994b *Mon. Not. R. Astron. Soc.* **268** 291–8

Whitworth A P, Boffin H, Watkins S and Francis N 1998 *Astron. Geophys.* **39** 10–13

Whitworth A P, Chapman S J, Bhattal A S, Disney M J, Pongracic H and Turner J A 1995 *Mon. Not. R. Astron. Soc.* **277** 727–46

Williams I P and Cremin A W 1969 *Mon. Not. R. Astron. Soc.* **144** 359–73

Williams J G and Benson G S 1971 *Bull. Am. Astron. Soc.* **2** 253

Williams S and Woolfson M M 1983 *Mon. Not. R. Astron. Soc.* **204** 853–63

Wise D U, Golombek M P and McGill G E 1979 *Icarus* **38** 456–72

Woolfson M M 1964 *Proc. R. Soc.* A **282** 485–507

——1979 *Phil. Trans. R. Soc.* A **291** 219–52

——1999 *Mon. Not. R. Astron. Soc.* **304** 195–8

Woosley S E, Fowler W A, Holmes J A and Zimmerman B A 1978 *At. Data Nucl. Data Tables* **22** 371–441

Yoder C F 1979 *Nature* **279** 767–70

Zel,dovich Ya B and Raizer Yu P 1966 *Physics of Shock Waves and High-temperature Hydrodynamic Phenomena* (New York: Academic)

Zinner E, Tang M and Anders E 1989 *Geochim. Cosmochim. Acta* **53** 3273–90

Printed and bound by CPI Group (UK) Ltd, Croydon, CR0 4YY

22/10/2024

01777333-0003

Index